Nondestructive Monitoring of Materials Properties

Nondestructive Monitoring of Materials Properties

Symposium held November 28-30, 1988, Boston, Massachusetts, U.S.A.

EDITORS:

John Holbrook
Battelle Columbus, Columbus, Ohio, U.S.A.

Jean Bussière
National Research Council of Canada, Boucherville, Quebec, Canada

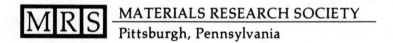

MATERIALS RESEARCH SOCIETY
Pittsburgh, Pennsylvania

CODEN: MRSPDH

Published by:

Materials Research Society
9800 McKnight Road
Pittsburgh, Pennsylvania 15237
Telephone (412) 367-3003

Library of Congress Cataloging in Publication Data

Nondestructive monitoring of materials properties : symposium held November 28-30, 1988, Boston, Massachusets, U.S.A. / editors, John Holbrook, Jean Bussière.

    p.   cm. — (Materials Research Society symposium proceedings : ISSN 0272-9172 ; v. 142)
    Symposium held in conjunction with the Fall Meeting of the Materials Research Society.
    Includes bibliographical references.
    ISBN 1-55899-015-1
    1. Non-destructive testing—Congresses.   2. Materials—Testing—Congresses.
I. Holbrook, John (John M.)   II. Bussière, Jean F.   III. Materials Research Society, Meeting (1989 : Boston, Mass.)   IV. Series: Materials Research Society symposium proceedings : v. 142.

TA417.2.N6725   1989   89-28073
620.1′127—dc20   CIP

Manufactured in the United States of America

Contents

*Invited Paper

PART III: DEGRADATION (CREEP, FATIGUE, ETC.)

*Invited Paper

PART IV: POLYMERS AND COMPOSITES

PART V: CERAMICS

*Invited Paper

*Invited Paper

Preface

The nondestructive monitoring of materials properties is a rapidly growing research area which faces the unique challenge of providing nondestructive, and preferably noninvasive, techniques and methodologies for assessing microstructure, morphology, physical properties, mechanical properties, etc., in the industrial environment from the early stages of materials processing and design to the final phases of a component's life.

Applications of these techniques, which at present are still largely in the early stages of laboratory development, range from process development and process control to assessment and extension of the useful life of existing structures.

The present volume contains the edited papers of the first MRS symposium devoted entirely to this area, and is certainly manifestation of the explosive growth of nondestructive evaluation in recent years and the extraordinary capacity for integration and increase of scope of Materials Science and Engineering. The symposium was held during the MRS Fall Meeting, Boston, November 28 - December 3, 1988. Sponsorship of the symposium was provided by the Electric Power Research Institute, the International Center for Diffraction Data, Battelle, and the Industrial Materials Research Institute of the National Research Council Canada.

In organizing the papers in this highly multidisciplinary area, one is always faced with the difficult choice between emphasizing techniques, materials, or the type of property to be measured. The major headings of these proceedings represent an eclectic--we hope--mixture of all these.

The first section, "Overview and New Techniques," reflects the concern for improving and making available new techniques for probing materials nondestructively, with applications ranging from microelectronics to steel processing. As described in the first paper, the availability of nondestructive evaluation tools for materials and flaw characterization is an important element for the realization of new concepts such as "materials by design" and "unified life cycle engineering," where they will serve not only process and quality control but will be totally integrated, even at the design phase of manufacturing.

The next section, on "degradation monitoring," is one of the focal areas of the symposium in response to rapidly growing interest, especially in the utilities industries, in extending the life of existing structures. Papers describe the use of ultrasonic, electromagnetic, NMR, microradiography and position annihilation techniques for monitoring damage due to creep, fatigue, hydrogen, and other environmental influences, primarily in metallic structures.

Monitoring of "texture and stress," using x-ray and neutron diffraction and ultrasonic velocity measurements for process evaluation and process control is the main theme of the next section.

The remaining sections are devoted to "polymers and composites," "ceramics" and "metals and metallic bonded structures," reflecting considerable interest in improving processing and quality control in those rapidly growing areas.

The editors are grateful to G. Cyr, Charlotte Roseberry, and Virginia Miller for participating in the organization of the symposium and in the preparation of the proceedings.

John Holbrook
Jean Bussière

August 1989

MATERIALS RESEARCH SOCIETY SYMPOSIUM PROCEEDINGS

ISSN 0272 - 9172

Volume 1—Laser and Electron-Beam Solid Interactions and Materials Processing, J. F. Gibbons, L. D. Hess, T. W. Sigmon, 1981, ISBN 0-444-00595-1

Volume 2—Defects in Semiconductors, J. Narayan, T. Y. Tan, 1981, ISBN 0-444-00596-X

Volume 3—Nuclear and Electron Resonance Spectroscopies Applied to Materials Science, E. N. Kaufmann, G. K. Shenoy, 1981, ISBN 0-444-00597-8

Volume 4—Laser and Electron-Beam Interactions with Solids, B. R. Appleton, G. K. Celler, 1982, ISBN 0-444-00693-1

Volume 5—Grain Boundaries in Semiconductors, H. J. Leamy, G. E. Pike, C. H. Seager, 1982, ISBN 0-444-00697-4

Volume 6—Scientific Basis for Nuclear Waste Management IV, S. V. Topp, 1982, ISBN 0-444-00699-0

Volume 7—Metastable Materials Formation by Ion Implantation, S. T. Picraux, W. J. Choyke, 1982, ISBN 0-444-00692-3

Volume 8—Rapidly Solidified Amorphous and Crystalline Alloys, B. H. Kear, B. C. Giessen, M. Cohen, 1982, ISBN 0-444-00698-2

Volume 9—Materials Processing in the Reduced Gravity Environment of Space, G. E. Rindone, 1982, ISBN 0-444-00691-5

Volume 10—Thin Films and Interfaces, P. S. Ho, K.-N. Tu, 1982, ISBN 0-444-00774-1

Volume 11—Scientific Basis for Nuclear Waste Management V, W. Lutze, 1982, ISBN 0-444-00725-3

Volume 12—In Situ Composites IV, F. D. Lemkey, H. E. Cline, M. McLean, 1982, ISBN 0-444-00726-1

Volume 13—Laser-Solid Interactions and Transient Thermal Processing of Materials, J. Narayan, W. L. Brown, R. A. Lemons, 1983, ISBN 0-444-00788-1

Volume 14—Defects in Semiconductors II, S. Mahajan, J. W. Corbett, 1983, ISBN 0-444-00812-8

Volume 15—Scientific Basis for Nuclear Waste Management VI, D. G. Brookins, 1983, ISBN 0-444-00780-6

Volume 16—Nuclear Radiation Detector Materials, E. E. Haller, H. W. Kraner, W. A. Higinbotham, 1983, ISBN 0-444-00787-3

Volume 17—Laser Diagnostics and Photochemical Processing for Semiconductor Devices, R. M. Osgood, S. R. J. Brueck, H. R. Schlossberg, 1983, ISBN 0-444-00782-2

Volume 18—Interfaces and Contacts, R. Ludeke, K. Rose, 1983, ISBN 0-444-00820-9

Volume 19—Alloy Phase Diagrams, L. H. Bennett, T. B. Massalski, B. C. Giessen, 1983, ISBN 0-444-00809-8

Volume 20—Intercalated Graphite, M. S. Dresselhaus, G. Dresselhaus, J. E. Fischer, M. J. Moran, 1983, ISBN 0-444-00781-4

Volume 21—Phase Transformations in Solids, T. Tsakalakos, 1984, ISBN 0-444-00901-9

Volume 22—High Pressure in Science and Technology, C. Homan, R. K. MacCrone, E. Whalley, 1984, ISBN 0-444-00932-9 (3 part set)

Volume 23—Energy Beam-Solid Interactions and Transient Thermal Processing, J. C. C. Fan, N. M. Johnson, 1984, ISBN 0-444-00903-5

Volume 24—Defect Properties and Processing of High-Technology Nonmetallic Materials, J. H. Crawford, Jr., Y. Chen, W. A. Sibley, 1984, ISBN 0-444-00904-3

Volume 25—Thin Films and Interfaces II, J. E. E. Baglin, D. R. Campbell, W. K. Chu, 1984, ISBN 0-444-00905-1

Volume 26—Scientific Basis for Nuclear Waste Management VII, G. L. McVay, 1984, ISBN 0-444-00906-X

Volume 27—Ion Implantation and Ion Beam Processing of Materials, G. K. Hubler, O. W. Holland, C. R. Clayton, C. W. White, 1984, ISBN 0-444-00869-1

Volume 28—Rapidly Solidified Metastable Materials, B. H. Kear, B. C. Giessen, 1984, ISBN 0-444-00935-3

Volume 29—Laser-Controlled Chemical Processing of Surfaces, A. W. Johnson, D. J. Ehrlich, H. R. Schlossberg, 1984, ISBN 0-444-00894-2

Volume 30—Plasma Processing and Synthesis of Materials, J. Szekely, D. Apelian, 1984, ISBN 0-444-00895-0

Volume 31—Electron Microscopy of Materials, W. Krakow, D. A. Smith, L. W. Hobbs, 1984, ISBN 0-444-00898-7

Volume 32—Better Ceramics Through Chemistry, C. J. Brinker, D. E. Clark, D. R. Ulrich, 1984, ISBN 0-444-00898-5

Volume 33—Comparison of Thin Film Transistor and SOI Technologies, H. W. Lam, M. J. Thompson, 1984, ISBN 0-444-00899-3

Volume 34—Physical Metallurgy of Cast Iron, H. Fredriksson, M. Hillerts, 1985, ISBN 0-444-00938-8

Volume 35—Energy Beam-Solid Interactions and Transient Thermal Processing/1984, D. K. Biegelsen, G. A. Rozgonyi, C. V. Shank, 1985, ISBN 0-931837-00-6

Volume 36—Impurity Diffusion and Gettering in Silicon, R. B. Fair, C. W. Pearce, J. Washburn, 1985, ISBN 0-931837-01-4

Volume 37—Layered Structures, Epitaxy, and Interfaces, J. M. Gibson, L. R. Dawson, 1985, ISBN 0-931837-02-2

Volume 38—Plasma Synthesis and Etching of Electronic Materials, R. P. H. Chang, B. Abeles, 1985, ISBN 0-931837-03-0

Volume 39—High-Temperature Ordered Intermetallic Alloys, C. C. Koch, C. T. Liu, N. S. Stoloff, 1985, ISBN 0-931837-04-9

Volume 40—Electronic Packaging Materials Science, E. A. Giess, K.-N. Tu, D. R. Uhlmann, 1985, ISBN 0-931837-05-7

Volume 41—Advanced Photon and Particle Techniques for the Characterization of Defects in Solids, J. B. Roberto, R. W. Carpenter, M. C. Wittels, 1985, ISBN 0-931837-06-5

Volume 42—Very High Strength Cement-Based Materials, J. F. Young, 1985, ISBN 0-931837-07-3

Volume 43—Fly Ash and Coal Conversion By-Products: Characterization, Utilization, and Disposal I, G. J. McCarthy, R. J. Lauf, 1985, ISBN 0-931837-08-1

Volume 44—Scientific Basis for Nuclear Waste Management VIII, C. M. Jantzen, J. A. Stone, R. C. Ewing, 1985, ISBN 0-931837-09-X

Volume 45—Ion Beam Processes in Advanced Electronic Materials and Device Technology, B. R. Appleton, F. H. Eisen, T. W. Sigmon, 1985, ISBN 0-931837-10-3

Volume 46—Microscopic Identification of Electronic Defects in Semiconductors, N. M. Johnson, S. G. Bishop, G. D. Watkins, 1985, ISBN 0-931837-11-1

Volume 47—Thin Films: The Relationship of Structure to Properties, C. R. Aita, K. S. SreeHarsha, 1985, ISBN 0-931837-12-X

Volume 48—Applied Materials Characterization, W. Katz, P. Williams, 1985, ISBN 0-931837-13-8

Volume 49—Materials Issues in Applications of Amorphous Silicon Technology, D. Adler, A. Madan, M. J. Thompson, 1985, ISBN 0-931837-14-6

Volume 50—Scientific Basis for Nuclear Waste Management IX, L. O. Werme, 1986, ISBN 0-931837-15-4

Volume 51—Beam-Solid Interactions and Phase Transformations, H. Kurz, G. L. Olson, J. M. Poate, 1986, ISBN 0-931837-16-2

Volume 52—Rapid Thermal Processing, T. O. Sedgwick, T. E. Seidel, B.-Y. Tsaur, 1986, ISBN 0-931837-17-0

Volume 53—Semiconductor-on-Insulator and Thin Film Transistor Technology, A. Chiang. M. W. Geis, L. Pfeiffer, 1986, ISBN 0-931837-18-9

Volume 54—Thin Films—Interfaces and Phenomena, R. J. Nemanich, P. S. Ho, S. S. Lau, 1986, ISBN 0-931837-19-7

Volume 55—Biomedical Materials, J. M. Williams, M. F. Nichols, W. Zingg, 1986, ISBN 0-931837-20-0

Volume 56—Layered Structures and Epitaxy, J. M. Gibson, G. C. Osbourn, R. M. Tromp, 1986, ISBN 0-931837-21-9

Volume 57—Phase Transitions in Condensed Systems—Experiments and Theory, G. S. Cargill III, F. Spaepen, K.-N. Tu, 1987, ISBN 0-931837-22-7

Volume 58—Rapidly Solidified Alloys and Their Mechanical and Magnetic Properties, B. C. Giessen, D. E. Polk, A. I. Taub, 1986, ISBN 0-931837-23-5

Volume 59—Oxygen, Carbon, Hydrogen, and Nitrogen in Crystalline Silicon, J. C. Mikkelsen, Jr., S. J. Pearton, J. W. Corbett, S. J. Pennycook, 1986, ISBN 0-931837-24-3

Volume 60—Defect Properties and Processing of High-Technology Nonmetallic Materials, Y. Chen, W. D. Kingery, R. J. Stokes, 1986, ISBN 0-931837-25-1

Volume 61—Defects in Glasses, F. L. Galeener, D. L. Griscom, M. J. Weber, 1986, ISBN 0-931837-26-X

Volume 62—Materials Problem Solving with the Transmission Electron Microscope, L. W. Hobbs, K. H. Westmacott, D. B. Williams, 1986, ISBN 0-931837-27-8

Volume 63—Computer-Based Microscopic Description of the Structure and Properties of Materials, J. Broughton, W. Krakow, S. T. Pantelides, 1986, ISBN 0-931837-28-6

Volume 64—Cement-Based Composites: Strain Rate Effects on Fracture, S. Mindess, S. P. Shah, 1986, ISBN 0-931837-29-4

Volume 65—Fly Ash and Coal Conversion By-Products: Characterization, Utilization and Disposal II, G. J. McCarthy, F. P. Glasser, D. M. Roy, 1986, ISBN 0-931837-30-8

Volume 66—Frontiers in Materials Education, L. W. Hobbs, G. L. Liedl, 1986, ISBN 0-931837-31-6

Volume 67—Heteroepitaxy on Silicon, J. C. C. Fan, J. M. Poate, 1986, ISBN 0-931837-33-2

Volume 68—Plasma Processing, J. W. Coburn, R. A. Gottscho, D. W. Hess, 1986, ISBN 0-931837-34-0

Volume 69—Materials Characterization, N. W. Cheung, M.-A. Nicolet, 1986, ISBN 0-931837-35-9

Volume 70—Materials Issues in Amorphous-Semiconductor Technology, D. Adler, Y. Hamakawa, A. Madan, 1986, ISBN 0-931837-36-7

Volume 71—Materials Issues in Silicon Integrated Circuit Processing, M. Wittmer, J. Stimmell, M. Strathman, 1986, ISBN 0-931837-37-5

Volume 72—Electronic Packaging Materials Science II, K. A. Jackson, R. C. Pohanka, D. R. Uhlmann, D. R. Ulrich, 1986, ISBN 0-931837-38-3

Volume 73—Better Ceramics Through Chemistry II, C. J. Brinker, D. E. Clark, D. R. Ulrich, 1986, ISBN 0-931837-39-1

Volume 74—Beam-Solid Interactions and Transient Processes, M. O. Thompson, S. T. Picraux, J. S. Williams, 1987, ISBN 0-931837-40-5

Volume 75—Photon, Beam and Plasma Stimulated Chemical Processes at Surfaces, V. M. Donnelly, I. P. Herman, M. Hirose, 1987, ISBN 0-931837-41-3

Volume 76—Science and Technology of Microfabrication, R. E. Howard, E. L. Hu, S. Namba, S. Pang, 1987, ISBN 0-931837-42-1

Volume 77—Interfaces, Superlattices, and Thin Films, J. D. Dow, I. K. Schuller, 1987, ISBN 0-931837-56-1

Volume 78—Advances in Structural Ceramics, P. F. Becher, M. V. Swain, S. Sōmiya, 1987, ISBN 0-931837-43-X

Volume 79—Scattering, Deformation and Fracture in Polymers, G. D. Wignall, B. Crist, T. P. Russell, E. L. Thomas, 1987, ISBN 0-931837-44-8

Volume 80—Science and Technology of Rapidly Quenched Alloys, M. Tenhover, W. L. Johnson, L. E. Tanner, 1987, ISBN 0-931837-45-6

Volume 81—High-Temperature Ordered Intermetallic Alloys, II, N. S. Stoloff, C. C. Koch, C. T. Liu, O. Izumi, 1987, ISBN 0-931837-46-4

Volume 82—Characterization of Defects in Materials, R. W. Siegel, J. R. Weertman, R. Sinclair, 1987, ISBN 0-931837-47-2

Volume 83—Physical and Chemical Properties of Thin Metal Overlayers and Alloy Surfaces, D. M. Zehner, D. W. Goodman, 1987, ISBN 0-931837-48-0

Volume 84—Scientific Basis for Nuclear Waste Management X, J. K. Bates, W. B. Seefeldt, 1987, ISBN 0-931837-49-9

Volume 85—Microstructural Development During the Hydration of Cement, L. Struble, P. Brown, 1987, ISBN 0-931837-50-2

Volume 86—Fly Ash and Coal Conversion By-Products Characterization, Utilization and Disposal III, G. J. McCarthy, F. P. Glasser, D. M. Roy, S. Diamond, 1987, ISBN 0-931837-51-0

Volume 87—Materials Processing in the Reduced Gravity Environment of Space, R. H. Doremus, P. C. Nordine, 1987, ISBN 0-931837-52-9

Volume 88—Optical Fiber Materials and Properties, S. R. Nagel, J. W. Fleming, G. Sigel, D. A. Thompson, 1987, ISBN 0-931837-53-7

Volume 89—Diluted Magnetic (Semimagnetic) Semiconductors, R. L. Aggarwal, J. K. Furdyna, S. von Molnar, 1987, ISBN 0-931837-54-5

Volume 90—Materials for Infrared Detectors and Sources, R. F. C. Farrow, J. F. Schetzina, J. T. Cheung, 1987, ISBN 0-931837-55-3

Volume 91—Heteroepitaxy on Silicon II, J. C. C. Fan, J. M. Phillips, B.-Y. Tsaur, 1987, ISBN 0-931837-58-8

Volume 92—Rapid Thermal Processing of Electronic Materials, S. R. Wilson, R. A. Powell, D. E. Davies, 1987, ISBN 0-931837-59-6

Volume 93—Materials Modification and Growth Using Ion Beams, U. Gibson, A. E. White, P. P. Pronko, 1987, ISBN 0-931837-60-X

Volume 94—Initial Stages of Epitaxial Growth, R. Hull, J. M. Gibson, David A. Smith, 1987, ISBN 0-931837-61-8

Volume 95—Amorphous Silicon Semiconductors—Pure and Hydrogenated, A. Madan, M. Thompson, D. Adler, Y. Hamakawa, 1987, ISBN 0-931837-62-6

Volume 96—Permanent Magnet Materials, S. G. Sankar, J. F. Herbst, N. C. Koon, 1987, ISBN 0-931837-63-4

Volume 97—Novel Refractory Semiconductors, D. Emin, T. Aselage, C. Wood, 1987, ISBN 0-931837-64-2

Volume 98—Plasma Processing and Synthesis of Materials, D. Apelian, J. Szekely, 1987, ISBN 0-931837-65-0

Volume 123—Materials Issues in Art and Archaeology, E.V. Sayre, P. Vandiver, J. Druzik, C. Stevenson, 1988, ISBN: 0-931837-93-6

Volume 124—Microwave-Processing of Materials, M.H. Brooks, I.J. Chabinsky, W.H. Sutton, 1988, ISBN: 0-931837-94-4

Volume 125—Materials Stability and Environmental Degradation, A. Barkatt, L.R. Smith, E. Verink, 1988, ISBN: 0-931837-95-2

Volume 126—Advanced Surface Processes for Optoelectronics, S. Bernasek, T. Venkatesan, H. Temkin, 1988, ISBN: 0-931837-96-0

Volume 127—Scientific Basis for Nuclear Waste Management XII, W. Lutze, R.C. Ewing, 1989, ISBN: 0-931837-97-9

Volume 128—Processing and Characterization of Materials Using Ion Beams, L.E. Rehn, J. Greene, F.A. Smidt, 1989, ISBN: 1-55899-001-1

Volume 129—Laser and Particle-Beam Chemical Processes on Surfaces, G.L. Loper, A.W. Johnson, T.W. Sigmon, 1989, ISBN: 1-55899-002-X

Volume 130—Thin Films: Stresses and Mechanical Properties, J.C. Bravman, W.D. Nix, D.M. Barnett, D.A. Smith, 1989, ISBN: 0-55899-003-8

Volume 131—Chemical Perspectives of Microelectronic Materials, M.E. Gross, J. Jasinski, J.T. Yates, Jr., 1989, ISBN: 0-55899-004-6

Volume 132—Multicomponent Ultrafine Microstructures, B.H. Kear, D.E. Polk, L.E. McCandlish, 1989, ISBN: 1-55899-005-4

Volume 133—High Temperature Ordered Intermetallic Alloys III, C.T. Liu, A.I. Taub, N.S. Stoloff, C.C. Koch, 1989, ISBN: 1-55899-006-2

Volume 134—The Materials Science and Engineering of Rigid-Rod Polymers, W.W. Adams, R.K. Eby, D.E. McLemore, 1989, ISBN: 1-55899-007-0

Volume 135—Solid State Ionics, G. Nazri, R.A. Huggins, D.F. Shriver, 1989, ISBN: 1-55899-008-9

Volume 136—Fly Ash and Coal Conversion By-Products: Characterization, Utilization, and Disposal V, R.T. Hemmings, E.E. Berry, G.J. McCarthy, F.P. Glasser, 1989, ISBN: 1-55899-009-7

Volume 137—Pore Structure and Permeability of Cementitious Materials, L.R. Roberts, J.P. Skalny, 1989, ISBN: 1-55899-010-0

Volume 138—Characterization of the Structure and Chemistry of Defects in Materials, B.C. Larson, M. Ruhle, D.N. Seidman, 1989, ISBN: 1-55899-011-9

Volume 139—High Resolution Microscopy of Materials, W. Krakow, F.A. Ponce, D.J. Smith, 1989, ISBN: 1-55899-012-7

Volume 140—New Materials Approaches to Tribology: Theory and Applications, L.E. Pope, L. Fehrenbacher, W.O. Winer, 1989, ISBN: 1-55899-013-5

Volume 141—Atomic Scale Calculations in Materials Science, J. Tersoff, D. Vanderbilt, V. Vitek, 1989, ISBN: 1-55899-014-3

Volume 142—Nondestructive Monitoring of Materials Properties, J. Holbrook, J. Bussiere, 1989, ISBN: 1-55899-015-1

Volume 143—Synchrotron Radiation in Materials Research, R. Clarke, J.H. Weaver, J. Gland, 1989, ISBN: 1-55899-016-X

Volume 144—Advances in Materials, Processing and Devices in III-V Compound Semiconductors, D.K. Sadana, L. Eastman, R. Dupuis, 1989, ISBN: 1-55899-017-8

Tungsten and Other Refractory Metals for VLSI Applications, R. S. Blewer, 1986; ISSN 0886-7860; ISBN 0-931837-32-4

Tungsten and Other Refractory Metals for VLSI Applications II, E.K. Broadbent, 1987; ISSN 0886-7860; ISBN 0-931837-66-9

Ternary and Multinary Compounds, S. Deb, A. Zunger, 1987; ISBN 0-931837-57-x

Tungsten and Other Refractory Metals for VLSI Applications III, Victor A. Wells, 1988; ISSN 0886-7860; ISBN 0-931837-84-7

Atomic and Molecular Processing of Electronic and Ceramic Materials: Preparation, Characterization and Properties, Ilhan A. Aksay, Gary L. McVay, Thomas G. Stoebe, 1988; ISBN 0-931837-85-5

Materials Futures: Strategies and Opportunities, R. Byron Pipes, U.S. Organizing Committee, Rune Lagneborg, Swedish Organizing Committee, 1988; ISBN 0-55899-000-3

Tungsten and Other Refractory Metals for VLSI Applications IV, Robert S. Blewer, Carol M. McConica, 1989; ISSN: 0886-7860; ISBN: 0-931837-98-7

PART I

Overviews: Applications to Manufacturing, New Techniques

THE ROLE OF NDE IN GLOBAL STRATEGIES
FOR MATERIALS SYNTHESIS AND MANUFACTURING

D. O. THOMPSON AND T. A. GRAY
Center for NDE, 303 Wilhelm Hall, Iowa State University, Ames, IA 50010

INTRODUCTION

During the past several years a number of design-centered "global" approaches have been offered that deal both with materials synthesis and with the manufacture of materials into finished products. Although the structure and constituent components of these approaches differ because of their different end purposes, they share major dependencies upon design, theoretical modeling, extensive computations, and confirming measurements using various NDE techniques. Materials-by-Design (MBD) is an example of these approaches that is focused on the synthesis of materials. As noted by Eberhard[1], a principal purpose of Materials-by-Design is to produce materials with prescribed macroscopic material properties by designing and controlling material structures at the atomic and molecular levels. At the other end of the spectrum, Unified Life Cycle Engineering (ULCE)[2] is an example of a "global" model for manufacturing. In this case, emphasis is placed upon the development of ways to predict the total set of important properties of a product--performance, quality, reliability, maintainability, and life cycle costs--at the designer's board. Taken together, these approaches offer the opportunity for designer-controlled materials with specified material properties to be fabricated into components of specified performance, quality, reliability, and cost. Even though this combination represents an idealistic vision which may never be perfectly realized, the potential payoff is so large that even imperfect realization may be worth a significant investment. Efforts in these directions can now be made because of the convergence of theoretical, instrumental, and computational techniques.

It is the purpose of this paper to describe some developing concepts in quantitative NDE and provide some examples of their potential use in the evolution of these global approaches. Due to limitations of space and time, emphasis in this paper will be placed primarily upon the role of NDE in the development of the ULCE concept; the conceptual extension of the role of NDE to Materials-by-Design concepts is straightforward.

UNIFIED LIFE CYCLE ENGINEERING (ULCE) AND PROBABILITY OF DETECTION (POD)

It is useful to refer to Fig. 1 in order to describe the ULCE concept and to examine the role of NDE in it. This figure shows a schematic diagram of a possible model of an integrated manufacturing system that is design centered and that contains the kinds of engineering functions commonly associated with a high-technology manufacturing enterprise. These functions are shown in circles and are connected by various links. The solid link connecting the Design and Forming Operations functions represents current CAD/CAM technology, i.e., the technology that has been developed in recent years that enables the designer to communicate and to interact with material forming operations (e.g. machining, casting, forging, etc.). The dotted lines indicate analogous, but

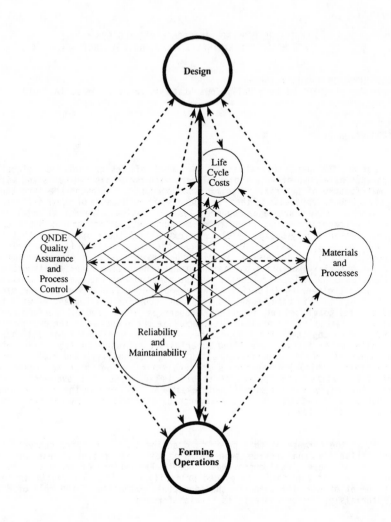

Fig. 1 Schematic representation of possible linkages needed for Unified Life Cycle Engineering (ULCE).

as yet, non-existent links that would permit the designer to interact equally well during preparation of the design with other engineering functions of the manufacturing operations, i.e., functions such as those listed in the other circles. These functions, which have a support character, are necessary in order to realize the ULCE goals. Activation of all these links would permit the designer to be interactively and simultaneously coupled with all the engineering functions in a way that is comparable to the current CAD/CAM link. Successful development and operation of this network would then allow the designer to incorporate all trade-off considerations at the time of product design, would reduce the need for design retrofitting, and would undoubtedly increase the efficiency of the manufacturing process remarkably. Successful development of the ULCE concept could provide a new paradigm in manufacturing.

PROBABILITY OF DETECTION (POD)

With the above model of a ULCE factory in mind, it is necessary to examine approaches that are available to develop the linkages shown in Fig. 1. In this paper, only the QNDE/Design linkage will be considered. Clearly, the approach needed to develop and implement this coupling is vastly different and more advanced than any encountered in current NDE practice. For example, the QNDE/Design link needs to provide the designer with several pieces of quantitative information related to the design. These include:

1. A figure-of-merit that can be calculated at each spatial point of a design that quantifies its inspectability for "critical" flaws (or values of material properties) and for various QNDE measurement techniques.

2. Feedback information that will guide the designer in altering design characteristics (e.g. shape, size, material, etc.) as needed in order to improve inspectability while simultaneously meeting other design constraints.

3. Ways to calculate a component scan plan for automated production inspections that will assure the realization of calculated design inspectabilities.

A logical approach for the development of the QNDE/Design link is based upon the probability of detection (POD) concept. This concept possesses the necessary features to fulfill the above requirements. By way of background, it is a broadly used NDE measure which measures the probability that a specified flaw will be found in a given sample using a specific inspection technique. Figure 2 shows an idealized POD curve in which the probability of a flaw's detection is shown plotted as a function of flaw size for both a real and an ideal inspection technique. For the ideal technique the POD of flaws smaller than a critical size is zero whereas the POD of any flaw greater than a critical size is unity. In this case there are neither false rejects (FR) of good parts nor false accepts (FA) of defective ones. However, real NDE techniques are seldom, if ever, as sharp and as discriminatory as that indicated by the ideal curve, with the result that there are regions of uncertainty shown by the false reject and false accept areas. Without going into detail, it is evident that various features related to these uncertain regions that are essentially defined by the quality of the QNDE inspection technology include information that can be utilized in the development of other linkages shown in Fig. 1.

To date essentially all applications of the above-described POD concept have been empirical, i.e., a statistically significant number of samples is prepared with artificial flaws, and then experiments are made by various operators utilizing specific NDE techniques. POD, or confidence level, results are then derived from these data. It is evident that such empirically derived results represent insufficient bases for the development of the QNDE/Designer linkage of Fig. 1, even if coupled with expert systems or other artificial intelligence approaches. First, the empirically determined POD curves represent a "convolution" of operator and instrumental capabilities. It is not possible to isolate these two sets of variables on the basis of empirical results only; hence, the degree to which any empirically determined POD curve represents the true POD determined by physical principles only (part shape, materials, details of measurement system, etc) is unknown. Secondly, empirically determined POD's cannot be used, with confidence, to predict POD values for other sets of measurement conditions. For example, there is no way to predict a POD at a given location in a part from an empirically determined POD measured at a different location with different part geometries. Without this predictive property, the utilization of POD during the evaluation of a design is impossible without readily available, extensive empirical histories of previous cases

Critical Flaw Size

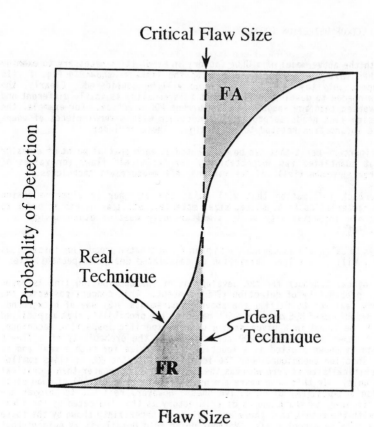

Fig. 2   Probability of detection (POD) curve as function of flaw size for both
ideal and real NDE techniques.

perhaps combined in an appropriate expert system. Even if this were possible
technically, it would more than likely be economically prohibitive to develop
such an inventory.

THEORETICAL MODELS OF POD AND APPLICATION TO DESIGN DECISIONS

In recent years, early versions of theoretical models that permit calculations of POD's to be made have appeared for three major QNDE
technologies--ultrasonics[3],   eddy   currents[4,5],   and   microfocus
radiography[6]. In contrast to empirically determined POD's, these are first
principle engineering models that can be used as a basis for the QNDE/Designer
linkage shown in Fig. 1. Without going into detail here, these first principle
models are analytical models of the QNDE measurement process and depend upon
details of the measurement setup for each inspection technology. For example,

the details include the geometry of the component being inspected, relative inspection configuration of probe and part, characterization of the generation, propagation, and reception of the interrogating energy (e.g. in the ultrasonic case, this characterization depends upon knowledge of the transducer radiation pattern, refraction of the beam at the part's surface, beam propagation characteristics in the host material including material anisotropy, attenuation, diffraction losses, etc.), critical flaw information that is obtainable from materials engineering, detailed models of field-flaw interactions from which flaw responses can be calculated for a known interrogating field, and finally a knowledge of noise conditions that adds uncertainty to the measurement results. Thus, the first-principle engineering POD models require the kind of fundamental results, e.g. ultrasonic scattering models, that have been obtained in various QNDE research efforts over the past 10-12 years.

Figure 3 shows an example of a POD calculation for μ-focus radiography using a film detector and for three different accelerating voltages in which the expected POD curve shape is realized[6]. In this case, the POD model consists of five parts that describe the generation of the x-ray beam from specific machine geometries, the energy-dependent interactions of the beam with the sample, the experimental configuration, details of the detector, and a description of the detectability criteria. They are based on eqns. (1) and (2), i.e.

$$I(x,y,E) = I_o(E) \int_{source} \frac{e^{-\mu(x,y,E)\cdot\rho} dA}{r^2(x,y)} \qquad (1)$$

$$D(E) = D_o(1 - e^{-\sigma[(1+\eta)I(E)t+\delta]}) \qquad (2)$$

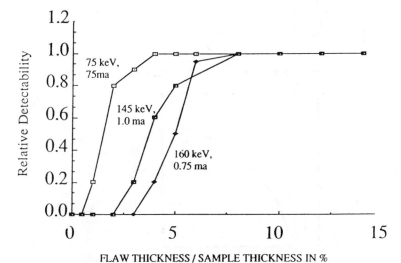

FLAW THICKNESS / SAMPLE THICKNESS IN %

Fig. 3   Theoretical POD curve for μ-focus radiography.

in which I is the intensity immediately above the detector, I_0 is the initial intensity produced by the x-ray generator, μ is the energy dependent linear absorption coefficient, ρ is the x-ray path length through the sample, r is the distance from the source to the detector, and x, y are the coordinates at the detector surface. In eqn (2), D is the film density, σ is the interaction cross section of an x-ray with a film grain, η is the coefficient of the x-ray scattering, δ is the natural film fog density, and D_0 is the maximum film density. As can be seen, the sensitivity drops as the hardness of the beam increases, a well known result.

POD modeling for ultrasonics is somewhat more advanced than the previously described x-ray case. Limited applications can now be shown for simple cases that demonstrate the way in which the models can be used to develop the QNDE/Designer link and which show that the quantitative information required by a designer can be obtained. Figure 4 shows a case in which a POD calculation

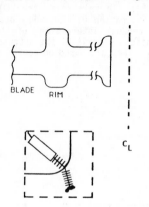

MATERIAL: IN100
GRAIN SIZE: 50μm
FILLET RADIUS: 0.635 cm
WATER PATH: 1.27 cm
FREQUENCY: 10 MHz
PROBE RADIUS: 0.3175 cm
PROBE FOCAL LENGTH: 1.27 cm
REFRACTED ANGLE 45° SHEAR
SCAN: IN-PLANE: 0.025 cm
            OUT-OF-PLANE: 0.025 cm

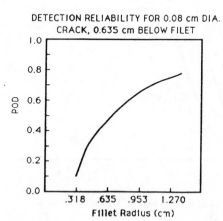

Fig. 4   Suggested design improvement using POD model for ultrasonics.

has been combined with a critical flaw specification to show how a design could be improved to enhance inspectability[3]. In this case, the component is a turbine engine blade as shown in the upper left hand corner of the diagram, and immediately below it, the particular area of interest (the fillet region of the blade including the location of a critical flaw) is shown in a magnified form. The inspection geometry is implied in this magnified section, and details of the ultrasonics and blade materials are given in the upper right. Results of POD calculations for this inspection geometry, material, and critical flaw are given in the lower part of the figure as a function of the fillet radius. It can be seen that, for the postulated inspection capabilities, the POD is less than 0.5 at the design fillet radius of 0.635 mm, but that it improves as the fillet radius is increased. The designer could obviously use this information to select a final fillet radius that would be both inspectable and compatible with other requirements. Most importantly, this information is available at the design stage before any final commitments of the design are made and before any added-value is accrued. Such information also provides a research guide for improved inspection capabilities. Design options could be extended with new variations in the inspection methodologies.

Figure 5 demonstrates another important capability of the POD model that can be used by the designer. In this case, the model is used to pre-determine the parameters of an automated scan plan that can be used to assure the designed-in inspectability[3]. An assumed circular part shape with a complex cross section is shown in the upper left hand corner together with the material and ultrasonic specifications shown in the upper right. The bottom part of the figure shows POD calculations as a function of flaw (crack) radius placed at 3 different depths in the part (cf code at bottom of figure). It is seen that scan plan #1 with its set of scan parameters produces a widely divergent set of POD results for the 3 different flaw depths whereas scan plan #2 with its selected scan parameters produces a uniform POD pattern for the same 3 different flaw depths. The designer could thus extract information from these calculations that can be used as inspection specifications for automated inspections.

Figures 6-10 show a sequence of preliminary results in which the ultrasonic POD models for both inspectability and optimal scan plan have been combined with other tools of the designer[7] to produce a design that is optimized for inspectability and thereby reliability. Figure 6 shows a perspective view of an axially symmetric component, a disk, that was produced using a wireframe and solid model design technique. As indicated in the figure, the fillet region is a region of particular interest for inspectability. Figure 7 shows schematically the next step in the analysis that brings in materials, fracture mechanics, and performance information. In this step, a finite element mesh is incorporated into the initial design, and stress analyses consistent with design performance profiles are then made. From this information together with known material properties, critical flaw information is developed. Critical flaws are then placed upon the mesh nodal points, and POD values are then calculated at the nodal points for the critical flaws assuming specific ultrasonic inspection parameters. The results of the POD calculations for two different fillet radius design values and scan inspection parameters are shown in Figs. 8-10. Values of the POD in these figures range from 0 to 1, the lightest end of the grey scale being 0 and the darkest end corresponding to a unity POD. Legends in the figures give the appropriate parametric values. The scan mesh (Δx) is a scan parameter and is measured radially along the disk diameter, and r is the fillet radius. It is assumed in these calculations that the axis of the inspecting transducer remains normal to the fillet surface while inspecting through the fillet region. It is evident, as shown in Fig. 10, that significant improvements in inspectability, and hence component

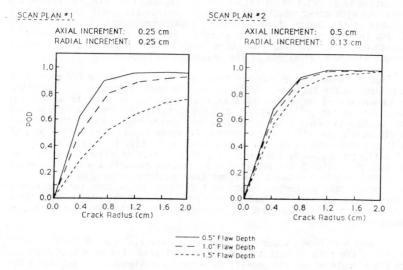

Fig. 5 Application of POD model to specification of automated inspection plan.

reliability, can be made at the design level while requiring only fairly insignificant changes in the design itself and in the specification of inspection routines.

SUMMARY

The development and utilization of computationally intensive "global" models to predict the properties of materials and manufactured products would appear to offer new paradigms in materials and product designs. In the first case, materials would be designed on the atomic level to produce desirable macroscopic properties, while in the second, products would be interactively and quantitatively designed for performance quality, reliability, maintainability, and estimates of life cycle costs. The realization of these new paradigms requires substantial new tools in QNDE measurements that can be applied both to material property and flaw measurements. One of the most promising approaches to provide the designer with key quantitative information in the manufacturing and processing case is the theoretical POD models that are becoming available for various NDE techniques. These models are no more and no less than quantitative physical models of the measurement process evaluated for specific design problems. When utilized and combined with other tools available to a designer, they form a key link in the development of the ULCE concept.

<u>Wireframe Design Model of Disk</u>

- Solid Model contains geometric and material properties of component

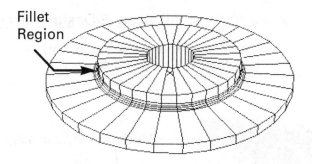

Fillet
Region

Perspective View

Fig. 6   Perspective of disk generated using wireframe and solid model techniques.

Finite Element Model

- Nodes provide points in object
  at which POD can be calculated

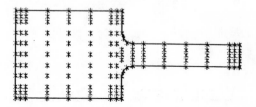

Fig. 7   Specification of finite element mesh in cross section of disk (cf.
Fig. 6)

POD Contours - "Nominal" Scan & Design Parameters

- Course scan mesh (Δx = 0.25 cm)
- Tight fillet radius (r = 0.50 cm)

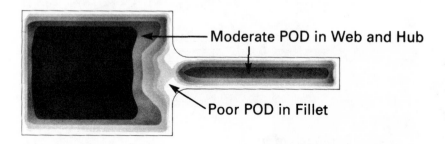

Fig. 8   Cross-section of disk showing POD contours for "Nominal" Scan and
Design Parameters for a specific set of ultrasonic tools. Note poor
POD in fillet region

POD Contours - "Optimal" Scan, "Nominal" Design

- Fine scan mesh ($\Delta x = 0.10$ cm)
- Tight fillet radius (r = 0.50 cm)

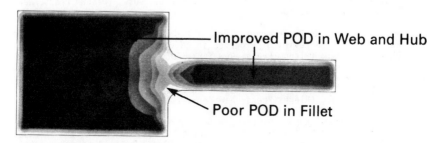

Fig. 9   Cross-section of disk showing improved ultrasonic POD contours that result from a scan mesh that is smaller than that used in Fig. 8.

POD Contours - "Optimal" Scan & Design Parameters

- Fine scan mesh ($\Delta x = 0.10$ cm)
- Moderate fillet radius (r = 1.0 cm)

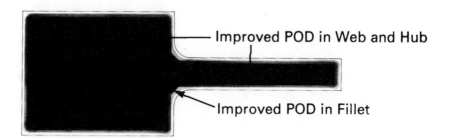

Fig. 10   Cross-section of disk showing nearly uniform ultrasonic POD contours that result from the smaller scan mesh (cf. Fig. 9) and an increased fillet radius (cf. Figs. 8 and 9).

REFERENCES

1. J. J. Eberhard, "Materials-by-Design," ECUT Program Bulletin, June, 1988, USDOE.

2. H. Burte and D. Chimenti, "Unified Life Cycle Engineering: An Emerging Design Concept," in Review of Progress in Quantitative Nondestructive Evaluation 6, D. O. Thompson and D. E. Chimenti, eds., (Plenum Press, New York, 1987), p 1797-1809.

3. T. A. Gray and R. B. Thompson, "Use of Models to Predict Ultrasonic NDE Reliability," in Review of Progress in QNDE 5, D. O. Thompson and D. E. Chimenti, eds., (Plenum Press, New York, 1986), p 911-918.

4. A. J. Bahr and D. W. Cooley, "Analysis and Design of Eddy Current Measurement Systems," in Review of Progress in QNDE 2, D. O. Thompson and D. E. chimenti, eds., (Plenum Press, New York, 1983), p 225-244.

5. R. E. Beissner, K. A. Bartels, and J. L. Fisher, "Prediction of the Probability of Eddy Current Flaw Detection," in Review of Progress in QNDE 7, D. O. Thompson and D. E. Chimenti, eds., (Plenum Press, New York, 1988), p 1753-1760.

6. J. N. Gray and F. Inanc, "Three Dimensional Modeling of Projection X-Ray Radiography," in Review of Progress in QNDE 8, D. O. Thompson and D. E. Chimenti, eds., (Plenum Press, New York, in press).

7. CAD Simulations were performed using I-DEAS from Structural Dynamics Research Corp.

NONDESTRUCTIVE EVALUATION FOR MATERIALS CHARACTERIZATION

ROBERT E. GREEN, JR.
Center for Nondestructive Evaluation
The Johns Hopkins University, Baltimore, MD 21218

ABSTRACT

In recent years classical nondestructive testing techniques for detecting macroscopic defects have been augmented by more sophisticated nondestructive evaluation methods for characterizing the microstructure and associated physical and chemical properties of materials. This paper will briefly describe several such nondestructive evaluation methods developed in the Center for Nondestructive Evaluation (CNDE) at The Johns Hopkins University.

MODERN NONDESTRUCTIVE EVALUATION

Traditionally the vast majority of material property characterization techniques have been destructive, e.g. chemical compositional analysis, metallographic determination of microstructure, tensile test measurement of mechanical properties, etc. The present unfavorable status of a large segment of American industry, coupled with premature failure of structures and devices, show that our traditional approaches must be dramatically modified if we are to be able to meet future needs. The development of high quality low porosity ceramic materials capable of performing reliably at high temperatures, the optimization of metal-matrix and polymeric composites, and the improvement of metallic superalloys can only result from proper application of nondestructive materials characterization techniques to monitor and control as many stages in the production process as possible.

THE JOHNS HOPKINS UNIVERSITY
CENTER FOR NONDESTRUCTIVE EVALUATION

In order to address this expanded role for NDE The Johns Hopkins University Center for Nondestructive Evaluation (CNDE) was established to exploit and integrate the resources and talent of the School of Engineering, the Applied Physics Laboratory, the School of Medicine, and the School of Arts and Sciences. The CNDE is dedicated to research which will lead to more sophisticated and accurate methods for the nondestructive evaluation of materials and systems and to the education of talented students who will enter the NDE field [1]. Brief descriptions of several of the research efforts underway in the CNDE as applicable to nondestructive evaluation for materials characterization follow.

ULTRASONIC CORRELATION

Unfortunately, the conventional ultrasonic techniques which have proven successful with relatively homogeneous, isotropic metals cannot penetrate thick composite materials which are both inhomogeneous and anisotropic. Therefore, a

Mat. Res. Soc. Symp. Proc. Vol. 142. ⌐1989 Materials Research Society

new ultrasonic correlation technique was developed in order
to overcome the problem of acoustic energy loss in these
materials [2]. In the correlation technique a reference
signal is multiplied by a delayed signal and the result is
integrated. Only the frequency and/or phase components
common to both waveforms are contained in the resulting
output. Utilizing the correlation scheme in conjunction
with broadband ultrasonic transducers, a method has been
developed that can acquire time-of-flight, frequency, and
phase response of relatively thick sections of highly
attenuating materials (Fig. 1). Test are also being
conducted on solid rocket motors from NASA to see if
delaminations, which occur inside the casing behind highly
attenuating insulators, can be detected.

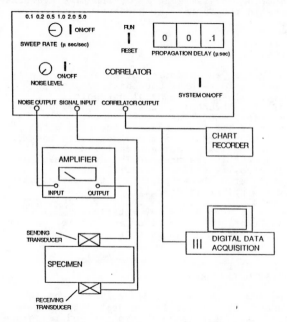

Figure 1. Schematic of ultrasonic correlator experimental
arrangement (after Lindgren [2]).

LASER GENERATION/DETECTION OF ULTRASOUND

A major problem associated with the conventional use of
piezoelectric transducers is the requirement that they must
be acoustically bonded to the test material with an
acoustical impedance matching coupling medium such as water,
oil, or grease. In addition to modification of the
ultrasonic signal and potential harm to the test structure
by the coupling medium, the requirement of physical contact
between transducer and test structure places limitations on
ultrasonic testing in structural configurations which
possess geometries with difficult to reach areas, as well as
on testing materials at elevated temperatures or in the
environment of outer space.

A number of techniques are available for non-contact generation and detection of ultrasound, including capacitive pick-ups and electromagnetic acoustic transducers (EMAT'S). However, only laser beam probes afford the opportunity to make truly non-contact ultrasonic measurements in both electrically conducting and non-conducting materials, in materials at elevated temperatures, in corrosive and other hostile environments, in geometrically difficult to reachlocations, in outer space, and do all of this at relatively large distances, i.e. meters, from the test structure surface. Incorporation of scanning techniques or full-field imaging greatly increase the capability of testing large structures without the present necessity of either immersing the test object in a water tank of using water squirter coupling. Figure 2 shows a schematic of a system which uses a pulsed laser to generate ultrasonic waves at the surface of a slab of steel and a laser beam optical interferometer for detection of the reflected waves. The pulsed laser beams are delivered to the metal surface through a linear array of optical fibers and the reflected probe laser beams are detected in the same fashion. An optical multiplexer permits a single pulsed laser to be used for generation and a single continuous wave laser to be used for detection [3].

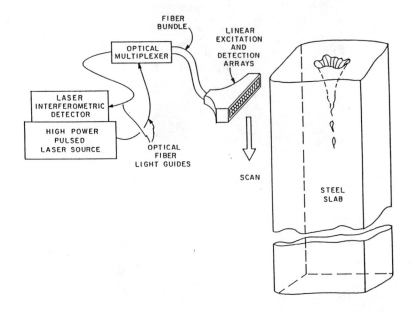

Figure 2. Schematic of a non-contact ultrasonic scanning system using laser generation and laser detection incorporating fiber optics (after Rosen and Green [3]).

Figure 3 shows the results of a recently developed heterodyne holographic interferometry technique which permits full field imaging of surface displacements due to ultrasonic wave propagation [4]. A laser pulse was used to record a holographic image of a graphite/epoxy composite plate. Subsequently, a second laser pulse was incident on the center (opposite side than shown in figure) of the plate and after sufficient time for the resulting ultrasonic wave to travel to the opposite surface of the plate (side shown in figure) a third laser pulse was used to re-expose the holographic plate. The resulting interference pattern shows the ultrasonic wavefront traveling outward from the source with the influence of the anisotropic character of the plate clearly evident. An alteration in the holographic image is also evident in the left lobe of the figure probably caused by a delamination between one of the internal composite layers.

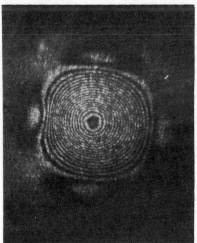

Figure 3. Heterodyne holographic interferometric full-field imaging of ultrasonic waves on surface of graphite/epoxy composite (after Wagner [4]).

OPTICAL ACOUSTIC EMISSION DETECTION

The importance of acoustic emission monitoring is that proper detection and analysis of acoustic emission signals can permit remote identification of source mechanisms and the associated microstructural alteration of the material. Several non-contact optical interferometric detectors of acoustic emission signals were developed to record emissions from a variety of metal specimens during tensile elongation. [5] The dimensions of the specimens, particularly the gauge section, were chosen to be very small so that any microstructural alteration would be visible on the gauge section surface and could be examined in detail using both optical and scanning electron microscopy. The small laser beam from the optical probe was focused on the tensile specimen just above the gauge section. Because of the flat large frequency bandwidth (0 - 60MHz) of the optical probe acoustic emission signals were obtained in a frequency regime not normally detected with conventional transducers. Tests run on a series of stainless steel specimens revealed a large number of acoustic emission signals at 9 - 10 MHz prior to fracture. Scanning electron microscopic examination of the fracture surface resulted in a one-to-one correspondence between these high frequency signals and the fracture of intermetallic particles included in the steel.

THERMAL- AND ELECTRON-ACOUSTIC IMAGING

As initially developed thermal-wave imaging, called photoacoustic microscopy or photothermal imaging, used laser beam scanning of a test object placed in a closed gas-filled container to cause changes in the gas pressure in direct proportion to thermal property changes in the surface layers of the test object. More recent developments have permitted elimination of the gas-filled container by use of a second probe laser beam which either detects surface displacements of the test object due to localized thermal expansion or changes in the refractive index of the air just above the sample surface. Thermal-acoustic imaging describes a modification where the laser beam introduces heat locally and periodically onto one surface of a specimen and the elastic wave resulting from local thermoelastic expansion of the sample is measured, usually with a piezoelectric transducer coupled to the other surface of the specimen.

Another modification uses a chopped electron beam in a scanning electron microscope to excite elastic and thermal waves at the top surface of a test object. These waves are detected by either a piezoelectric transducer coupled to the bottom surface of the test object or an optical interferometric displacement probe. By displaying and recording the output of the detector as a function of position of the scanning electron beam an "electron-acoustic" image of the test object can be obtained. Alteration of the energy of the electron beam or chopping frequency permits different layers in the test object to be imaged as desired. Figure 4, which shows four images of the same portion of an integrated circuit, serves to illustrate this [6]. Figure 4(a) is the conventional scanning electron microscope image recorded with a 5 keV electron beam and Fig. 4(b) the electron-acoustic image recorded simultaneously. Both of these images show the surface layer of the integrated circuit. Figure 4(c) is the conventional scanning electron microscope image recorded with a 30 keV electron beam and Fig. 4(d) the electron-acoustic image recorded simultaneously. Note that although this higher voltage scanning electron microscope image only shows more of the surface layer of the integrated circuit, the electron-acoustic image reveals details of the second layer of the circuit. These emerging imaging techniques will have a marked impact on nondestructive material property characterization.

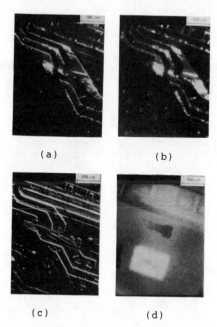

(a)   (b)

(c)   (d)

Figure 4. SEM and electron-acoustic images of integrated circuit (after Murphy et al. [6]):
  5keV (a) SEM, (b) electron-acoustic
  30 keV (c) SEM, (d) electron-acoustic

ELECTRON PARAMAGNETIC RESONANCE IMAGING

Magnetic resonance imaging techniques, although well established in biology and medicine, are being explored as possible nondestructive material property characterization techniques for structural and electronic materials. Researchers in the CNDE have found that there is a great deal of promise in investigating the electron paramagnetic resonance (EPR) that arises when a polymeric solid is damaged. The fact that the EPR signals can be enhanced and controlled by adding dopants to the polymers opens up the possiblity of solving problems impossible by other means. When polymer based composite materials are damaged, for example by impact, the extent of damage or even its existence often cannot be accessed by examination of the outer surface. By doping each layer of a polymer based composite prior to lay-up with a different dopant, subsequent damage can be quantitatively determined as a function of depth into the composite by means of scanning EPR. Figure 5 shows an example of an EPR line scan across a polymer composite which had organometallic dopant applied at four positions along the line as indicated in the upper left inset of the figure. Note that the EPR technique is also sensitive to the quantity of dopant [7].

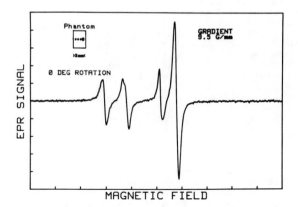

Figure 5. Electron paramagnetic image of dopant enhanced damage sites in polymer composite (after Bryden and Poehler [7]).

MAGNETIC DETECTION OF CORROSION

Research is in progress to develop a novel NDE method for remote non-contact detection and evaluation of corrosion in metallic materials [8]. Measurements on 100 ft. sections of underground gas pipelines, under varied conditions of coatings and soil coverings, have been made using Electrochemical Impedance Spectroscopy (EIS) instrumentation with magnetometer sensing of pipe and earth current. Multiple holidays have been individually analyzed and their impedance evaluated despite the presence of pipe coating impedance. Once optimized this methodology will allow the survey of underground pipelines by surface measurements, locate corroding regions in base pipelines, determine defective regions in coated pipelines including defect size and corrosion rate, and help assure the integrity and safe operation of the pipeline system. The system developed will also prove extremely useful for other applications such as corrosion evaluation of steel reinforcing bars in concrete structures.

RAPID X-RAY DIFFRACTION IMAGING

Electro-optical systems optimized for rapid x-ray diffraction imaging have been used in the CNDE to orient single crystals, to study crystal lattice rotation accompanying plastic deformation, to measure the rate of grain boundary migration during recrystallization annealing of cold-worked metals, to determine the physical state of exploding metals, to monitor the amorphous to crystalline phase transformation of rapidly solidified metals, to rapidly measure residual stress (strain), to study the dynamics of structural phase transitions in ferroelectric crystals, and to record topographic images of lattice defects in quartz, nickel, cadmium and zinc telluride, and gallium arsenide crystals [9-11].

Although these techniques are unique in that they can yield information about the defect structure, down to the size of individual dislocations, throughout the volume of fairly thick crystals, the exposure times required when using conventional x-ray generators and techniques often run to hours or even days for recording a single image. Recent use of beam line X-19C at the National Synchrotron Light Source, Brookhaven National Laboratory, has permitted x-ray topographic film exposure times to be reduced to seconds and replacement of film with state-of-the-art electro-optical detectors has permitted x-ray diffraction images to be recorded continuously at millisecond (television) framing rates.

Quartz Crystal Resonators

If variations in quartz crystal resonator parameters due to manufacturing processes are to be separated from those due to inherent material properties then the defect state of such resonators must be measured at as many stages of the manufacturing process as possible. Synchrotron white beam topographic images have been used to characterize the defect morphology in natural and cultured quartz as obtained from different vendors, bars as thick as one-half inch, surface defects induced by blank manufacturing processes, mounting strains, strains induced by evaporating electrodes, blanks before and after sweeping (electrodiffusion) to reduce level of impurities, artificially induced strains, and mode shapes of oscillating resonators.

Nickel Single Crystal Turbine Blades

The conventional technique for inspection of nickel base alloy single crystal turbine blades is to use the Laue back-reflection x-ray diffraction technique to determine the orientation of one small localized point on the turbine blade at one time. Examination of the entire blade using this point probe method requires an inordinately long time. An asymmetric crystal topographic (ACT) x-ray diffraction technique permits imaging of a large portion of the single crystal blade at one time, while incorporation of an x-ray sensitive electro-optical detector permits this to be done in real-time.

Telluride Infrared Detecting Crystals

Extensive research is being conducted on zinc telluride, cadmium telluride, zinc cadmium telluride, and mercury cadmium telluride crystals motivated by the desirability to develop superior infrared imaging systems. However, the yield of infrared imaging arrays obtained by epitaxial growth is limited by the poor quality of the zinc cadmium telluride crystal substrate. Unresolved material problems with the substrates are impurities, inclusions, dislocation tangles, subgrains, and often even more severe structural defects. The capability of obtaining synchrotron transmission x-ray topographs from relative thick telluride crystals has exhibited the superiority of synchrotron radiation over conventional x-ray sources which can only sample surface layers of such high linear absorption coefficient materials.

Gallium Arsenide Crystals

Among the most important new materials for manufacture of microelectronic devices is gallium arsenide and aluminum gallium arsenide. Characterization techniques for arsenide crystals based on white beam synchrotron x-ray topography possess the capability of real-time full-field imaging of the spatial distribution of dislocation networks over the entire area of the wafer. Topographic images can be made on cut and polished wafers prior to further processing, on wafers both prior to and immediately following ion implantation, and during annealing of ion implanted wafers.

Figure 6 shows a white beam synchrotron transmission x-ray topograph obtained by scanning half of a 3 inch diameter gallium arsenide wafer. Initially a large number of topographic images were recorded simultaneously on 8 x 10 film. In this case, since the gallium arsenide wafer was much larger than the size of the incident x-ray beam, the size and shape of each topographic image is essentially a "footprint" of the incident x-ray beam on the specimen. After recording a large number of Laue images, a particular one from a single set of crystallographic planes was selected for scanning. This Laue image was isolated from its neighbors by moving the film sufficiently far from the specimen to provide enough room to scan the specimen without generating overlapping images. There appear to be two basically distinct defect structures apparent in the gallium arsenide topographs examined to date. The first is a fairly uniform cellular network covering the entire wafer. Second are the more striking highly linear arrays which are most pronounced at the edges and the center of the wafer. It was observed that synchrotron topographic images could be obtained from gallium arsenide wafers which were left in plastic containers without any degradation of the recorded image. This permits synchrotron topographic images to be recorded from gallium arsenide and other semiconductor crystals which are in different stages of processing without the undesirable feature of environmental contamination.

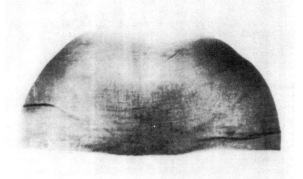

Figure 6. Continuous scan synchrotron white beam transmission x-ray topograph of one Laue image from a 3 inch diameter GaAs crystal wafer (after Winter and Green [11]).

Flash X-ray Diffraction

Flash, or pulsed, x-ray generators, which are normally used for radiographic applications, produce x-ray tube currents of the order of thousands of amperes as opposed to the tens of milliamperes in conventional tubes. However, the burst of emitted x-radiation only lasts a few tens to perhaps a hundred nanoseconds. Flash x-ray systems have been used to record x-ray diffraction patterns from exploding metal foils, shock wave compression of pyrolytic boron nitride, a shaped charge jet, and structural phase transformations in ferroelectric crystals caused by polarity switching and electrically initiated temperature jumps.

OTHER NDE TECHNIQUES FOR MATERIALS CHARACTERIZATION

In addition to those described above other techniques are in various stages of development for nondestructive material property characterization. Among these techniques are high resolution real-time microradiographic imaging, computerized axial tomography, flash x-ray radiography and diffraction imaging, EXAFS (Extended X-ray Absorption Fine Structure), laser speckle interferometry, fiber optic sensor technology, acoustic microscopy, scanning tunneling microscopy, scanned force probe imaging, infrared thermographic imaging, vibrothermography, and nuclear magnetic resonance imaging.

REFERENCES

[1] 1988 CNDE Annual Report, Center for Nondestructive Evaluation, The Johns Hopkins University, Maryland Hall 102, Baltimore, MD 21218

[2] Eric A. Lindgren, "The Use of Ultrasonic Correlation Methods for the Investigation of Phase Transitions in Polytetrafluorethylene and the Precipitation Hardening Process in 2024 Aluminum Alloy", Masters Essay, Materials Science & Engineering Department, The Johns Hopkins University, Baltimore, MD 21218 (May, 1988).

[3] M. Rosen and R.E. Green, "Non-Contact Laser Ultrasonic System for Inspection of Hot Steel Bodies During Processing", Report of NBS Workshop on Internal Discontinuity Sensor Needs for Steel (June, 1988).

[4] J.W. Wagner, "Full Field Mapping of Transient Surface Acoustic Waves Using Heterodyne Holographic Interferometry", pp. 159-164, Proceedings of Ultrasonics International 87, Butterworths Scientific, London (1987).

[5] J.T. Glass, S. Majerowicz, and R.E. Green, Jr., "Acoustic Emission Determination of Deformation Mechanisms Leading to Failure of Naval Alloys", Final Report, Volumes I & II), DTNSRDC-SME-CR-18-83 AND DTNSRDE-SME-CR-19-83 (May, 1983).

[6] J.C. Murphy, J.W. Maclachlan, and L.C. Aamodt, "Image Contrast Processes in Thermal and Thermoacoustic Imaging", IEEE Transactions on Ultrasonics, Ferroelectrics, and Frequency Control, Vol. UFFC-33, pp. 529-541 (1986).

[7] W.A. Bryden and T.O. Poehler, "NDE of Polymer Composites Using Magnetic Resonance Techniques", in Review of Progress in Quantitative Nondestructive Evaluation, D.O. Thompson and D.E. Chimenti (Eds.), pp. 441-447, (1987).

[8] J.C. Murphy, G. Hartong, R.F. Cohn, P.J. Moran, K. Bundy, and J.R. Scully, "Magnetic Field Measurement of Corrosion Processes", Journal of the Electrochemical Society, 135, pp. 310-313, (1988).

[9] K.A. Green and R.E. Green, Jr., "Application of X-ray Topography to Improved Nondestructive Inspection of Single Crystal Turbine Blades", Proceedings of 16th NDE Symposium, pp. 13-22, Southwest Research Institute, San Antonio, Texas (April, 1987).

[10] R.E. Green, Jr., "Real-Time X-ray Diffraction for Materials Process Control", Materials Research Society Bulletin, pp. 44-48 (April, 1988).

[11] J.M. Winter, Jr. and R.E. Green, Jr., "Characterization of Industrially Important Materials Using X-ray Diffraction Imaging Methods", to be published in Proceedings of 3rd International Symposium on Nondestructive Characterization of Materials, Saarbrucken, Germany (October, 1988).

LASER-ULTRASONICS FOR MATERIALS CHARACTERIZATION

J.-P. MONCHALIN, J.-D. AUSSEL, R. HÉON, J.F. BUSSIÈRE AND P. BOUCHARD*
National Research Council Canada, Industrial Materials Research Institute
75 De Mortagne Blvd., Boucherville, Québec, Canada   J4B 6Y4
* Tecrad Inc., 1000 Ave. St. Jean-Baptiste, Québec, Québec, Canada   G2E 5G5

ABSTRACT

Several material properties and microstructural features can be determined or monitored by measuring ultrasonic velocity and/or ultrasonic
attenuation. Conventional techniques which use piezoelectric transducers
for generation and reception have several limitations, in particular in the
case of materials at elevated temperature, of samples of complex shapes,
and in regard to the detection bandwidth. These limitations are eliminated
by laser-ultrasonics, a technique which uses lasers for generation and
detection of ultrasound. Following a review of the various principles and
methods used for generation and detection, we discuss the use of laser-
ultrasonics for velocity and attenuation measurement. Examples of application to various materials are presented.

INTRODUCTION

Increased requirements for quality and process control are creating
new needs for nondestructive techniques which can be used to characterize
materials after and during production. Ultrasonic techniques are widely
used for the determination of the elastic properties and the characterization of the microstructure of materials. The elastic constants are determined by the measurement of the velocities of the longitudinal and shear
ultrasonic waves [1], while the microstructure (grain size, porosity ...)
is generally evaluated by measuring their attenuation [2]. Ultrasonics
also offers the possibility of measuring texture [3], phase changes and
residual stresses [4]. Ultrasonic velocity or attenuation has been shown
also to correlate in certain cases to fracture properties such as fracture
toughness and fatigue damage [5].
Present ultrasonic technology has several limitations which render its
application to material characterization difficult and even impossible in
some cases. The need for contact or coupling fluid prevents measurements
at elevated temperatures which represent an important area of application
since materials like metals and ceramics are manufactured at high temperatures. EMAT transducers are limited to metals and have the drawback to
require close proximity of the sensor to the hot specimen. Laser and
optical ultrasonic techniques (noted below as laser-ultrasonics), which are
based on the generation of ultrasound by a strong laser pulse [6, 7] and
the detection of the resulting ultrasonic displacement pulses by another
laser coupled to an optical interferometer [8] have no limitation of this
kind. Another limitation of present technology is the requirement of
orienting precisely the transducer to follow a complex surface contour.
Laser-ultrasonics, on the other hand, generate ultrasound with a wavefront
matched to the surface and receive ultrasound directly from the surface and
can readily be used on samples of awkward shapes. A third limitation of
present technology, unless special techniques are used, is its limited
bandwidth at emission and reception. By opposition, lasers enable to
produce broadband ultrasonic pulses extending from zero frequency to 50 MHz
and even more, and interferometric receivers can be made with a bandwidth
extending from a minimum value of 10 to 50 KHz limited by ambient vibrations to a maximum value given by the cutoff frequency of the detector (50,
100 MHz and above).

However, laser-ultrasonics is not generally as sensitive as conventional piezoelectric techniques, essentially because of detection. Nevertheless, as demonstrated by the results shown below, it is often possible to reach adequate sensitivity to perform various measurements. Before describing the use of laser-ultrasonics for velocity and attenuation measurements, we review first the principles of generation and detection.

GENERATION OF ULTRASOUND WITH LASERS

The generation of ultrasound, following the absorption of a high power laser pulse, can proceed essentially from two mechanisms [6, 7]. At low laser power density, there is no phase change at the surface and only transient surface heating which produces essentially tangential stresses (thermoelastic regime) [9]. At higher laser power density there is surface melting and surface vaporization giving near the surface a hot expanding plasma (ablation regime). The ultrasonic stress in this case is produced by the recoil effect following material ejection and is essentially normal. A longitudinal wave is then emitted normally to the surface. Ultrasonic stresses in this case are comparable in magnitude to the ones produced by conventional piezoelectric transducers using peak voltage excitation of a few hundred volts, whereas they are weaker in the thermoelastic regime. The ablation regime has the drawback of vaporizing a small quantity of material at the surface, but in many cases, this small surface damage of the order of a micron per hundred shots is acceptable. It is also acceptable on steel products at elevated temperature on a production line. Strong longitudinal pulses are also produced in the damage-free thermoelastic regime by covering the surface with a transparent layer [10].The acoustic source is in this case distributed underneath the surface and the stresses are essentially normal to the surface as in the ablation regime. The same effect is also observed without additional coating when the material weakly absorbs laser light [11].
One important advantage of the generation of ultrasound with lasers for material characterization is the generation of shear waves as well, simultaneously to longitudinal waves. The amplitude and characteristics of the displacements associated with these two kinds of waves (step-like, monopolar or bipolar pulse) depend on the generation mechanism (thermoelastic or ablation), on the penetration of light through the material, on the size of the illuminated zone and if detection is performed on-axis (at epicenter) or off-axis [6, 7, 9]. In all cases, the generated ultrasonic wavefront follows the surface curvature, thus permitting to probe readily parts of complex shape.
Rayleigh surface waves and plate waves can be generated also with magnitude comparable or exceeding traditional means. Good directivity can be obtained by focusing the beam with a cylindrical lens in order to obtain a line source [12]. Large signal magnification has been demonstrated by generating a circular wave with a conical lens (axicon) and detecting with an interferometer at the center of convergence [13]. This setup minimizes heat loading on the surface, which is important for some materials.

OPTICAL DETECTION OF ULTRASOUND

The various techniques for optical detection of ultrasound have been recently reviewed [8], except for the new methods described below. Generally, in the case of laser generated ultrasound, even for shear wave generation, the ultrasonic displacements on the surface of generation and any other surface have a non-vanishing normal component. Therefore, it is generally sufficient to detect this component, although in-plane motion can be detected as well by optical techniques [8]. For detecting normal

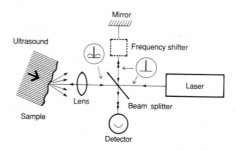

Figure 1: Basic configuration for optical heterodyning (simple interfero-metric detection). A frequency shifter (e.g. Bragg cell) can be introduced in either arm (heterodyne Michelson interferometer). The inserts in circles indicate the optical frequency spectrum at various locations.

displacement, two interferometric methods can essentially be used.

The first one, which we call optical heterodyning or simple interfero-metric detection, consists in making the wave scattered by the surface interfere with a reference wave directly derived from the laser [8], and is sketched in Fig. 1. Such a technique is sensitive to optical speckle and the best sensitivity is obtained when one speckle is effectively detected. This means that the mean speckle size on the focusing lens has to be about the size of the incoming beam and that this beam should be focused onto the surface. Therefore, this technique generally permits to measure the ultra-sonic displacement over a very small spot, which, except at high ultrasonic frequencies, can be considered as giving point-like detection. Compensa-tion for vibrations can be performed by an electromechanical feedback loop which uses a piezoelectric pusher for pathlength compensation. For more severe vibration environments, an heterodyne configuration is preferred. In this scheme, the frequency in one arm is shifted by a RF frequency and the detector receives a signal at this shift frequency, phase modulated by ultrasound and vibrations. Electronic circuits can be devised to retrieve the ultrasonic displacement independently of vibrations.

The second detection method, called velocity interferometry or time-delay interferometry [8] is based on the Doppler frequency shift produced by the surface motion and its demodulation by an interferometer having a filter-like response (see fig. 2). This technique is primarily sensitive to the velocity of the surface and is therefore very insensitive to low frequencies. The filter-like response is obtained by giving a path delay

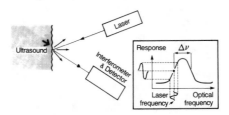

Figure 2: Ultrasound detection with a velocity or time-delay interfero-meter. The insert indicates the principle of detection.

between the interfering waves within the interferometer. Two-wave interferometers (Michelson, Mach-Zehnder) or multiple-wave interferometers (Fabry-Pérot) can be used. Unlike optical heterodyning, this technique permits to receive many speckles and allows a large detecting spot (several mm and even more), especially when known modifications which increase étendue and field of view are used (field-widened Michelson interferometer, confocal Fabry-Pérot interferometer) [8].

Velocity interferometry, being based on a filter-like principle, does not have a flat detection response, but has generally a large étendue (or throughput) corresponding to its ability of detecting over a large spot. Optical heterodyning or simple interferometric detection, on the other hand, has a flat response, limited by the detector cut-off frequency or the Bragg frequency, but a small étendue ($\approx \lambda^2$, λ being the optical wavelength), corresponding to its restriction to small spot detection. This limitation has been circumvented by two recent variants of the technique. One of them makes use of nonlinear optics to generate with beams derived from the detection laser a phase-conjugating mirror. This mirror makes

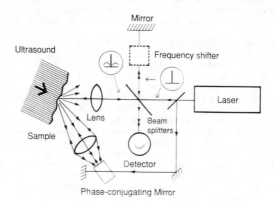

Figure 3: Optical heterodyning using a phase-conjugating mirror, which is produced by two counter-propagating waves derived from the detection laser.

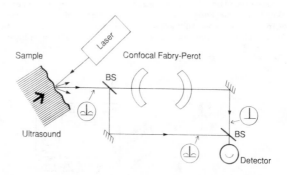

Figure 4: Optical heterodyning providing a large étendue by sidebands stripping.

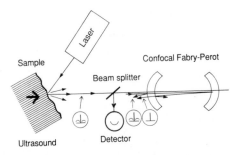

Figure 5:   Optical heterodyning by sideband stripping using
a confocal cavity in reflection mode.

light to retrace its path back onto the surface and then into the interferometer as a speckle-free wave which interferes with the reference wave
[14], (see Fig. 3). The second variant uses a confocal cavity (confocal
Fabry-Pérot) to strip the scattered light from its sidebands [15] and is
sketched in Fig. 4. In this detection scheme, the cavity is tuned to the
laser frequency and the Mach-Zehnder configuration is stabilized at the
zero fringe crossing for sensitive and linear detection. A simpler
embodiment, sketched in Fig. 5, is obtained by noting that in the confocal
Fabry-Pérot the reflected light actually includes the interference of a
beam with sidebands directly reflected from the surface and a beam stripped
from its sidebands leaked by the cavity. In this case, the laser frequency
is tuned on the slope of a resonance peak. For both schemes based on
sideband stripping a large étendue is obtained and the ultrasonic frequency
response is flat from $\Delta\nu$ to $\Delta\nu_0 - \Delta\nu$, $\Delta\nu$ being the optical bandwidth and
$\Delta\nu_0$ being the free spectral range (e.g. from 5 MHz to 145 MHz for a 50 cm
long cavity with mirror reflectivities chosen to give a bandwidth of 5
MHz).

ULTRASONIC VELOCITY DETERMINATION

We have seen that longitudinal and shear waves can be generated at the
same time by laser illumination. This feature is particularly useful since
it permits, from the measurement of the corresponding velocities, to deduce
at once the elastic constants of the material. The measurement of ultrasonic velocities can also be used to monitor or evaluate phase changes,
anisotropy, concentration and shape of inclusions, porosity ... Typical
laser-ultrasonic data are presented in Fig. 6 which shows the arrival of
longitudinal and shear waves [16]. These data were obtained by generating
ultrasound in the slight ablation regime on one side of the specimen with a
Q-switched Nd-YAG laser and detecting on the other side at epicenter (opposite to the source) with a point-like probe based on optical heterodyning.
In order to perform the same kind of measurement at elevated temperature,
the sample was located inside an evacuated tube fitted inside a tubular
oven. These data show that the technique produces a complicated waveform
from which wave identification and wave arrival times may be difficult to
determine. Wave identification is made easier by theoretical models when
such models have been developed and are applicable [17]. Approximate
values of velocities are also very useful to identify proper wave arrivals,
in particular in cases where no theoretical model exist (e.g. anisotropic
materials). The accurate measurement of the arrival times appears also
difficult and different approaches have been considered recently. These
include the use of ablation in order to get strong spike pulses [18, 19],

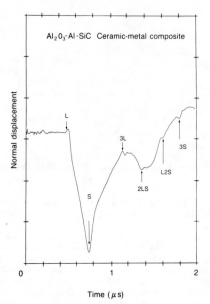

Figure 6: Displacement measured at epicenter with a point-like detecting probe on a 3.25 mm thick Al_2O_3-Al-SiC ceramic-metal composite plate. The generation laser pulse has 10 ns duration and its spot size is ≈ 0.5 mm in diameter [16].

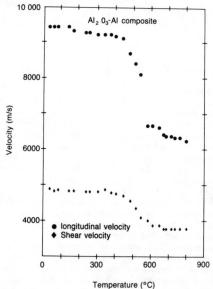

Figure 7: Ultrasonic velocities vs temperature for an Al_2O_3-Al ceramic-metal composite. The step around 500-600°C corresponds to the aluminum melting temperature [16].

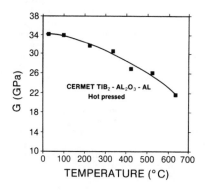

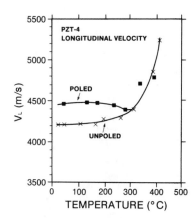

Figure 8: Shear modulus of a hot pressed cermet made of $T_i B_2$, Al_2O_3 and Al [19].

Figure 9: Velocity of poled and unpoled PZT-4 samples versus temperature [19].

off-epicenter probing [19] to enhance shear features and modelling of the source [20]. In a recent work, we have shown that satisfactory results, providing an accuracy similar to conventional ultrasonics, can be obtained by generation in the slight ablation regime, detection at epicenter and determination of the time delay by a cross-correlation technique applied between consecutive echoes [16]. In this work, we analyzed the various limitations to the accuracy of such a measurement. These include noise in the signal, uncertainty of alignment between generation and detection spots, ultrasonic diffraction and material velocity dispersion. In particular, we showed that velocity dispersion can be measured if the signal-to-noise ratio is sufficient.

We present below a few examples of application. The first example (Fig. 7) shows the plot of the longitudinal and shear velocities of an Al_2O_3-Al ceramic-metal composite [16]. A discontinuity of velocities is noted at the melting point of aluminum. Below this point the elastic constants can be readily determined from the values of velocities using standard formulas [1]. Above this point the material consists of a solid filled with liquid and a propagation model including the liquid phase would be needed to relate elastic constants and velocities. Figure 8 shows a result obtained on a hot pressed cermet composed of titanium diboride, alumina and aluminum [19]. This figure shows the large decrease of the shear modulus when the material is heated up to the melting point of aluminum ($\approx$ 660°C). Above this temperature, this modulus vanishes and the material loses cohesion. Figure 9 shows the behavior of the longitudinal velocity versus temperature of two PZT-4 specimens [19]. One of them was unpoled and the other one was poled under high electric field for piezo-electric ultrasonic transduction. It is noted, as expected, that poling stiffens the material and increases its velocity. The Curie temperature is revealed by the point where the velocities of the poled and the unpoled materials become identical ($\approx$ 300°C). Above this temperature, one notices a large change of velocity. Later work has shown that this change actually levels off at higher temperature and that it corresponds to a phase transition from a tetragonal to a cubic structure [21].

ULTRASONIC ATTENUATION DETERMINATION

The conventional measurement of ultrasonic attenuation with piezoelectric transducers [22] presents several requirements which are often
difficult to satisfy. In particular, the samples should be machined with
precisely parallel surfaces. In the case of a measurement made in immersion, the transducer should be precisely oriented, whereas, in the case of
a measurement made by contact, a smooth surface and an uniform bond are
required. Furthermore, in order to be able to apply diffraction corrections, the transducer should verify in good approximation the assumption of
a piston source and of an uniform baffled receiver [22]. Therefore such
measurements have been so far restricted to the laboratory.

Laser-ultrasonics, on the other hand, while requiring samples of
sufficiently uniform thickness, is free of the orientation and bond
problems since the source and the receiver are located on the surface of
the specimen. The last requirement appears more difficult to satisfy since
the intensity distributions of most lasers are not well defined, are very
sensitive to cavity mirrors alignment and cannot generally be represented
by analytical functions. This is further aggravated by local variations of
surface absorption and reflectivity. Therefore, in general, it is not
possible to calculate the diffraction correction. This seems to explain
why the application of the laser-ultrasonic technique to ultrasonic attenuation measurement has not been so far reported, although the decreasing and
widening of the successive echoes has been discussed [23] and several
attenuation measurements have been performed with laser generation and
conventional piezoelectric reception [24]. We have shown recently [25],
that there are at least two cases where the laser beam distributions and
surface variations have negligible effects and precise ultrasonic attenuation measurement can be performed [25, 26]. These two cases correspond to
the spherical wave approximation and the plane wave approximation.

The spherical wave approximation is valid when the source and the
receiver can be considered as point-like compared to the ultrasonic wave-

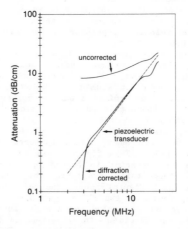

Figure 10: Longitudinal ultrasonic attenuation measured by laser-ultrasonics on a PZT ceramic illustrating the spherical wave limit. The
detecting sensor is a laser-heterodyne interferometer. Also shown is the
variation obtained with a piezoelectric transducer after diffraction
correction and smoothing [25].

lengths. These conditions are more easily met by focusing the lasers, using thick samples and low frequencies. In this case the diffraction losses are inversely proportional to the ultrasonic propagation length and the diffraction correction is a constant which can be exactly calculated. An example of measurement in this case is presented in Fig. 10, which shows the attenuation spectra of a PZT ceramic sample. In this example, ultrasound was generated in the ablation regime by a Q-switched Nd:YAG laser focused to a spot of the order of 0.5 mm in diameter. Detection was performed at epicenter by a laser probe of the optical heterodyning type detecting displacements over a small spot of the order of $\approx$ 0.1 mm in diameter. Figure 10 shows good agreement between the measurements by laser-ultrasonics and by classical piezoelectric techniques.

The plane wave approximation on the other hand applies to large spot sizes, thin samples and high frequencies. In this case, the diffraction losses are negligible. We have recently shown that, in order to minimize diffraction effects, the generation and detection spots should have similar sizes [25]. Since the detecting spot has to be large, a measurement in this limit case can only be performed in practice with a receiver of the velocity interferometric type or using one of the two variants of optical heterodyning mentioned above.

This case is illustrated by data shown in Fig. 11 taken on a 4.19 mm thick hot-rolled carbon-steel plate with 14 μm average grain size. Ultrasound is generated in the slight ablation regime with a frequency doubled Nd:YAG laser giving a generating spot on the surface estimated to 8 mm in diameter. The receiving laser is a Nd:YAG long-pulse laser focused onto the surface to a spot estimated to 10 mm in diameter on top of the generating zone. The scattered light is coupled to a Fabry-Pérot velocity interferometer. The results uncorrected for the diffraction effect are plotted in Fig. 11 and it can be seen that they are in good agreement with a diffraction-corrected measurement obtained with a piezoelectric trans-

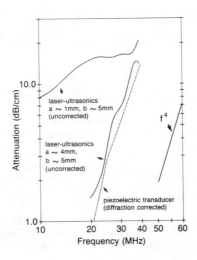

Figure 11: Longitudinal ultrasonic attenuation of a hot-rolled steel plate measured by a conventional piezoelectric transducer and by laser-ultrasonics using a Fabry-Pérot velocity interferometric receiver.

ducer and that they roughly follow a variation in f^4, which is characteristic of Rayleigh scattering. Figure 11 also shows the uncorrected attenuation obtained with the generation laser focused to a diameter of $\approx$ 1.5 mm. It is clear that, in this case, the diffraction correction is not negligible any more. This result shows the importance of having beams of large and approximately equal sizes. In a more recent development, using higher ultrasonic frequencies and the sideband stripping technique mentioned above, we were able to obtain an excellent correlation between measured attenuation and grain size, for sizes ranging from 4 μm to 15 μm.

SUMMARY AND CONCLUSION

We have reviewed the principles and the techniques for generating and detecting ultrasound with lasers. We have seen that laser—ultrasonics has several distinct advantages over classical piezoelectric ultrasonics, including in particular, its ability to perform measurements at elevated temperatures, on specimen of complex shapes, and its broad bandwidth at detection and generation. We have presented our recent developments concerning the application of laser—ultrasonics to ultrasonic velocity and attenuation measurements. We have analyzed for both types of measurement the various limitations to accuracy, originating in particular from diffraction effects. We have shown that laser—ultrasonics, after taking appropriate precautions, permits to obtain velocity and attenuation results as accurate as classical piezoelectric ultrasonics. In conclusion, there are good prospects that, whenever classical ultrasonics has been demonstrated to be useful for materials characterization, laser—ultrasonics could be used as well and, furthermore, laser—ultrasonics being free of the limitations associated to sample temperature and shape, material characterization in conditions of industrial interest appears now feasible.

ACKNOWLEDGEMENTS

We gratefully acknowledge the collaboration of Dr. C.K. Jen of IMRI and B. Farahbakhsh of Alcan Int. Ltd for several results reported here. We wish to also thank Dr. A.J. Gesing and G. Burger of Alcan Int. Ltd and J. G Thomson of Stelco Inc. for providing several samples.

REFERENCES

1. E. Schreiber, O. Anderson, N. Soga, Elastic constants and their measurements, McGraw Hill, New York (1973).

2. E.P. Papadakis, "Absolute measurements of ultrasonic attenuation using damped nondestructive testing transducers", J. Testing Eval. Vol. 12, 1984, pp. 273-279.

3. D. Daniel, K. Sakata, J.J. Jonas, I. Makarow and J.F. Bussière, "Acoustoelastic determination of the fourth order ODF coefficients and application to R—value prediction" and Y. Li and R.B. Thompson "Ultrasonic characterization of texture" in these proceedings.

4. D.R. Allen, W.H.B. Cooper, C.M. Sayers and M.G. Silk, "The use of ultrasonics to measure residual stresses", in Research Techniques in Nondestructive Testing, Vol VI, R.S. Sharpe ed., Academic Press (1982), pp. 151-209.

5. A. Vary, "Ultrasonic measurement of material properties", in Research Techniques on Nondestructive Testing, Vol. IV, R.S. Sharpe ed., Academic Press (1980), pp. 159-204.

6. C.B. Scruby, R.J. Dewhurst, D.A. Hutchins and S.B. Palmer, "Laser generation of ultrasound in metals" in Research Techniques in Nondestructive Testing, Vol. V, R.S. Sharpe ed., Academic Press (1982), pp. 281-327.

7. D.A. Hutchins, "Mechanisms of pulsed photoacoustic generation", Can. J. Physics, Vol. 64, (1986), pp. 1247-1264.

8. J.-P. Monchalin, "Optical detection of ultrasound", IEEE Trans. on Ultrasonics, Ferr. and Frequency Control, Vol. UFFC-33, (1986), pp. 485-499.

9. U. Schleichert, K.J. Langenberg, W. Arnold, S. Fassbender, "A quantitative theory of laser-generated ultrasound", Rev. of Progress in Quantitative Nondestructive Evaluation, D.O. Thompson and D.E. Chimenti ed., Plenum Press, N.Y., Vol. 8, 1989, in press.

10. R.J. von Gutfeld, "20 MHz acoustic waves from pulse thermoelastic expansions of constrained surfaces", Appl. Phys. Lett., Vol. 30, 1977, pp. 257-259.

11. R.J. Conant and K.L. Telschow, "Longitudinal wave precursor signal from an optically penetrating thermoelastic laser source", Review of Progress in Quantitative Nondestructive Evaluation, D.O. Thompson and D.E. Chimenti ed., Plenum Press, Vol. 8, 1989, in press.

12. A.M. Aindow, R.J. Dewhurst and S.B. Palmer, "Laser-generation of directional surface acoustic wave pulses in metals", Optics Com., Vol. 42, (1982) pp. 116-120.

13. P. Cielo, F. Nadeau, M. Lamontagne, "Laser generation of convergent acoustic waves for material inspection", Ultrasonics, Vol. 23, 1985, pp. 55-62.

14. M. Paul, B. Betz and W. Arnold, "Interferometric detection of ultrasound at rough surfaces using optical phase conjugation", Appl. Phys. Lett., Vol. 50, 1987, pp. 1569-1571.

15. J.-P. Monchalin, unpublished.

16. J.-D. Aussel and J.-P. Monchalin, "Precision laser-ultrasonic velocity measurement and elastic constant determination", Ultrasonics, 1989, in press.

17. N.H. Hsu, Dynamic Green's function of an infinite plate, A computer program", Report NBSIR 85-3234, National Bureau of Standards, Gaithersburg, MD (1985).

18. R.J. Dewhurst, C. Edwards, A.D.W. McKie, S.B. Palmer, "A remote laser system for ultrasonic velocity measurement at high temperatures", J. Appl. Phys., Vol. 63 (1988) pp. 1225-1227.

19. J.-P. Monchalin, R. Héon, J.F. Bussière, B. Farahbakhsh, "Laser-ultrasonic determination of elastic constants at ambient and elevated temperatures", in Nondestructive Characterization of Materials II, ed. by J.F. Bussière, J.-P. Monchalin, C.O. Ruud and R.E. Green, Jr., Plenum Press, 1987, pp. 717-723.

20. L.F. Bresse, D.A. Hutchins, K.Lundgren, "Elastic constants determination using ultrasonic generation by pulsed lasers", J. Acoust. Soc. Am., Vol. 84, 1988, pp. 1751-1757.

21. B. Jaffe, W.R. Cook and H. Jaffe, Piezoelectric ceramics, Academic Press, New York, 1971, see chap. 7.

22. R. Truell, C. Elbaum and B.B. Chick, Ultrasonic methods in solid state physics, Academic Press, 1969, Chap. 2.

23. C.B. Scruby, R.C. Smith and B.C. Moss, "Microstructural monitoring by laser-ultrasonic attenuation and forward scattering", NDT Int. Vol. 19, 1986, pp. 307-313.

24. K. Telschow, "Microstructural characterization with a pulsed laser-ultrasonic source", Review of Progress in Quantitative Nondestructive Evaluation, D.O. Thompson and D.E. Chimenti ed., Plenum Press, New York, Vol. 7b, 1988, pp. 1211-1218.

25. J.-D. Aussel and J.-P. Monchalin "Measurement of ultrasound attenuation by laser-ultrasonics" to be published.

26. J.-P. Monchalin, J.-D. Aussel, R. Héon, J.F. Bussière, P. Bouchard, J. Guèvremont, C. Padioleau, "Laser-ultrasonics developments towards industrial applications", IEEE Ultrasonics symposium, Proceedings, 1988, in press.

NONDESTRUCTIVE EVALUATION OF THIN FILM MICROSTRUCTURES BY PICOSECOND ULTRASONICS

HUMPHREY J. MARIS, HOLGER T. GRAHN, AND JAN TAUC

Department of Physics and Division of Engineering, Brown University, Providence, RI 02912

ABSTRACT

We describe a technique by which ultrasonic measurements can be made in the picosecond time domain. A light pulse (duration of the order of 0.1 psec) is absorbed at a surface, thereby setting up an elastic stress. This stress launches an elastic pulse into the interior. The propagation of this strain, including its reflection at interfaces within a microstructure, is monitored through measurements of the time-dependent changes of the optical reflectivity. These measurements are made using a time-delayed probe pulse. In these experiments the spatial length of the elastic pulses can be as short as 50 $\mathring{A}$. We can therefore use this technique to perform a nondestructive ultrasonic evaluation of thin-film microstructures. We describe here results we have obtained which demonstrate the application of the method to the study of the mechanical properties of thin films, the geometry of microstructures, and the quality of bonding at interfaces.

INTRODUCTION

In this paper we review experiments we have performed using picosecond light pulses to generate and detect ultrasonic waves. This technique makes possible the study of microstructures by nondestructive ultrasonic techniques.

In traditional pulse-echo ultrasonic experiments piezoelectric transducers are used to produce and detect the ultrasonic waves. The generating transducer is typically driven by an rf pulse of duration 0.1 to a few μsec. Experiments with pulses significantly shorter than 0.1 μsec are difficult because the rf pulses used to drive the generating transducer are normally produced by conventional electronics. In a solid with sound velocity 5 x 10^5 cm sec^{-1} the length L of an acoustic pulse of time duration 0.1 μsec is 0.05 cm. Thus, in order to avoid the overlap of echoes of different order it is necessary to study samples of a thickness appreciably larger than $L/2$, such as 0.1 cm. This requirement in turn leads to a restriction on the highest frequency that can be studied. If one wants successive echoes to be attenuated by no more than 20 dB, the attenuation α per unit length in a 0.1 cm sample must be no more than 100 dB cm^{-1}. In dielectric crystals at room temperature this condition restricts the highest frequency that can be studied to the order of 3000 MHz; in other materials (e.g., glasses) the attenuation is higher and so the maximum frequency is less.

To avoid these problems it is clearly necessary to devise a technique with better time resolution. It is now possible to generate light pulses with a duration as short as 6 fsec [1], and so it is natural to try to develop a high time-resolution ultrasonic technique based on this technology.

GENERATION

To generate a strain pulse we direct the light pulse at the surface of an optically absorbing material. The simplest picture of the generation process is as follows. The light is absorbed within a characteristic absorption length ς of the surface. For visible light this distance is typically 100 to 300 $\mathring{A}$ in a metal; in a semiconductor ς undergoes a much wider variation depending particularly on the amount by which the photon energy exceeds the bandgap. If the energy of the light pulse is Q and the area excited is A, the energy deposited per unit volume at depth z below the surface is

$$(1-R)\frac{Q}{A\varsigma}e^{-z/\varsigma} \tag{1}$$

where R is the optical reflectivity. This heating produces a temperature rise which sets up a stress tensor which has components

$$\sigma_{\alpha\beta} = \frac{-3B\beta Q}{AC\varsigma}e^{-z/\varsigma}\delta_{\alpha\beta} \tag{2}$$

where β is the thermal expansion coefficient, B is the bulk modulus, and C is the specific heat. Immediately after the light is absorbed the elastic strain is zero. The stress (2) produces an elastic strain pulse propagating into the material. If we assume that the illuminated area is very large the only nonzero component of the propagating strain pulse is [3]

$$\eta_{33} = -\frac{(1-R)Q\beta}{2A\varsigma C}\frac{1+\nu}{1-\nu}e^{-|z-vt|}sgn(z-vt) \tag{3}$$

where v is the speed of sound and ν is Poisson's ratio. The spatial shape of the generated pulse thus has the form shown in Fig. 1.

In our first experiments we used a passively mode-locked colliding-pulse ring dye laser[2], which gave pulses of duration 0.2 psec, photon energy 2 eV, and energy per pulse of 0.2 nJ. The repetition rate of the laser is 110 MHz. The light is focussed onto a spot on the sample of diameter $\sim 20\mu$. In a typical case this gives a temperature rise in the heated layer of a few K, and an elastic pulse in which the strain has a magnitude $\sim 10^{-4}$. In more

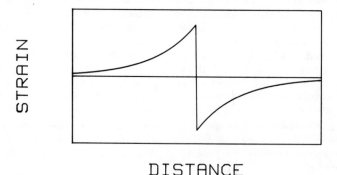

Fig. 1. Shape of the generated strain pulse.

recent experiments we have also used a synchrously-pumped system with 0.5 nJ per pulse and pulse widths between 0.5 and 5 psecs. As we will describe later, there are some experimental situations in which longer pulses have an advantage.

This description of the generation process we have just given is highly over-simplified. A more complete picture should include the following effects:

1) The light energy is always intially given to the electrons, and then transferred to the thermal phonon system. The change in the distribution of the electrons makes a contribution to the elastic stress. This contribution is comparable in magnitude to the thermal expansion effect we have considered, but may have the opposite sign [3]. In addition, as the electrons relax this electronic stress can change with time.

2) The spatial form of the strain pulse can be significantly modified by diffusion. Initially, energy is deposited over a distance ς. The time it takes the strain pulse to leave this region is ςv. If the thermal diffusivity is D, in this time heat diffuses a distance $\varsigma* = (D\varsigma/v)^{1/2}$. Thus, diffusion is important as a mechanism for changing the pulse shape if $\varsigma*$ is comparable to ς, i.e., if D is appreciable compared to $v\varsigma$. In metals one generally has $D \gtrsim v\varsigma$ and so diffusion considerably broadens the generated strain pulses. One also has to consider the possibility that the electrons diffuse a significant distance before they lose their energy to the thermal phonon bath. This last effect should increase as the energy of the photons in the light pulse is increased, since this gives a larger initial energy of the excited electrons. We have very recently seen evidence for this effect in experiments in which we compare the pulse shape produced by 2 eV and 4 eV photons [4]. A longer strain pulse is produced by 4 eV photons even though ς at 4 eV is only half of ς at 2 eV.

It is interesting that the duration τ_L of the light pulse does not play a significant role in the calculation of the shape of the generated strain pulse. In the time domain the length of the strain pulse is ς/v which is in almost all materials at least a few psecs. Provided that τ_L is less than this, there is a negligible effect of τ_L on the shape of the strain pulse.

In all of the experiments we have performed so far, the diameter D of the region illuminated has been very large compared to the absorption length ς. Under these conditions it is a good approximation to treat the generation problem as one-dimensional as we have done. Because of the slow variation of the light intensity across the source there will only be small amounts of shear wave and Rayleigh surface waves generated. It may be possible to enhance the generation of surface waves by depositing onto a surface a structure similar to the interdigital transducers used to generate surface waves at low frequencies [5] (Fig. 2). This type of structure generates surface waves of frequency v_s/d (v_s = surface wave velocity), and would also act as a grating source for bulk longitudinal and transverse waves.

DETECTION

Elastic strain produces a change in the optical constants n and α (n = refractive index and α = absorption coefficient). This provides a means to detect the strain pulses. The simplest version of this is shown in Fig. 3. In this experiment the sample is a thin film of absorbing material deposited onto a substrate. The strain pulse is generated as we have just described by a light pulse (the "pump"). After the strain pulse propagates through the sample film it is partially reflected at the film-substrate interface. When the reflected part returns to the free surface of the film it changes the optical reflectivity. This change ΔR is detected by measuring the reflected intensity of a probe light pulse. (This pulse is

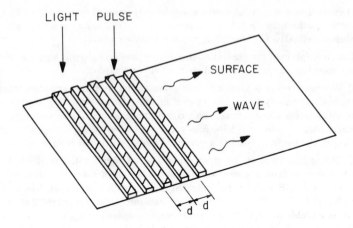

Fig. 2. Possible transducer design for the generation of surface waves. The shaded regions denote narrow thin films deposited onto the surface of the sample. These strips have a different optical absorption from the sample.

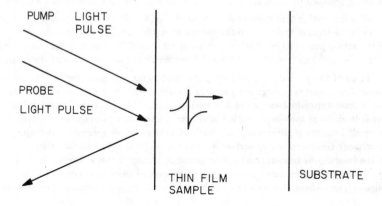

Fig. 3. Simplest version of the experiment.

derived from the same source as the pump, but is time-delayed with respect to it). An example of some early data [3] we obtained in this way is shown in Fig. 4. The sample was a nickel film of thickness ~ 1200 Å and two echoes are clearly visible.

The measured change in reflection can be expressed as

$$\Delta R(t) = \int f(z)\eta(z,t)dz \qquad (4)$$

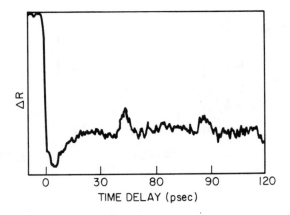

Fig. 4. Acoustic echoes observed in a 1200 Å nickel film.

It is straightforward to calculate the "sensitivity function" $f(z)$ from the optical properties of the sample [3]. The range of $f(z)$ (i.e., the distance from the surface over which the strain pulse produces a significant change in the optical reflectivity) is roughly equal to the absorption length ς. By combining Eq. (4) with the result (3) for the strain one can calculate the expected shape of the echoes observed by the measurement of $\Delta R(t)$. This calculation comes out in reasonably good agreement with experiment [3] in the cases that have been investigated.

APPLICATIONS

Measurements of Velocity and Attenuation

The technique as we have described so far enables pulse-echo experiments to be performed on thin films deposited onto substrates. The film has to absorb light in order for the strain pulse to be generated, and the optical constants (n and α) must have some dependence on strain so that the echoes can be detected. From the arrival times of the echoes the sound velocity can be found. From the relative heights of a sequence of echoes the attenuation can be calculated [3]. It is possible to make attenuation measurements over a frequency range by taking the Fourier transforms of each echo and determining how the magnitude of a given Fourier component deceases for the higher echoes [3].

To make measurements on transparent materials one approach is to add a thin optically-absorbing transducer film (Fig. 5). One wants a transducer with a high optical absorption, and optical constants which depend strongly on strain. The effectiveness of a transducer depends on the photon energy used; at 2 eV we have had success with films of As_2Te_3, InSb, In_2Te_3, a-Ge, and Al. A transducer with a high sound velocity (such as Al) generates short strain pulses with Fourier components extending up to high frequencies. We have been able to measure ultrasonic attenuation in a-SiO_2 at frequencies up to 230 GHz using transducer techniques [6]. At this frequency the attenuation has the very large value of

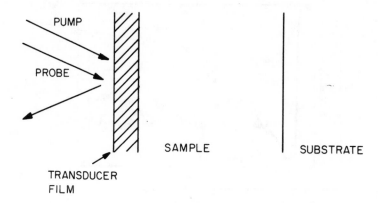

Fig. 5. Transducer film used to generate and detect strain pulses in transparent sample films.

7×10^5 dB cm^{-1}. Thus, the energy in the wave drops by a factor of e in a distance of only 600Å, which at this frequency corresponds to only 2.4 wavelengths. More recently, we have used an As$_2$Te$_3$ transducer to study the sound velocity in a series of silicon oxynitride films with different nitrogen content [7]. These films were prepared by reactive sputtering of a silicon target in a mixed atmosphere of oxygen and nitrogen, and had thicknesses beween 2500 and 6100 Å.

We have also developed an interesting variation of this technique which, while restricted in frequency range, makes possible measurements of velocity and attenuation with very high accuracy [8]. A transducer is deposited onto a film of the material to be studied (thickness preferably $> 10\mu$). The pump and the probe are now directed at the transducer from the substrate side as shown in Fig. 6. The strain pulse generated by the pump light propagates into the sample film as before, but the detection mechanism is different. There is a strong reflection (a) of the probe pulse at the interface between the transducer and the sample. In addition, a small part of the probe is reflected at the propagating strain pulse, either before (b) or after (c) this light has been reflected at the transducer. The interference between the beams (a), (b), (c) causes the total intensity of the reflected light to oscillate as the strain pulse moves further into the sample film. The period of this oscillation is

$$\tau = \frac{\lambda}{2nvcos\theta} \qquad (5)$$

where λ is the light wavelength in free space, n is the refractive index, v is the sound velocity, and θ gives the propagation direction of the light inside the film. These oscillations continue as long as the strain pulse remains in the film. For a thick film there are therefore a large number of oscillations, and so the period τ and the sound velocity v can be determined with high accuracy. (Typically in a 10μ film there will be roughly 50 oscillations). The method can also be used to give the sound velocity in a bulk material in which case the transducer is simply deposited on the surface. Some recent results [9] of this type are shown in Fig. 7. The transducer is a-Ge and the sample is a slab of borosilicate glass. It

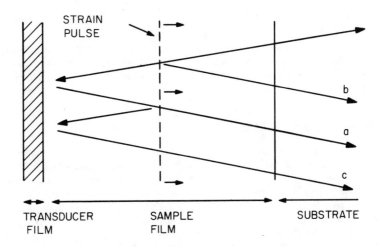

Fig. 6. Technique for measurement of sound velocity and attenuation in thick films or bulk materials.

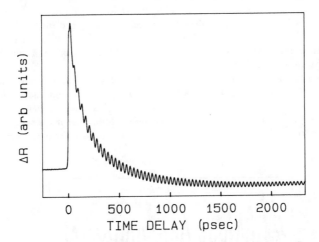

Fig. 7. Data obtained for borosilicate glass using an a-Ge transducer by the method shown in Fig. 6.

is possible to determine the sound velocity with an accuracy of about 1 part in 10^4 in this way. The rate of decay of the oscillations can also be used to give an accurate measurement of the attenuation. One can show [8] that the measured quantity is the attenuation of a sound wave of frequency $1/\tau$, which is typically in the range 20 to 30 GHz. In materials which have had some special surface treatment (such as ion implantation) one should be able to use this method to study the sound velocity as a function of distance from the surface, simply by measuring how the oscillation period changes as the pulse moves into the material.

In the method we have just described it is important that the light pulses not be too short, since interference is required between beams (a), (b) and (c) which travel substantially different path lengths. For this reason the experiment is best done with pulses of length ≥ 1 psec.

Studies of Complex Microstructures

In a structure composed of several films the pump light pulse may be absorbed partially in more than one of the films. Thus, strain pulses can be launched simultaneously from several points in the structure. Similarly, there may be a number of contributions to the observed change in reflectivity. From the analysis of $\Delta R(t)$ it may be possible to determine the sound velocity of the different films, or if these velocities are reliably known one can use the data to make precise measurements of the film thickness. In a recent letter [10] we describe results obtained for a structure consisting of an AlAs epitaxial film grown on a GaAs substrate. On top of the AlAs film was a capping layer of GaAs and a thin transducer of InSb. From the measured $\Delta R(t)$ we were able to determine the sound velocity in AlAs and also the refractive index.

We have also studied amorphous multilayers using this technique [11]. These structures consisted of alternating layers of amorphous hydrogenated silicon and germanium. The thickness of a double layer in the different samples studied ranged from 190 to 1140 Å. For a 2 eV light source the optical absorption in a-Ge:H is much larger than in a-Si:H. Thus, the intial stress is almost entirely set up in the Ge layers. Instead of considering the initial stress to generate a large number of strain pulses propagating in different directions, it turns out to be more fruitful to consider that the initial condition excites the various normal modes of the structure. (Of course, these two views must be formally equivalent). A surprising result was obtained in these experiments. The main part of the response $\Delta R(t)$ consisted of an oscillation which was only weakly damped. For the sample of repeat distance 405Å this is shown in Fig. 8. The existence of this persistent oscillation in

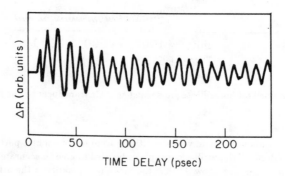

Fig. 8. Measured change in reflectivity for an a-Ge:H/a-Si:H multilayer structure with repeat distance 405 Å.

$\Delta R(t)$ indicates that for some reason the vibrations of the structure are trapped near the free surface and do not diffuse throughout the multilayer. This is because the reflectivity measurement is sensitive only to the oscillating strain near to the free surface. We were able to show that these oscillations arise from a surface mode of the structure which is localized near to the free surface. The calculated dispersion relation for elastic waves propagating in the structure is shown in Fig. 9. The surface mode has a frequency in the first zone-boundary gap. (There may also be other surface modes in the higher gaps). It is a longitudinal mode, i.e., the displacement is along the normal to the surface. This mode is very sensitive to the detailed geometry of the multilayer. For example, it only exists if the multilayer begins with a silcon layer which has a thickness in a certain range. In addition, the frequency of the mode and the distance over which it is localized depend sensitively on the thickness of this layer. Thus, the mode may be a useful diagnostic of multilayer structures.

Studies of Bonding of Films to Substrates

According to the laws of continuum mechanics, the reflection coefficient r for a strain pulse incident on an interface between two elastic materials is

$$r = \frac{\rho_2 v_2 - \rho_1 v_1}{\rho_2 v_2 + \rho_1 v_1} \tag{6}$$

where ρ_1, ρ_2 and v_1, v_2 are the densities and sound velocities in the two media. A larger reflection coefficient indicates that the bonding between the two materials is imperfect.

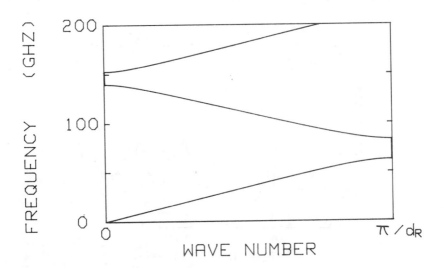

Fig. 9. Calculated dispersion relation for elastic waves propagating in a silicon-germanium multilayer structure. The repeat distance d_R is 405 Å.

Typically, this may occur because the interface is dirty, e.g., there is an intermediate layer of "elastically soft" material which reduces the transmission across the boundary. It is easy to show that, regardless of its elastic properties, an intermediate layer has no effect on r if its thickness is negligible compared to the acoustic wavelengths making up the strain pulse. (In our experiments these wavelengths are roughly the same as the pulse length). Thus, ordinary ultrasonic experiments conducted in the MHz frequency range are sensitive only to thick intermediate layers, i.e., thicknesses in the micron range and up.

The very short strain pulses that we can produce thus make possible much more sensitive tests of the quality of interfaces. We have confirmed that the measured value of r is sensitive to even very small amounts of surface contamination. As an example we show in Fig. 10 results obtained for a film of nickel which is approximately 200 $\mathring{A}$ thick. The nickel was deposited onto a silicon substrate. For a very thin metal film like this the pump light sets up an initial stress which is nearly uniform throughout the film thickness. One can show that the expected response $\Delta R(t)$ is then a triangular waveform. The ratio of the amplitudes of successive maxima should equal r, which for the nickel-silicon interface is 0.44. The experiment gives a much larger reflection coefficient ($r \sim 0.85$) and there is a noticeable distortion of the waveform from triangular as the oscillation proceeds. If we take the intermediate layer to have a density 1 gm cm^{-3} and a sound velocity of 2 x 10^5 cm sec^{-1} (values typical for an organic material) we find that to explain a reflection coefficient of 0.85 a layer thickness of 4 $\mathring{A}$ is needed. The origin of the distortion of the waveform is not so clear; it may arise from the thin oxide layer on the silicon surface.

SUMMARY

We have reviewed recent progress towards the development of picosecond ultrasonics as

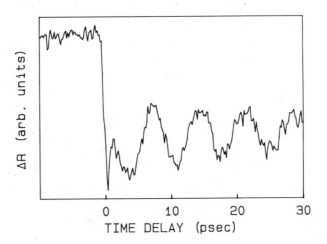

Fig. 10. Measured change in reflectivity for a 200 $\mathring{A}$ Ni film on a Si substrate. The oscillations correspond to thickness vibrations of the Ni film.

a useful tool for the nondestructive evaluation of thin films and multifilm microstructures. The technique makes possible pulse-echo ultrasonic measurements at frequencies much higher than previously possible and because of the excellent time resolution can be used to study films whose thickness is as small as 100 Å.

We thank T. R. Kirst for technical assistance and R. I. Devlen, K. S. Hatton, J. M. Hong, H.-N. Lin, T. P. Smith, C. Thomsen, Z. Vardeny, D. A. Young, and C.-D. Zhu for their contributions to these experiments. This work was supported in part by the Department of Energy Grant DE-FG02-86ER45267.

REFERENCES

1. R. L. Fork, C. H. Brito Cruz, P. C. Becker, C. V. Shank, Opt. Lett. 12, 483 (1987).

2. R. L. Fork, B. I. Greene, and C. V. Shank, Appl. Phys. Lett. 38, 671 (1981).

3. C. Thomsen, H. T. Grahn, H. J. Maris, and J. Tauc, Phys. Rev. B 34, 4129 (1986).

4. Unpublished results of R. I. Devlen.

5. E. Salzmann, T. Plieninger, and K. Dransfeld, Appl. Phys. Lett. 13, 14 (1968).

6. H. J. Maris, C. Thomsen, and J. Tauc, in Phonon Scattering in Condensed Matter V, edited by A. C. Anderson and J. P. Wolfe (Springer, Berlin, 1986), p. 374.

7. H. T. Grahn, H. J. Maris, J. Tauc, and K. S. Hatton, Appl. Phys. Lett., to be published November, 1988.

8. C. Thomsen, H. T. Grahn, H. J. Maris, and J. Tauc, Opt. Comm. 60, 55 (1986).

9. Unpublished data of H.-N. Lin.

10. H. T. Grahn, D. A. Young, H. J. Maris, J. Tauc, J. M. Hong, and T. P. Smith, Appl. Phys. Lett. to appear November, 1988.

11. H. T. Grahn, H. J. Maris, J. Tauc, and B. Abeles, Phys. B 38, 6066 (1988).

PART. II

Texture, Stress

X-RAY DETERMINATION OF STRAIN DISTRIBUTION IN INCONEL ALLOY 600 C-RING

C.F. Lo*, H. Kamide**, G. Feng*, W.E. Mayo* and S. Weissmann*
*Department of Mechanics and Materials Science, Rutgers University, Piscataway, NJ 08855-0909
**Muroran Institute of Technology, 27-1 Mizumoto, Muroran, Japan.

ABSTRACT

An Inconel Alloy 600 C-ring was subjected to various strain levels and the deformation process was monitored by a Computer Aided Rocking Curve Analyzer (CARCA). A large grain population was sampled, and the calibration curve of average rocking curve halfwidth of the individual grains relating to the nominal strain was established. The strain distribution as a function of the angular position along the peripheral surface layer and layers at different depth distance was obtained. Up to C-ring closure at the nominal strain of 3.3% at the apex, the induced plastic strains were confined to a surface layer of 30-40 μm in depth.The largest strain and strain gradients below the surface occurred at the apex and near apex region. The extent and the spread of microdeformation inhomogeneity increased with applied strain. At ring closure some grains exhibited large plastic strains while others exhibited only small plastic strains or were not affected by the deformation process at all. These experimental results were not in agreement with the current theoretical understanding of the deformation of C-ring since these theories did not take shape changes into account. When such changes were included, good agreement on the angular strain dependence for the apex and near apex region were achieved between experiment and theory. It was concluded that the CARCA X-ray method can be a useful research tool in aiding and guiding mathematical modeling of non-linear inelastic behavior of solids by disclosing important microstructural and micromechanical aspects.

INTRODUCTION

Local stress concentration in inherently ductile materials give rise to plastic deformation which must be taken into account in continuum mechanics calculations, in assessing stress concentration. Such calculations, however, pertain to the domain of macromechanics and depend upon taking an average over a region which contains a sufficiently large number of grains to justify the assumption of macroscopic isotropy. If closed form solutions are used, they ignore local variations of microplasticity associated with severe localization of stresses. The assessment of the latter falls into the domain of micromechanics.

The objective of the present study is to focus attention on the plastic deformation rather than the elastic stresses measured by the usual X-ray residual stress method. The latter measures a uniform macrostrain which is deduced from the relative shift of the X-ray line. It is assumed then that the strain is uniform over relatively large distances and that the spacings of the lattice planes change from their strain-free value to a new value governed by the magnitude of the strain. Based on isotropic elasticity theory, the macrostrain is then converted into an average stress value. By contrast, the present method (CARCA) [1] assesses the accrued damage in a material by measuring and mapping directly the workhardened plastic regions. The latter give rise to lattice misalignment and latice curvature which are determined by the X-ray rocking curve measurements taken of a large grain population.

It is the purpose of this X-ray investigation to establish a link between the micro- and macromechanic approach in elucidating the stress-strain distribution of Inconel 600 C-rings when subjected to applied external stresses. This alloy has been the subject of many studies for steam generator tubing in pressurized water reactors [2-5] and recent investigations have concentrated on its stress corrosion cracking resistance. The C-ring are widely used as convenient models for simulation studies of tubes and pipes subjected to stresses with or without a hostile environment.

In addition to the important technological implications, the objective of this investigation is to compare the results of micromechanics with those obtained from closed form calculations of continuum mechanics and to explore the common ground and divergence between the micro- and macromechanic approach.

EXPERIMENTAL PROCEDURE

C-ring samples were prepared from mill annealed Inconel 600 tubes (0.028 wt% C, 0.27 wt% Mn, 9.42 wt% Fe, 15.84 wt% Cr, 73.94 wt% Ni, 0.24wt% Al, 0.22 wt% Ti, 0.23 wt% Si, 0.27 wt% Cu) in accordance with ASTM G38-73. They were then annealed at 704°C for 15 h in argon. Prior to deformation, all samples were electropolished to remove any nonuniform or damaged surface layers (~ 100 μm). Samples were deformed to controlled strain levels by tightening a restraining bolt and monitoring the total strain with a precision resistance strain gauge mounted onto the apex position of the C-ring.

For the structure characterization mapping and analysis of the microplasticity induced by the deformation process, the Computer-Aided Rocking Curve Analyzer (CARCA) was employed as the principal research tool. Sampling a large grain population of 100-500 grains, this method sensitively and rapidly measures the microdeformation range of the individual grains through its rocking curve halfwidth. The sites of maxima in microplasticity can thus be mapped, and on the basis of a calibration curve relating rocking curve halfwidth to the macrostrain, the strain distribution in stressed C-ring can be experimentally measured. This method is most sensitive when individual diffraction spots are clearly imaged either on photographic plates or electronic detectors. When the strain level is too high, or the starting grain size too small, the number of diffraction spots is very large thus obscuring the observation of any individual grain reflection. This effect occurs at about 5% normal plastic strain or at initial grain size of about 2 μm.

RESULTS

Deformation Calibration

Tests were performed on control samples to relate the macroscopic strain levels with the average X-ray rocking curve halfwidth, β, of the grain population as analyzed by CARCA. Flat samples were machined and heat treated at 704°C for 15 h before testing. Sample with resistance strain gauges mounted parallel to the tensile axis were strained to various strain levels(ε)and removed for X-ray investigation by CARCA. Additional tests were performed to relate the β / ε calibration to the local dislocation density(ρ).Direct observations of the dislocation structure were made at several different strain levels between 0.3 to 2.4% plastic strain with the aid of a JOEL 100CX transmission electron microscope. Samples for these observations were prepared by conventional jet polishing techniques in a solution of 7% perchloric solution at -27°C. The deformed microstructure observed in the TEM consisted of duplex distribution of dislocations. At low strain levels (< 0.7%), only planar arrays adjacent to the grain boundary were observed, consistent with the planar slip expected of this alloy, while at higher strains dislocation entanglements in the grain interior began to appear. Quantitative measurements of the dislocation density were made by a series of tilting experiments and use of a linear intercept method. The results of this calibration are shown in Fig. 1 in which the interrelationships between the three principal quantities are shown. By use of this figure, the experimentally measured rocking curve halfwidth, β, can be used to determine the local strain, ε, or dislocation density, ρ .

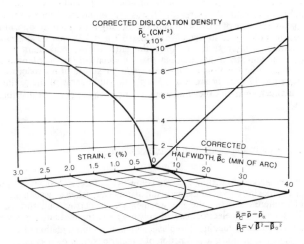

Figure 1.   Relationship between the average X-ray rocking curve halfwidth, β, the average macroscopic strain, ε, and the average dislocation density, ρ.

Surface Mapping of Strain- Induced Microplasticity

Using CARCA, the average halfwidth (β) of the grain population was measured as a function of angular position on the C-ring, for 0.3% plastic strain and for 3.1% (ring gap closure) plastic strain as shown in Fig. 2.

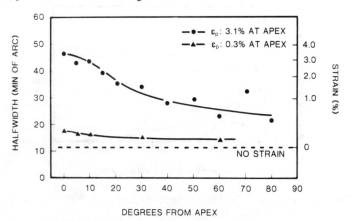

Figure 2.   Dependence of rocking curve halfwidth on angular position along the periphrery of a C-ring for small deformation (0.3 % plastic strain) and for large deformation (3.1 % plastic strain).

For the unstrained C-ring, β was 12 minutes of arc and did not reveal any angular position dependence. It will be noted that for plastic strains as small as 0.3% , β has increased significantly above the reference level of the unstrained ring and most conspicuously in the apex region. For gap closure the increase of β was very pronounced for all angular positions, reaching a maximum value of about 45 minutes of arc, slightly off the apex position.

The distribution curves for the unstrained, 0.3% and 3.1% plastic strains are shown in Fig.3. Of particular interest is not only the shift of the maximum toward greater microdeformation with strain increase, but also the spread of the distribution curves which became significantly larger. Whereas for zero strain the spread of microdeformation is very narrow, centered at 12 min. of arc, for 0.3% plastic strain some of the grains had already undergone microdeformation exceeding even the mean value, β, of the distribution of 3.1% plastic strain.

Surface-Depth Deformation Gradient of Deformed C-Ring

To determine the deformation gradient below the surface, a C-ring was strained to 3.1% plastic strain and rocking curves of the grain population were taken at several angular positions along the circumference. After the measurements were carried out, surface layers were incrementally removed by electrochemical polishing and the measurements were repeated. From the calibration curve relating β to plastic strain, ε, a representation of the plastic strain gradients in depth was obtained as a function of angular position. As may be seen from Fig. 4 the strain gradient is steepest in the apex region, dropping sharply between 15 and 20 degrees from the apex.

Comparison of Experimental Observations with Calculations of Continuum Mechanics

Since the X-ray method employed in this study was able to map the induced microplasticity along the peripheral surface of the strained C-rings and also measure the

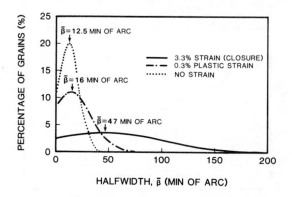

Figure 3.   Distribution of grains with various β values at 0%, 0.3% plastic strain and 3.3% total strain.

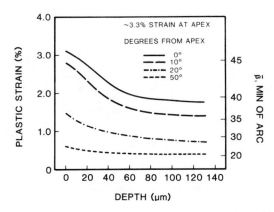

Figure 4.   Dependence of microplastic strain on depth and angular position of a
C-ring

effective strain gradient in depth, it was felt that sufficient experimental data were generated to
enable a comparison between continuum mechanics and experimentally observed data.

Blake's analysis [6] was applied as a basis of comparison to the present microplasticity
studies by X-rays. From his model the relationship of the force moment to the yield stress can
be used to calculate the strain distribution along the periphery of the C-ring as a function of the

angle θ. The results are shown in Fig.5. Poor agreement was obtained on the angular strain
dependence between experimental observations and theoretical calculations, if induced shape
changes were not taken into account. Good agreement, however, was achieved between
experiment and theory, when shape changes were allowed for by partitioning the C-ring
periphery into sectors with varying radii and inserting the experimentally determined values
into the existing theory. Details of these calculations can be found in Ref. 7.

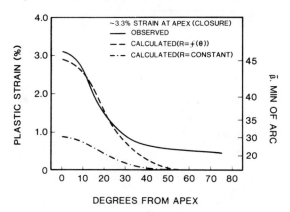

Figure 5.   Comparison of experimentally determined strain distribution with existing
theory (R = constant) and modified theory (R = $f(\theta)$).

DISCUSSION

The inhomogeneous nature of the deformation process is clearly demonstrated in Fig. 3 where even a small macroscopic plastic deformation of 0.3% produces a broad microscopic distribution of strain. Grains at the extreme end of the strain distribution experience a strain at least one order of magnitude above the mean value. It is principally due to this inhomogeneity of plastic deformation that discrepancies arise between experimental observation and theoretical calculations, based on closed form solutions. When calculations were carried out on C-ring deformation based on the simple model of a an elastic core surround by plastic skins [6], no agreement between theory and experiment was obtained (Fig.5). Fairly good agreement, however, for the apex and near apex region was obtained when small shape changes in the C-ring, induced by the deformation, were taken into account. At large peripheral angles of the C-ring, all theoretical calculations were in disagreement with the experimental measurements. This discrepancy resulted from the simple assumption of induced plasticity on which the theory is based.

SUMMARY AND CONCLUSIONS

An Inconel 600 C-ring was subjected to various strain levels and the deformation process was monitored by a computer-aided X-ray rocking curve analysis (CARCA) of individual grains, sampling a large grain population. With the aid of tensile specimens and direct observations on a TEM, calibration curves were established, relating the average rocking curve halfwidth of the sampled grain population to the prevailing strain and the local dislocation density. The strain analysis was carried out as a function of the angular position along the periphery of the C-ring for surface layer and layers at different depth distances.

It was concluded that the CARCA X-ray method can be a very useful research tool in aiding and guiding mathematical modeling of nonlinear behavior of solids by disclosing important microstructural and micromechanical aspects.

ACKNOWLEDGMENT

The authers gratefully acknowledge the financial support of the Electric Power Research Institute (EPRI) under contract 2163-3. The encouragement of C. Shoemaker of EPRI is deeply appreciated.

REFERENCES

1. R. Yazici, W.E. Mayo, T. Takemoto and S. Weissmann, J. appl. Crystallogr. 16, p 89 (1983).

2. D. Van Rooyen, Corrosion J., 1975, 31, pp. 327-337.

3. M. Cornet et al., Metall. Trans. A., 1982, 1, pp. 141-144.

4. D. Lee and D.A. Vermilyea, Metall. Trans. A., 1971, 2, pp. 2565-2571.

5. G.P. Airey, EPRI Project S303-3, September 1987, NP-5282.

6. A. Blake, Deflection of Arches and Rings, M.Sc. Thesis, London University, 1955.

7. C.F. Lo, H. Kamide, G. Feng, W.E. Mayo and S. Weissmann, Acta Metall., Vol. 36, No. 12, pp. 3069-3076 (1988).

NEUTRON DIFFRACTION MEASUREMENTS OF STRAIN AND TEXTURE IN WELDED Zr 2.5 wt.% Nb TUBE

T.M. HOLDEN, R.R. HOSBONS, J.H. ROOT AND E.F. IBRAHIM
Atomic Energy of Canada Limited, Chalk River Nuclear Laboratories,
Chalk River, Ontario, K0J 1J0, Canada

ABSTRACT

The axial and tangential components of residual strain have been measured through the region of an electron beam weld in Zr 2.5 wt.% Nb tube. Marked changes in grain orientation in the tube occur as far away as 10 mm from the weld-centre. In particular a large fraction of the grains which had an [0002] direction aligned tangentially in the original tube were reoriented with this direction aligned axially. These grains showed high (5×10^{-4}) tensile strain. Compressive strains $(-5$ to $-10 \times 10^{-4})$ were measured in the region 5 - 10 mm from the weld-centre in those grains with $[10\bar{1}0]$ directions aligned axially. The hoop strains vary from tensile (maximum 5×10^{-4}) to compressive (maximum -5×10^{-4}) as a function of distance from the weld centre.

INTRODUCTION

Neutron diffraction offers the possibility of making residual strain measurements within welded components. The technique is here applied to a study of the residual strain and the grain orientation in the neighbourhood of electron beam welds in Zr 2.5 wt.% Nb tubes. This material is used extensively in the nuclear industry and in the petrochemical industry because of its good corrosion resistance. The economically favoured mode of joining tubing is by welding. Measurements were made in the as-welded condition, after a cosmetic weld pass and after a stress relief treatment.

One cause of failure of welded tubes of Zr 2.5 wt.% Nb is delayed hydride cracking at room temperature. Our measurements show why this can occur in a guillotine fashion. Neutron diffraction offers the materials engineer a diagnostic tool to help select starting material and develop welding procedures that will minimize the likelihood of failure.

EXPERIMENTAL

Sample

The Zr 2.5 wt.% Nb tubing, supplied by NuTech (Arnprior), had an outside diameter of 108 mm and wall thickness of 4 mm. In preparation for welding, the ends of the tubes were machined so that one had a 0.38 mm (0.015 in) deep groove and the other a 0.38 mm projection to align the pieces. The welds were made by electron beam welding in a vacuum. A draw-bolt was used to hold the pieces together during the initial seal pass. The draw-bolt was removed for the full penetration pass which was made at a speed of 62 mm sec^{-1} with a voltage of 26 kV and a beam current of 210 mA.

One of the welds was given a cosmetic pass with a defocussed beam at a voltage of 26 kV. The welds were between 4 and 5 mm wide at the base and the surface of the tube was oxidized over a width of about 11 mm. The stress relief treatment given to the weldment was at a temperature of 673K for 24 hours.

The Zr 2.5 wt.% Nb tubes had a strong texture with the majority of grains having a $[10\bar{1}0]$ direction along the tube axis, a $[2\bar{1}\bar{1}0]$ direction along the radius and an $[0002]$ axis along a tangent to the tube. The FWHM of the distribution of $[0002]$ axes in the plane containing the tube axis and tube radius is about 6°, while the distribution in the radial-tangential plane is about 60°. This tube texture is strongly modified near the weld and will be discussed below.

Neutron diffraction

The neutron diffraction measurements were made with the L3 spectrometer at the NRU reactor, Chalk River. The wavelength of the monochromatic beam was found to be 2.6124 Å from a calibration experiment with germanium powder. The monochromator was germanium (113 planes) and the collimation of the incident and scattered beams was 0.3°. The incident and scattered beams were defined for axial strain measurements by vertical slits in absorbing cadmium sheet 1 mm wide and 10 mm high. The incident beam first passed through the first wall of the tube, but the scattered neutrons from this wall were prevented from reaching the counter by shielding. The gauge volume, where the incident and scattered beams intersect over the centre of the spectrometer, was arranged to lie at the centre of the second wall of the tube. The tube was carried on an X-Y translator under computer control so that any point in the tube could be placed in the gauge volume to a precision of ± 0.1 mm. The arrangement of the tube on the spectrometer for a measurement of the axial strain is shown in Fig. 1. In the measurement of hoop strain a horizontal slit 10 mm wide and 1 mm high was employed. In this case the strain measured is a through-wall average rather than the result for the middle of the tube. Previous unpublished work, however, indicated that there is very little through-wall variation.

The measurements of interplanar spacing, d_{hkil}, were made by scanning the detector over the expected angular ranges for the $(10\bar{1}0)$ and (0002) Bragg reflections. When the diffraction condition

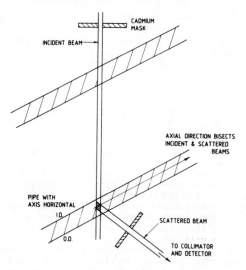

Fig. 1. Arrangement of the tube on the spectrometer for measurement of the axial strain.

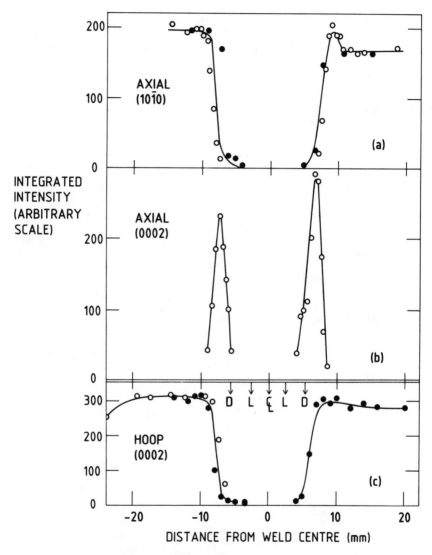

Fig. 2 – Integrated intensities of Bragg peaks determined in the measurement of axial, (a) and (b), and hoop strains, (c), in a Zr 2.5 wt%Nb welded tube with a cosmetic pass (open circles) and after strain relief (closed circles). The (10$\bar{1}$0) intensity in Fig. 2(a) far away from the weld is characteristic of the as-produced tube, but grain reorientation occurs about 10 mm from the weld-centre. The (0002) intensity in Fig. 2(b) corresponds to grains which have been 'rotated' from a tangential orientation. The (0002) intensity in Fig. 2(c) far from the weld is characteristic of the as-produced tube, but grain reorientation occurs about 10 mm from the weld centre and grains are rotated into an axial orientation. L denotes the edge of the lap mark and D denotes the edge of the surface discolouration.

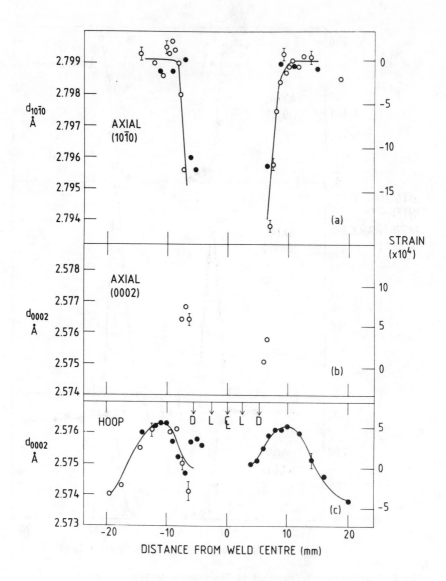

Fig. 3 — Measured residual strains in the axial direction, (a) and (b) and the hoop direction (c) in a Zr 2.5 wt%Nb welded tube with a cosmetic pass (open circles) and after strain relief (closed circles). The strain is close to zero in the grains with $\lfloor 10\bar{1}0 \rfloor$ along axis but becomes compressive near the weld. The reoriented $\lfloor 0002 \rfloor$ grains show large tensile effects. The hoop strains in $\lfloor 0002 \rfloor$ grains are tensile close to the weld but change to compressive as the distance from the weld increases.

$$\lambda = 2d_{hkil} \sin \theta_{hkil} \tag{1}$$

is fulfilled, a sharp diffraction peak is observed at a scattering angle $2\theta_{hkil}$. From the peak position, determined by fitting a Gaussian line-profile on a sloping background, the spacing may be found. The integrated intensity of the diffraction peak is proportional to the total volume of the grains in the gauge volume oriented to fulfil the condition (1). The integrated intensity was used to follow the texture modification caused by welding.

RESULTS AND DISCUSSION

Texture Variation

The variations of the integrated intensity of the $(10\bar{1}0)$ and (0002) Bragg peaks as a function of distance from the weld in measurements of the axial strain component are shown in Figs. 2(a) and 2(b). The results are given for the case of the cosmetic weld (open circles) and after strain relief treatment (closed circles). The strong $(10\bar{1}0)$ intensity characteristic of the original tube begins to fall rapidly to a few percent of its maximum value about 10 mm from the weld centre. This is well outside the edge, D, of the oxidized region shown in Fig. 2. The (0002) peak, on the other hand, increases sharply in this region to a maximum value about 7 mm from the weld centre but falls to zero 5 mm from the weld centre. The observations show that welding caused grain reorientation in the region 5-10 mm from the weld centre. It is interesting to note that a study of texture and residual strain in compressed Zircaloy rod samples showed marked grain reorientation with $\lfloor 0002 \rfloor$ directions effectively rotating from radial to axial orientations and $\lfloor 10\bar{1}0 \rfloor$ directions rotating from axial to radial as in the present case. Twinning was shown to create the equivalent of a 90° rotation of grains about the $\lfloor 11\bar{2}0 \rfloor$ direction. A possible explanation of our observation in the weld is that the compressive axial stress generated by the heating cycle of the weld causes a similar reorientation of the grains. If, however, the critical shear stress for twinning is greater in Zr2.5wt%Nb than Zircaloy (this appears to be the case, N. Christodoulou and A. Salinas-Rodriguez, private communication), it may be that combinations of several slip systems acting together can also generate 90° rotations. In any case, an explanation of this kind may not hold closer to the weld, particularly in the melted region. Texture measurements to find the grain orientation in the weld centre are planned. The variation of the number of grains with [0002] axes aligned along a tangent to the tube, as manifested in the (0002) peak intensity in a measurement of the hoop strain, is shown in Fig. 2(c). A rapid decrease of (0002) intensity occurs about 10 mm from the weld, corresponding to the re-orientation along the tube axis.

Strain variation

The variation of the $(10\bar{1}0)$ and (0002) spacings, and hence the axial residual strains, are shown in Figs. 3(a) and 3(b). Reference spacings were measured in the tube 100 mm from the weld centre. The strains induced by welding, indicated on the right hand scale, are therefore calculated with respect to the macroscopic strain state existing in a manufactured tube.

The axial strain state of grains with [10\bar{1}0] directions along the tube is close to zero until the position where grain reorientation begins. Thereafter, the remaining grains are strongly compressed. The axial strain, however, in the reoriented grains with [0002] directions along the tube axis is tensile and large. When gauge volume is centred on the region of maximum intensity for these grains (see Fig. 2(b)) the strain measured is an average over the whole region because of the finite width of the gauge volume. Note that different grain orientations have completely different strains so that a single stress or strain tensor would not describe the effects.

The highly strained reoriented grains may have high technological impact since the increased basal plane separation provides a natural location for zirconium hydride platelets to form in a hydrogen rich environment. Cracking of the hydride platelets (delayed hydride cracking) causes cracks to propagate along (0002) planes and in the case of the reoriented grains could give rise to a guillotine failure.

The hoop strain as determined from the main grain orientation, (0002), shows a maximum in tension (5×10^{-4}) just outside where grain reorientation begins. Beyond 15 mm from the weld centre, these grains are in compression and the amplitude of the strain gradually falls to zero, 100 mm from the weld.

It is interesting to note that in the study of compressed Zircaloy[1] the non-rotated ($10\bar{1}0$) grains were in compression and the non-rotated (0002) grains were in tension, just as in the present case.

The results for the as-welded section and for the section given a cosmetic pass are very similar and the effect of the stress relieving heat treatment appears to be negligible.

ACKNOWLEDGEMENTS

We wish to acknowledge the expert technical assistance of M.M. Potter, H.F. Nieman, A.H. Hewitt and E.M. Kirkus. Useful discussions of the results were held with N. Christodoulou and A. Salinas-Rodriguez.

REFERENCES

1. S.R. MacEwen, N. Christodoulou, C. Tome, J.A. Jackman, T.M. Holden, J. Faber Jr., and R.L. Hitterman. Proceedings of the International Conference on the Texture of Materials (ICOTOM-8), edited by J.S. Kallend and G. Gottstein (The Metallurgical Society) 1988, p. 825.

DETERMINATION OF THERMALLY AND MECHANICALLY INDUCED INTERNAL STRESSES
IN METAL-MATRIX COMPOSITES BY X-RAY METHODS

RAHMI YAZICI, K. E. BAGNOLI AND Y. BAE
Department of Materials and Metallurgical Engineering, Stevens Institute of
Technology, Hoboken, New Jersey 07030

ABSTRACT

In this study the progression of thermally and mechanically induced internal strains (stresses) in metal-matrix composites was investigated by X-ray methods. The materials
studied were whisker-reinforced 2124 Al-SiC(w) and 6061 Al-
SiC(w) composites. X-ray diffractometry was used to measure
thermally induced stresses on samples cycled from ambient to
280°C. Significant variations in residual stress values were
observed in the matrix depending on the location and direction of the measurements with respect to the whisker orientation. The determined stress states of the as-processed
and the thermally cycled samples were evaluated with continuum models.

The microstrains in composites induced during processing
and tensile loading were also investigated by nondestructive
means. Individual grains of the matrix were analyzed by
rocking-curve measurements using a modified X-ray double-
crystal diffractometer. The relationship between the plastic
deformation induced by applied loads and the progression of
the microstrain/excess-dislocation values was determined.

INTRODUCTION

The impetus for the development of metal matrix composites comes from
the need for light weight, high strength and high temperature resistant
structural materials. In order to satisfy these expectations, composites
usually constitute individual components that have strong differing properties such as coefficient of thermal expansion, elastic modulus and ductility. The mismatch in the coefficient of thermal expansion (CTE) of the
components causes large internal stresses when the composite is subject to
temperature changes during processing or service. Mechanical tests showed
that Al-SiC(w) composites exhibit tensile strength values higher than
expected according to the continuum models of discontinuous reinforcement
[1]. The relatively high dislocation density generated by thermal
stresses, even in the annealed composites was reasoned as the major
strengthening mechanism. The strength of the interface in the Al-SiC(w)
system exceeds that of the matrix [4,5] so that if high enough internal
stresses develop or under external loads, the matrix would yield before any
slippage at the interface [6,7]. It was also shown that thermal cycling
significantly reduces the tensile strength of 2124 Al-SiC(w) [2] and 6061
Al-SiC(w) [3] composites. In these studies the premature failure of the
matrix was attributed to the thermally induced high residual stresses. It
is technologically important, therefore, to evaluate the strain state in
these materials by nondestructive means.

EXPERIMENTAL

The two composites studied in this investigation were extruded 2124
Al-SiC(w) rod and 6061 Al-SiC(w) plate. Both contained 20 volume % β-SiC
whiskers. The samples prepared for thermal cycling were 6.3 mm x 6.3 mm x
6.3 mm cubes. The tensile specimens were 12.7 mm x 6.4 mm x 3.2 mm in
gage. All samples were cut with a low-speed diamond saw and ground with
diamond paste in order to avoid surface damage and roughness for the subse-

quent X-ray measurements. During thermal cycling the temperature excursion for each cycle was from 25°C to 280°C with a heating rate of 50°C/min. On each cycle samples were held for ten minutes and quenched in water.

The residual stresses were measured by a two-tilt method [8] using an X-ray powder diffractometer equipped with a stress goniometer. The position of the 2θ peaks were determined by step scanning in 0.1 degree intervals with the ψ angle set at 0 and 45 degrees. The corrections, curve fittings and anaylsis were done by an on-line computer. The instrumental error in the measurements was less than 10 MPa. Strain measurements were made on both the longitudinal and the transverse extrusion directions. With the Al matrix the (422) reflections were used in all measurements. With the SiC phase (422) and (333) reflections were utilized in measurements on the longitudinal and transverse surfaces, respectively. The elastic constants that were used were: for SiC_w, E = 427 GPa, ν = 0.17; for 2124 Al, E = 69 GPA, ν = 0.33; for 6061 Al, E = 47 GPa, ν = 0.33. In-depth stress gradients were determined in thick samples by sectioning at 1.2 mm intervals from the surface(s).

The defect structure of the matrix was studied by microstrain measurements on as-received and tensile-deformed samples. Samples which were subjected to predetermined plastic strains were intermittently evaluated by rocking-curve analysis on (200) and (111) planes. The experimental tool used to measure the microstrains and to determine the excess dislocation density of the matrix was based on double-crystal diffractometry. A thorough review of the technique was described elsewhere [9] and only a brief description will be given here.

A polycrystalline sample is held stationary while being irradiated by a crystal-monochromated X-ray beam. Several grains in the specimen will be in reflecting positions which result in individual microscopic spots and be recorded by a film or a position-sensitive detector. The intensity of each spot and its location are then stored for numerical analysis. Subsequently, a small discrete change in the angular setting of the sample is made and the X-ray measurements are repeated. The individual crystallites are step-wise rotated through the range of reflecting positions. The reflected intensities, recorded as a function of crystallite rotation, represent rocking curves. The half-width, β, of the rocking curve provides a measure of the angular lattice misalignment introduced during the deformation of the crystallite. The β values provide information about the local densities of the accumulated excess dislocations of one sign. Assuming that the excess dislocations are randomly located in the grains, then the excess dislocation density $D = β^2/9b^2$, where b is the magnitude of the Burgers vector.

RESULTS

In Figure 1 a montage of scanning electron micrographs of the etched 2124 alloy composite is shown. The preferential alignment of the whiskers in length with the extrusion direction was clearly evident. The average whisker diameter was 1 μm and the average length was 10 μm; setting the aspect ratio, l/d, at 10. The average matrix grain size was 2 μm. Very similar structural features were found in the 6061 Al composites [10]. The crystallographic texture in both of the composites was the same. In the Al matrix the (220) planes were aligned parallel to the longitudinal direction and the (111) planes parallel to the transverse direction. With the SiC whiskers, also, the (220) planes were aligned parallel to the longitudinal direction and (111) planes parallel to the transverse direction.

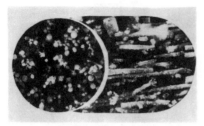

Fig. 1. Whiskerous microstructure of 2124 Al 20% vol
SiC(w) composite (HCl
etched)

In the as-received condition the average residual stress values in the 6061 Al matrix were -40 MPa in the transverse and +40 MPa in the longitudinal directions. In the 2124 Al matrix the as-received stress values were -80 MPa and +10 MPa in the transverse and longitudinal directions, respectively. The SiC whiskers were under an average compressive stress of -500 MPa.

Upon thermal cycling the compressive stresses in the matrics increased. After six thermal cycles, in 6061 Al-SiC(w) samples, the compressive stress in the matrix transverse direction increased and leveled at -140 MPa. In the longitudinal direction the stresses remained essentially constant at +40 MPa. With the 2124 Al-SiC(w) samples subject to thermal cycling the results were similar: after a single thermal cycle in the compressive stress in the matrix transverse direction increased to -130 MPa and remained constant after twelve cycles. In the longitudinal direction the tensile stresses remained essentially constant, +20 MPa after twelve cycles. At this stage the stress level in the SiC whiskers was -430 MPa. The stress state in the 2124 Al-SiC(w) composite after twelve cycles is shown schematically in Figure 2.

Some of the samples were heavily etched in order to remove most of the matrix except for the layers adherent to the SiC whiskers. The stress values in the remaining matrix layers were -30 MPa for the transverse and +20 MPa for the longitudinal directions.

The variation of the residual stresses as a function of depth in 2124 Al-SiC samples subjected to three cycles is schematically shown in Figure 3. In the transverse direction the matrix stress varied from -130 MPa at the surface to -230 MPa at the center, and the SiC(w) stress varied from -430 MPa at the surface to -690 MPa at the center. In the longitudinal direction a similar effect was observed and the matrix stress dropped from +20 MPa at the surface to -110 MPa at the center.

The results of the rocking-curve analysis of the 6061 Al-SiC composite showed that in the as-received condition the average excess dislocation density in the matrix was $6.5 \times 10^9/cm^2$. As the material was loaded to its yield strength (0.2% strain), the dislocation density level increased to $7.1 \times 10^9/cm^2$ and eventually reached to $1.1 \times 10^{10}/cm^2$ upon fracture (2.7% strain). The increase in the excess dislocation density was almost linear with respect to the macroscopic deformation. The two curves given in Figure 4 show the percent distribution of the rocking-curve half-width values from individual grains in as-received and fractured samples. All measurements were taken on the long transverse surface of the extruded material and away from the fracture site.

DISCUSSION

In discontinuously reinforced metal-matrix composites the directional variation of the residual stresses evidently depend on the volume fraction and the aspect ratio of the hard phase. When the aspect ratio approaches unity as in SiC particle reinforced Al composites, it was shown that the matrix is in the state of hydrostatic tension [4]. In composites with a SiC aspect ratio larger than unity, residual stresses were found to vary

68

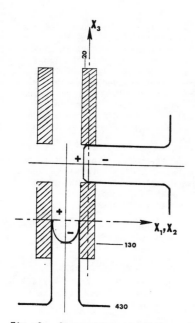

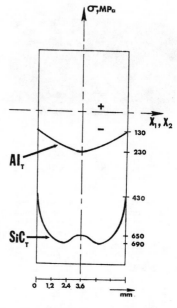

Fig. 2. Stress state in 2124 Al-SiC(w) composite after thermal cycling (transverse surface).

Fig. 3. Stress state as a function of depth in 2124 Al-SiC(w) after thermal cycling.

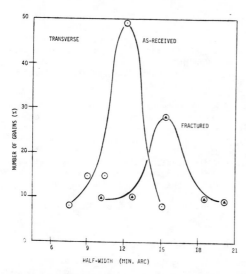

Fig. 4. Statistical distribution of rocking-curve half-width values from individual grains.

with respect to the direction of the whisker-alignment [11] and even into a compressive state [3] in the matrix . The presence of compressive stresses is in conflict with the mathematical model [3] adopted from Eshelby's concepts [12] for inclusions. In this model by Arsenault [3], the matrix would be under nonuniform tensile stresses. According to the experimental results of this study (Figure 2), upon thermal cycling, compressive stresses were set in the transverse and tensile stresses in the longitudinal directions. A similar stress state was also observed in the as-received materials, although these were heat treated to minimize the residual stresses. The observed stress state can qualitatively be explained by the orthotropic nature of the composite CTE. According to Schapery's approximation [13] for similar structures, the composite CTE in transverse direction, α_t, would be larger than in the longitudinal direction, α_l. This could create compressive stresses in the matrix transverse direction when a stress-free composite at high temperatures is cooled down. Moreover, according to Schapery, α_t could exceed even the matrix CTE, α_m, for low whisker contents. Therefore pockets of matrix could be set under compression during thermal cooling. The presence of such regions between the whiskers was explored in this study by etching the matrix in order to expose the remaining layers adherent to the whiskers for X-ray measurements. The experimental results were in support of the assumptions made and a substantial decrease of compressive stresses from -110 MPa to -30 MPa was observed in the transverse direction. Also with stresses in excess of -400 MPa in the whiskers, there have to be matrix regions near the interface that are under very high tensile stresses. These regions however are very small in size and cannot be resolved by conventional methods.

According to Figure 2, the stress gradients between the two components would be highest near the whisker tips, making this area highly susceptable to failure. In addition to the internal stresses the shear stresses generated in the matrix during load transfer are also expected to be maximum near the tips [7]. The TEM work done by Nutt [14] showed that under applied loads the dislocation density rapidly increase in these regions, eventually leading to fracture. In this study the increase in the average dislocation density from the as-received to the fractured state was limited which is a sign for the high localization of plastic deformation in the matrix. The salient feature of the rocking-curve analysis (the ability to study the defects in the individual matrix grains), has revealed the nonuniform distribution of the microplasticity. As shown in Figure 4, both of the as-processed and the deformed materials exhibit a range of half-width values. Especially in the deformed sample, the spread in the β values was very pronounced. While some of the grains remained in the as-processed state others were highly deformed. This was most likely caused by the location of the grains with respect to the whiskers.

The long-range stresses developed during thermal cycling were progressively compressive into the bulk. The values shown in Figure 3 were not corrected for material removal or relaxation and the actual values could be different. Such high stresses in the matrix could only be sustained by a state of triaxial stress. The matrix stresses measured near the surface on the other hand had approximated the yield strength of the matrix, indicating the yielding of the matrix in these regions.

In summary, both micro- and macro-strains were determined in Al-SiC(w) composites. Some of the values observed cannot be predicted with the present continuum models. Further study is required, experimentally as well as theoretically with more emphasis on the micromechanical aspects of materials.

REFERENCES

1. R. J. Arsenault and Y. Flom, Mat. Sci. & Eng., 77, 191 (1986).

2. W. G. Patterson and M. Taya, Proc. ICCM5, ed. W. Harrigan, San Diego (1985).

3. R. J. Arsenault and M. Taya, Poc. ICCM5, ed. W. Harrigan, p. 21, San Diego (1985).

4. H. M. Ledbetter and M. W. Austin, Mat. Sci. & Eng., 89, 53 (1987).

5. H. M. Ledbetter and M. W. Austin, Adv. in X-ray Analysis, 29, 71 (1986).

6. K. K. Chawla, Composite Materials, Springer Verlag (1987).

7. J. C. Le Flour and R. Locicero, Scripta Met., 21, 1071 (1987).

8. I. C. Noyan and J. B. Cohen, Residual Stress, Springer Verlag, NY (1987).

9. R. Yazici, W. Mayo, T. Takemoto and S. Weissmann, J. Appl. Cryst. 16, 89-95 (1983).

10. R. Yazici and S. B. Han, Proceedings of 1987 TMS/MRS Conference on High-Temperature Structural Composites, Hoboken.

11. Y. Flom and R. J. Arsenault, Deformation of SiC/Al Composites, J. of Metals, July 1986, 31.

12. J. D. Eshelby, Solid State Physics, 3, 79 (1956).

13. R. A. Schapery, J. Composites Materials, 2, 311, (1969).

14. S. R. Nutt and A. Needleman, Scripta Met., 21, 705 (1987).

X-RAY DIFFRACTION FOR NONDESTRUCTIVE CHARACTERIZATION
OF POLYCRYSTALLINE MATERIALS

C. O. RUUD AND S. D. WEEDMAN
The Pennsylvania State University, 159 Materials Research Laboratory,
University Park, PA 16802.

ABSTRACT

X-ray diffraction has long been the mainstay for materials
characterization in the laboratory. This characterization includes the
determination of phase composition, residual stress, microstrain, grain size,
and crystallographic texture of polycrystalline metals, ceramics, and
minerals. The analytical capabilities of XRD techniques have been expanded
recently by the application of computer control to data collection and
processing. These capabilities include the identification of irregularities in
metals and ceramics that are caused by processing and fatigue damage, as well
as the apriori prediction of processing anomolies. While the above
applications have been largely restricted to the laboratory, the possibility
for exploitation of the nondestructive nature of x-ray diffraction for in-
process evaluation of materials is now being realized.

The availability of computer-controlled position-sensitive x-ray
detectors can now provide rapid, non-contacting, in-process interrogation of
materials. The examples of nondestructive characterization illustrated in this
paper will be those that can be used for process control and/or damage
assessment.

BACKGROUND

In-process nondestructive characterization should be fast, non-
contacting and tolerant of detector to component distance variation. A
position-sensitive scintillation detector (PSSD), developed at The
Pennsylvania State University, is unique in its ability to satisfy these
requirements, and has been successful in determining phase composition,
residual stresses and microstrain, grain size and crystallographic texture in
ceramics, non-ferrous metals, and steel. The PSSD relies on the coherent
conversion of the diffracted x-ray pattern to an optical signal (light); the
amplification of this signal by electro-optical image intensification; the
electronic conversion of the signal; and the transfer of the electronic signal
to a computer for refinement and interpretation. It has been described
elsewhere [1], and its application to materials characterization discussed
[2,3].

With the PSSD, data collection times of less than one second are
possible using modern x-ray tubes and constant potential power supplies -- a
two order of magnitude improvement over conventional XRD instrumentation. A
compact version of this instrument has been developed that is capable of
making measurements on the inside of a pipe as small as 100 mm [4]. Also, the
geometry of its x-ray optics allows stress readings in confined areas such as
gear teeth, pipe, and turbine vane bases. The compact size, speed, and non-
contacting nature of the PSSD offers the potential for reliable in-process
materials characterization.

X-RAY DIFFRACTION PEAK ANALYSIS

X-ray diffraction techniques measure the interatomic crystal lattice
spacing of crystalline materials. Data are generated graphically as plots of
intensity of diffracted radiation versus angle of diffraction. Plastic

deformation, phase composition, residual stress, and crystalline texture in materials can be measured from the breadth, position, and intensity of the x-ray diffraction peaks. When metals are stressed during fabrication or in use, they can deform both plastically and elastically [5]. Plastic deformation results in the distortion of individual crystallite lattices, and is manifested in peak broadening. Elastic strain (residual stress) occurs uniformly over several tens of grains in both metals and ceramics, causing the d-spacing between elastically strained lattices to change, and the diffraction peak to shift position with respect to the diffraction angle, θ. In addition, a preferred orientation of crystallites, i.e. texture, occurs in many manufacturing processes, which results in a greater intensity, or peak height, of diffraction by a given set of crystallographic planes at different orientations to the surface of a sample. Phase composition measurements, critical to the evaluation of steel, are determined from both peak position and intensity.

The following discussion summarizes x-ray diffraction data collected on the PSSD for metals and ceramics for which there is a potential for in-process use.

Peak Position

When a metallic crystalline material is stressed either during fabrication or use, the elastic strains in the material are manifested in the crystal lattices of its grains. The stress applied externally or residual within the material, if below its yield strength, is taken up by uniform interatomic macro-strain that is spread over several tens of grains. The distance between lattice planes, d-spacing, is thus changed, as well as the angle (2θ) the x-ray beam is diffracted, i.e., there is a shift in the peak position. Because x-ray diffraction techniques measure interatomic spacing (peak position), the elastic macrostrain experienced by the specimen can be quantified [6]. Residual stresses have been studied in copper strip by the PSSD in preparation for development of an in-process instrument.

Residual stresses are induced in copper strip as a result of slitting in preparation for the fabrication of electrical switches and contacts. These stresses affect forming response. The presence of residual stresses in the copper strips can lead to warpage immediately after fabrication, or in later manufacturing operations. Residual stress measurements on copper alloy strips were performed by the single exposure technique (SET) [6], using the (420) plane and Cu K-alpha x-radiation. Preliminary results of residual stress measurements on the edge of a copper strip are shown in Fig. 1.

In addition, residual stress measurements were made by the PSSD on five samples of titanium diboride, a ceramic material, prepared by two different processes. The sponsor suspected that the electro-discharge machining (EDM) process had induced residual stresses in the surface of the samples, and that hand grinding helped reduce them. The results, in Table 1, show that hand grinding with a diamond paste does reduce residual stresses induced during processing, though the reduction is not directly controlled by the grain size of the abrasive.

Peak Broadening

Plastic deformation, or microstrain, results in a distortion of the crystal lattice of individual crystallites and, subsequently, variability of the d-spacing of the lattice planes. Consequently, the angle of x-ray diffraction for each plane will vary slightly, yielding a broader peak than would an undeformed crystallite. Preliminary measurements were made on Inconel 600 C-rings with the intention of applying the techniques to study the

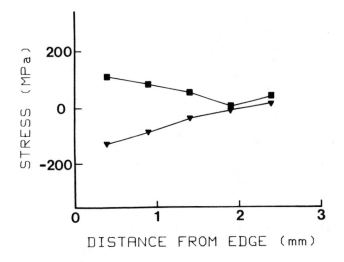

Figure 1. Plot of the residual stresses in copper alloy strip as
a function of distance from the edge of the strip. Measurements
were made on both concave (■) and convex (▼) surfaces.

susceptibility for stress corrosion cracking in Inconel U-bend tubes, used in
heat exchangers in nuclear reactors. Nine microstrain measurements made around
the curvature of a mill-annealed Inconel 600 C-ring. The results are plotted
in Figure 2 along with the macrostress resulting from elastic deformation.
Note that the greatest microstrain was detected at the site of greatest
curvature of the ring at 0 degrees, as expected. Measurements were made using
the methods of Delhez et al. [7] and Lo et al. [8].

TABLE I.

Residual Stresses in Titanium Diboride (TiB$_2$) Cut Surfaces
Machined by Electro-Discharge Machining (EDM) and
Hand Grinding

Sample	Residual stresses	MPa (KSI)[1]
1. EDM	345±83	(50±12)
2. EDM	558±83	(81±12)
3. EDM w/hand grinding 100 μm with diamond	-365±83	(-53±12)
4. EDM w/hand grinding 130 μm with diamond	76±83	(11±12)
5. EDM w/hand grinding 700 μm with diamond	-503±83	(-73±12)

[1]Residual stresses calculated by the SET method [6].

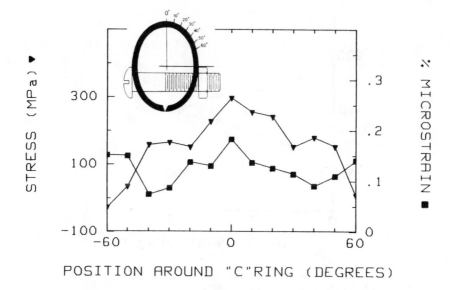

STRESS (MPa) ▼

% MICROSTRAIN ■

POSITION AROUND "C"RING (DEGREES)

Figure 2. Microstrain and microstress variation versus
position on an Inconel C-ring (upper left). The position of
greatest curvature, as well as greatest strain, is at 0°.

Fatigue damage can result in plastic deformation as well. Line broadening
has been shown to be a sensitive indicator of the magnitude of stress levels
in Inconel 718, when measured in fatigue tests, and has proven useful in
predicting the expended fraction of life of the Inconel samples studied [9].

In addition, the measurement of peak breadth has been used in the
evaluation of microstructural changes in steel upon quenching. If steels are
not quenched fast enough to attain a fully martensitic structure, a mixed
martensite - perlite structure remains [10]. The x-ray diffraction peak for
this material has a distinct sharp peak superimposed on a broad base. The
resulting peak shape, then, can be compared quantitatively to, and
distinguished from, the diffraction peak of a fully hardened steel [10].

Peak Height (Intensity)

Preferred orientation of crystallites is often produced in metals by
manufacturing methods such as wire drawing or sheet rolling. Such crystalline
texture is detected in the surface of the aluminum sheet by variation in peak
intensity when measured at two different orientations to the rolling
direction. A preferred orientation in aluminum can stock can affect its
forming response, and in many cases produce a distorted can, i.e. "earing"
texture. Bad stock can be anticipated in advance by an analysis of peak
breadth and intensities.

Data were collected with the PSSD on aluminum can stock from diffraction
patterns in the form of relative full width at half peak height (b/b_p),
relative intensity (I/I_p), and the peak intensity area function percent

$(b/b_p \times I/I_p \times 100)$, from diffracted x-ray peaks at several orientations to the surface. The b_p and I_p values were used to normalize the b and I values from the textured samples and were obtained from an aluminum powder specimen.

The data are plotted in Figure 3. A linear regression line is drawn through the points, with a correlation coefficient of 0.958. This is an excellent fit despite the fact that the samples came from two different manufacturers, were different thicknesses, and were irradiated by two different sources.

Peak intensity has also been utilized in the measurement of retained austenite in steels. Austenite is an interstitial solid solution of carbon and γ-iron. The hardening of steel requires that first it must be heated to a high temperature to form a homogeneous polycrystalline solid solution called austenite. The material is then cooled rapidly to transform the austenite to a hard metastable solid solution called martensite. Because the amount of retained austenite affects the properties of the steel, it is important to

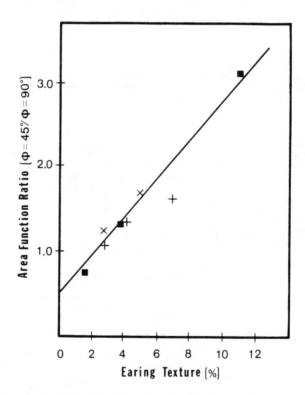

Figure 3. Plot of the area function ratio at two orientations to the rolling direction, 45° and 90°, for incident beam angle of 28° ± 2° to the surface versus percent earing in aluminum can stock. Sheet thickness varied from 0.012" (+), to 0.08 " (■), to 0.127" (x). The regression line drawn through the points has a slope of 0.22, with a correlation coefficient of 0.958.

quantitatively measure that amount. The x-ray diffraction of retained austenite requires that the intensity of the x-rays diffracted be compared with the intensity of the martensite phase. With a recent modification in the PSSD, three x-ray detection surfaces have be used to provide nearly simultaneous measurement of both retained austenite and residual stresses.

CONCLUSIONS

X-ray diffraction techniques have been shown to be effective in the measurement of plastic deformation, residual stress, and crystalline texture in both metals and ceramics caused by manufacturing process and fatigue damage. These material properties are evaluated by the analysis of x-ray diffraction peak position, breadth, and height. The position-sensitive scintillation detector offers the potential for the measurement of these properties of materials as they are being processed and used. It has been shown to be fast, non-contacting, and tolerant of detector to component distance variation -- the necessary requirements for cost-effective in-process inspection of materials. New modifications of the PSSD will allow the simultaneous measurement of phase compositions and residual stresses.

ACKNOWLEDGMENTS

Portions of the work described previously were funded by the Army Materials Technology Laboratory, contract number DAAG46-83-K-0036, and the National Science Foundation, grant number DMC-8615863.

REFERENCES

1. C. O. Ruud, Ind. Res. and Dev. 84-87, (January, 1983).
2. C. O. Ruud, P. S. DiMascio, and D. J. Snoha, in Adv. in X-Ray Anal. 27, (Plenum Press, New York, 1984), pp. 273-282.
3. C. O. Ruud, Nondestructive Methods for Material Property Determination, edited by C. O. Ruud and R. E. Green, Jr. (Plenum Press, New York, 1984), Vol. 1, pp. 21-37.
4. Personal communication, M. A Doxbeck, Benet Weapons Laboratory, Watervliet, New York, May, 1986.
5. B. D. Cullity, Elements of X-Ray Diffraction, Second Edition (Addison-Wesley, Reading, MA, 1978).
6. SAE, "Residual Stress Measurement by X-Ray Diffraction - J784a," Soc. of Auto. Eng., Warrendale, PA (1971).
7. R. Delhez, Th. H. de Keijser, and E. J. Mittemeijer, Surface Engineering 3, 331-342 (1987).
8. C. F. Lo, H. Kamide, W. E. Mayo, and S. Weissman, in press.
9. R. P. Khatri, R. N. Pangborn, T. S. Cook, and M. Roberts, Jour. of Mater. Sci. 21, 511-521 (1986).
10. M. Kurita, I. Ihara, M. Shinbo, and H. Koguchi, Jour. of Testing and Eval. 33-39 (January, 1986).

ACOUSTOELASTIC DETERMINATION OF THE FOURTH ORDER ODF COEFFICIENTS
AND APPLICATION TO R-VALUE PREDICTION

D. DANIEL*, K. SAKATA*, J. J. JONAS*, I. MAKAROW** AND J. F. BUSSIERE**
* McGill University, 3450 University Street, Montreal, Canada, H3A 2A7
** National Research Council Canada, Industrial Materials Research
Institute, 75 De Mortagne Blvd., Boucherville, Québec, Canada, J4B 6Y4

ABSTRACT

The fourth order orientation distribution function (ODF) coefficients
of textured low carbon steel sheets were determined nondestructively from
the anisotropy of the velocity of Lamb (S_0) and SH_0 plate waves measured
using electromagnetic acoustic transducers (EMATs). The three coefficients
(C_4^{11}, C_4^{12}, C_4^{13}) are calculated from five velocity measurements made in
three directions in the rolling plane of the sheet using the Hill
approximation by an iterative numerical method. The coefficients were also
determined from Young's modulus measurements based on a resonance technique
and are compared to those obtained ultrasonically. The comparison with
coefficients determined from X-ray diffraction pole figures permits
adjustment of the C_4^{11} coefficient and then very good agreement is obtained.
The plastic strain ratios (R-values) of the steel samples are predicted from
the adjusted coefficients using a series expansion method based on the
Taylor theory of crystal plasticity. These are compared with experimental
measurements and again good agreement is displayed.

INTRODUCTION

The preferred grain orientations produced by the thermomechanical
treatment of rolled steel sheet induces anisotropy of physical properties
such as Young's modulus, the ultrasonic velocities and the plastic strain
ratio or R-value. Because of the effect of the latter on deep drawability,
it is of industrial interest to perform "on-line" characterizations of the
texture. In this context, considerable research is being carried out
currently on the use of ultrasonic techniques for the determination of both
texture and R-value [1-6].
In this study the determination of the texture by such non-destructive
means is investigated in several types of low carbon steel. The method is
evaluated through a comparison with X-ray (or neutron) diffraction
measurements. With the aid of the X-ray data, the technique for measurement
of the C_4^{11} is "calibrated" by adjusting the single crystal elastic
constants. By this means, the plastic strain ratio can be accurately
predicted from information obtained ultrasonically.

DETERMINATION OF TEXTURE BY ULTRASONIC MEASUREMENTS

The theory of acoustoelasticity in unstressed anisotropic orthorhombic
materials has been reviewed by many authors [7,8]. These works have shown
how the plane wave phase velocities corresponding to different propagation
and polarization directions can be calculated from the 2nd order elastic
constants of the material. For use in thin rolled sheets, Thompson et al.[9]
introduced 'effective' elastic constants which permit the plane wave
solution to be adapted to the calculation of plate wave velocities.
Using the mathematical formalism of texture description (Bunge or Roe)
[10], one can express the polycrystalline elastic constants as functions of
the single crystal ones and of the 4th order orientation distribution
function coefficients assuming an averaging model. Here the Hill average is

used as it has been shown to give reasonably accurate results and to be within 2% of the more realistic (but time-consuming) Kröner type of calculation [11].

On the basis of single crystal values from the literature and employing the Hill averaging procedure, the planar distribution of ultrasonic velocity can be expressed as a function of the three 4th order ODF coefficients : C_4^{11}, C_4^{12} and C_4^{13} in Bunge's formalism [7,10]. Inversely, by measuring three independent ultrasonic quantities, these coefficients can be estimated, from which a partial description of texture can be obtained. In many practical applications such restricted information is sufficient for an evaluation of the physical properties, as shown later in this paper.

This procedure for determination of the 4th order ODF coefficients can be employed with any elastic data, such as Young's modulus measurements. The angular variation of Young's modulus in the rolling plane is described by the following equation :

$$E(\alpha) = [\, S_{1111} \cos^4 \alpha + S_{2222} \sin^4 \alpha + (S_{1212} + \frac{1}{2} S_{1122}) \sin^2 2\alpha \,]^{-1} \qquad (1)$$

where, on the basis of the Hill assumption, the polycrystalline elastic compliances S_{ijkl} are functions of the C_4^{11}, C_4^{12} and C_4^{13} coefficients and of the single crystal elastic constants [10].

NUMERICAL METHOD

Sayers [8] and Thompson et al. [9,12] have derived linear expressions linking the ODF coefficients and the longitudinal or shear wave velocities propagating in the rolling plane or through the thickness of a sheet. These expressions were obtained by neglecting the higher order terms in the Voigt approximation. Hirao et al. [13] have derived similar expressions on the basis of the Hill assumption. This method allows a resolution by a simple calculator.

The method used here is different in that it retains the original expressions of the ultrasonic quantities as functions of the texture coefficients. Due to the complexity of these expressions including higher order terms, the inverse resolution cannot be done analytically without making mathematical approximations. Instead, a numerical method is employed, which can nevertheless solve the problem in about 0.5s on a microcomputer. It employs and operates iteratively a system of equations (at least three independent ones) with three unknowns. If the input is a set of three independent ultrasonic quantities, the solution is exact. Alternatively, if more than three sets of elastic data are available, the method gives a "best fit" (least square type) solution. Comparison of this method with the approximative analytical one indicates a difference of less than 2%. A similar approach was used for the determination of the texture coefficients from Young's modulus data.

EXPERIMENTAL WORK

Fifteen commercial low carbon steel sheets of 4 different types were studied : HSLA, commercial grade rimmed, deep drawable Al-killed and extra deep drawable interstitial free steels. All these steels were cold rolled and annealed.

The plastic strain ratio R at 15% elongation was measured by tensile testing in samples whose longitudinal axes were inclined at 15 or 22.5° intervals with respect to the rolling direction. A grid technique was used to measure the width and longitudinal strains.

Young's modulus was measured by means of Modul-R equipment every 15°
in the rolling plane of the sheets. This apparatus determines the resonant
frequency, the square of which is proportional to the Young's modulus of the
specimen. It is commonly used in steel industry as a *destructive* way of
determining $R=(R_0+2R_{45}+R_{90})/4$ and $\Delta R=(R_0-2R_{45}+R_{90})/2$ from empirical
correlations derived by Stickels and Mould [14]. Nevertheless, texture
information can be obtained from these measurements by employing a variation
of the ultrasonic method described above. The latter has the advantage that
it is *nondestructive* ; also, it is not limited to sheet thicknesses of about
2mm.

The ODFs of the 15 steels were also determined from X-ray (or neutron
for 2 samples) diffraction measurements of pole figures ; the data were
converted into texture coefficients in the conventional way.

The longitudinal S_0 and quasishear SH_0 (plate) wave velocities were
measured for three directions of propagation (0,45,90°) in the rolling plane
using electromagnetic acoustic transducers (EMATs). The S_0 mode velocities,
measured at a frequency of 500kHz, were extrapolated to zero frequency to
correct for the dispersion effect associated with finite plate thickness
[15]. The velocities of bulk ultrasonic (longitudinal and transverse) waves
propagating through the thickness of the sheet were also determined. Because
of the errors inherent in thickness measurement, this method seems to be
less reliable than the in-plane ones.

COMPARISON BETWEEN THE X-RAY AND ULTRASONIC TEXTURE COEFFICIENTS

The major problem encountered in the determination of texture
coefficients from ultrasonic measurements arises from the inaccuracy
associated with the C_4^{11} coefficient which is calculated from the *absolute*
values of the velocities, while C_4^{12} and C_4^{13} are determined mainly from the
variations in the velocities [1]. The comparison of texture coefficient
predictions by ultrasonic measurements with those obtained from Young's
modulus data shows very good agreement. Thus the problem of the accurate
determination of C_4^{11} seems to be related to the calculation of the
polycrystalline elastic constants and not to absolute errors in velocity
measurements.

Two problems are involved in the calculation of these constants : the
limitations of the averaging model and the somewhat arbitrary choice of the
single crystal constants to be used. On the basis of the Hill approximation,
the present calculation was "calibrated" by adjusting the single crystal
data. From a review of literature data for single crystal elastic constants
of iron and mild steels [16], a first calculation was carried out with the
following set : $C_{11}=230GPa$, $C_{12}=136GPa$ and $C_{44}=116.5GPa$. The inaccuracies
associated with these values are of the order of 4%. By modifying these
constants within their intervals of reliability, it was found that C_4^{11}
results obtained from Young's modulus data are consistent with X-ray data
obtained with the following set : $C_{11}=236GPa$, $C_{12}=141GPa$ and $C_{44}=118.5GPa$.
As illustrated in Figure 1, the calibration only affects the prediction of
C_4^{11}. The spreads observed in Figure 1 around the x=y line are reasonable
compared to the errors involved in X-ray analysis ($|\Delta C_4^{\mu\nu}|\sim0.3$) and the
reproducibility of ultrasonic velocity measurements (0.2%).

APPLICATION TO R-VALUE PREDICTION

R-values were predicted in two different ways. The first involves the
use of the Taylor crystal plasticity model ; the second is based on
empirical correlations. The series expansion method of texture description
can be employed to calculate the angular dependence of plastic strain ratio
in the rolling plane [10]. It calculates the total plastic work in the

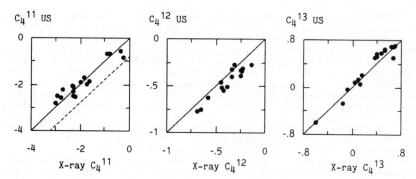

Figure 1 : Correlation between X-ray and ultrasonically determined 4th order ODF coefficients. The broken line indicates the correlation before calibration.

polycrystal by assuming that each crystallite undergoes the same macroscopic strain rate tensor as the sample as a whole. The "correct" R-value is assumed to be the one which minimizes the average plastic work. In the present calculation, the plastic deformation of the grains was described in terms of slip on both the {110} and {112} planes along the <111> direction. Although the higher order ODF coefficients are required for a rigorous description of the plastic properties [10,17], it is of interest that, as illustrated in Figure 2, the predictions made on the basis the 4th order coefficients alone are acceptable for the four types of steels studied.

Using the present experimental measurements on the 15 low carbon steels, some linear correlations have also been obtained between R and ΔR on the one hand and the ultrasonically measured C_4^{11} and C_4^{13} coefficients on the other. The correlations are illustrated in Figure 3 and the equations are listed below :

$$R = -0.57 \times C_4^{11} + 0.57 \qquad (2)$$

$$\Delta R = 0.77 \times C_4^{13} + 0.06 \qquad (3)$$

The respective correlation coefficients are 0.94 and 0.88. They are equivalent to those determined from Modul-R equipment (destructive way).

CONCLUSIONS

The 4th order ODF coefficients were determined acoustoelastically on 15 commercial low carbon steels. The calculations were carried out by a numerical iterative method on the basis of the Hill approximation. Good agreement can be obtained with the coefficients derived from X-ray measurements (even for the C_4^{11}) if the single crystal elastic constants employed are adjusted in the manner shown. It has been pointed out that the same information can be derived from Young's modulus measurements obtained destructively from Modul-R equipment.

The ultrasonically determined 4th order ODF coefficients lead to good predictions of R-value when the Taylor model is used. It is also of practical interest that a good empirical correlation can be obtained between

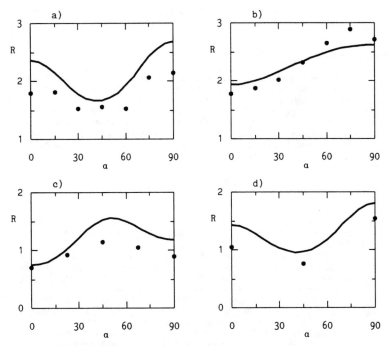

Figure 2 : Comparison between experimental (●) and ultrasonically predicted (-) R values in the rolling plane of a) Al-killed, b) interstitial free, c) HSLA and d) rimmed steels as a function of the angle α with the rolling direction.

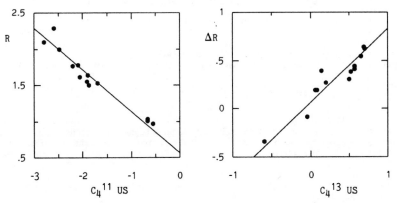

Figure 3 : Correlations between R and C_4^{11} and ΔR and C_4^{13} in the present low carbon steels.

the ultrasonic C_{44}^{11} and C_{44}^{13} coefficients on the one hand and the experimentally measured values of R and ΔR on the other.

ACKNOWLEDGEMENTS

The authors acknowledge with gratitude the financial support received from the Natural Sciences and Engineering Research Council of Canada, the Canadian Steel Industry Research Association and the Ministry of Education of Québec (FCAR program). They are indebted to M. McLean (Algoma), J. Thompson (Stelco) and A. Vigeant (Dofasco) for supplying steels. They also express their thanks to Atomic Energy of Canada Limited for X-ray and neutron diffraction measurements, to the Kawasaki Steel Corporation for supplying steels and texture measurements, and to Dr. C. K. Jen for the dispersion correction.

REFERENCES

[1] A. V. Clark, Jr, R. C. Reno, R. B. Thompson, J. F. Smith, G. V. Blessing, R. J. Fields, P. P. Delsanto and R. B. Mignogna, Ultrasonics 26, 189 (1988).

[2] D. R. Allen, R. Langman and C. M. sayers, Ultrasonics 23, 215 (1985)

[3] M. Hirao, N. Hara and H. Fukuoka, J. Acoust. Soc. Am. 84, 667 (1988)

[4] O. Cassier, C. Donadille and B. Bacroix, in Review in Progress of Quantitative N. D. E., San Diego, jul 31-aug 5 1988, in press.

[5] A. Wilbrand, W. Repplinger, G. Hübschen and H. J. Salzgurger, in 3th International Symposium on the Nondestructive Characterization of Materials, Saarbrucken, Germany, oct 3-6 1988, in press.

[6] J. F. Bussière, C. K. Jen, I. Makarow, B. Bacroix, P. Lequeu and J. J. Jonas, in Nondestructive Characterization of Materials II, edited by J. F. Bussière et al. (Plenum Press, New York, 1987) pp. 523-533.

[7] R. B. Thompson, S. S. Lee and J. F. smith, J. Acoust. Soc. Am. 80, 921 (1986).

[8] C. M. Sayers, J. Phys. D 15, 2157 (1982).

[9] R. B. Thompson, S. S. Lee and J. F. smith, Ultrasonics 25, 133 (1987).

[10] H. J. Bunge, Texture Analysis in Materials Science, (Butterworth, London, 1982).

[11] P. R. Morris, Int. J. Engng. Sci. 8, 49 (1970).

[12] S. S. Lee, J. F. Smith and R. B. Thompson, in Nondestructive Characterization of Materials II, edited by J. F. Bussière et al. (Plenum Press, New York, 1987) pp. 555-562.

[13] M. Hirao, K. Aoki and H. Fukuoka, J. Acoust. Soc. Am. 81, 1434 (1987).

[14] C. A. Stickels and P. R. Mould, Metall. Trans. 1, 1303 (1970).

[15] B. A. Auld, Acoustic Waves and Fields in Solids, (Wiley-Interscience, New York, 1973), Vol. II, Chap. 10.

[16] H. M. Ledbetter and R. P. Reed, J. Phys. Chem. Ref. Data 2, 531 (1973).

[17] D. Daniel and J. J. Jonas, in 8th International Conference on Textures of Materials, edited by J. S. Kallend and G. Gottstein (AIME Proc. 1988) pp. 1079-1084.

ULTRASONIC CHARACTERIZATION OF TEXTURE

Y. LI AND R. B. THOMPSON
Ames Laboratory, Iowa State University, Ames, IA 50011

ABSTRACT

This paper will propose a new technique to characterize texture of rolled plates of cubic crystallites. This technique uses information from ultrasonic velocities of high order plate mode to improve the estimation of orientation distribution coefficients (ODC's), especially W_{400}. Also discussed will be the generalization of this technique to the case of hexagonal crystallites.

INTRODUCTION

It is well know that the texture (preferred grain orientation) of a polycrystalline metal sheet or plate strongly influences the formability of the material. Traditionally, texture is determined destructively through X-ray or neutron diffraction techniques. Recently, ultrasonic techniques have shown the promise of being able to characterize texture quickly and nondestructively in polycrystals of cubic crystallites [1,2]. Present techniques are generally based on the information from velocities of S_0 and SH_0 plate waves. This paper presents two extensions to the current theory and techniques. In one extension, higher order plate modes are utilized to make better estimations of the orientation distribution coefficients. The second extension is made to polycrystals of hexagonal crystallites. Included are discussions on the similarities and differences in applying the present and newly proposed theory and techniques.

The texture or preferred grain orientation of a plate is often described by the crystallite orientation distribution function (CODF) expressed as a series of spherical harmonics with weightings W_{LMN} (Roe's notation) [3,4] or $C_{\ell}^{\mu \upsilon}$ (Bunge's notation) [5] known as the orientation distribution coefficients (ODC's). The texture is characterized quantitatively by the ODC's. Although the complete specification of texture requires knowledge of all W_{LMN} for $L \geq 0$, it has been shown that the formability of polycrystalline metals is most strongly influenced by the lowest order coefficients, W_{LMN} for $L \leq 4$. Of this order, W_{400}, W_{420}, and W_{440} are the only nonzero and independent coefficients for cubic materials such as Fe, Al, and Cu. For hexagonal materials such as Ti and Zr, W_{200} and W_{220} are also nonzero and independent [6,7].

PRESENT TECHNIQUE FOR CUBIC MATERIALS

The presence of texture gives rise to anisotropy in the polycrystalline metal and hence influences the ultrasonic wave speed. Ultrasonic measurement of texture relies on the information from ultrasonic wave speeds in different propagation directions. In a technique which has received considerable recent attention [8,9], use is made of S_0 and SH_0 modes of plate waves. Through the measurements of S_0 and SH_0 wave velocities at $0°$, $45°$, and $90°$ with respect to the rolling direction, W_{400}, W_{420}, and W_{440} for cubic materials can be calculated from following equations:

$$SH_0: \quad W_{400} = \frac{35 \sqrt{2} \rho}{16\pi^2 C} \left[V_{SH_0}^2 (45) + V_{SH_0}^2 (0) - \frac{2T}{\rho} \right] ;$$

$$W_{440} = \frac{\sqrt{35}\,\rho}{16\pi^2 C} \left[v_{SH_o}^2 (45) - v_{SH_o}^2 (0) \right] \quad ;$$

$$S_o: \qquad W_{400} = \frac{35\sqrt{2}\,\rho}{32\pi^2 C(3 + 8\,P/L + 8\,P^2/L^2)} \left(v_{S_o}^2 (0) + v_{S_o}^2 (90) + 2v_{S_o}^2 (45) - \right.$$

$$\left. 4\left(\frac{L - P^2/L}{\rho}\right) \right] \quad ; \tag{1}$$

$$W_{420} = \frac{7\sqrt{5}\,\rho}{32\pi^2 C(1 + 2P/L)} \left[v_{S_o}^2 (90) - v_{S_o}^2 (0) \right] \quad ;$$

$$W_{440} = \frac{\sqrt{35}\,\rho}{32\pi^2 C} \left[v_{S_o}^2 (0) + v_{S_o}^2 (90) - 2v_{S_o}^2 (45) \right] \quad .$$

In the above equations, T, P, and L are elastic moduli for the corresponding isotropic material (texture free), and C is the elastic anisotropy constant. Depending on the averaging scheme employed, these constants are related differently to the single crystal elastic constants and compliances [10].

Theoretically, the S_o velocities in Eqs. (1) should be the long wave length limit of the S_o mode velocities due to the dispersive nature of S_o plate wave. However, the difference between the two is in general small as long as the plate thickness is relatively small with respect to the wave length.

The above procedure makes consistently accurate prediction of W_{420} and W_{440}, which depend on relative velocities. However difficulties have been encountered in the prediction of W_{400} in aluminum. This is believed to be a consequence of the need for an absolute rather than a relative measurement, the strong dependence of the prediction on the isotropic moduli L, P, and T, and the small anisotropy of aluminum plates.

USE OF HIGHER ORDER PLATE MODES IN CUBIC POLYCRYSTALS

An improved technique for determining W_{400} takes advantage of higher order plate modes [11]. It is known that, for an isotropic free plate, there always exist infinite Lamé modes, occurring at $K = \frac{\pi}{b} n$. The lowest order Lamé mode occurs at the point where the S_o and SH_1 modes touch each other tangentially on the dispersion curves, as depicted in Fig. 1. At the tangency point, the S_o mode consists of pure SV partial waves polarized in the sagittal plane and propagating at ±45° with respect to the plate surfaces. The SH_1 mode is polarized perpendicularly to the sagittal plane and also consists of partial waves propagating at ±45° with respect to the plate surfaces. In the presence of texture (weak anisotropy), the dispersion curves of S_o and SH_1 modes have been shown to cross over one another or split at the Lamé point as shown in Fig. 2 [12]; therefore, information on texture can be inferred from the differences between these two wave speeds. In other words, W_{400}, W_{420} and W_{440} all can be determined through relative measurement of S_o and SH_1 wave velocities. Although computation of the W's from S_o and SH_1 velocities based on exact solution is not mathematically feasible, perturbation technique is readily available to accomplish this inversion process. Using the formula developed by Auld [13], we find the difference in wave numbers for SH_1 and S_o modes are:

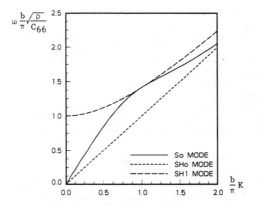

Fig. 1: SH_o, S_o, and SH_1 mode dispersion curves in an isotropic plate

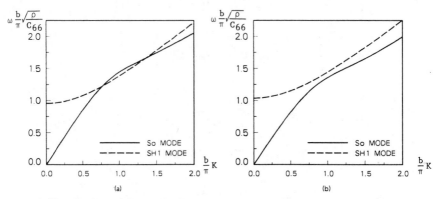

Fig. 2: S_o and SH_1 mode dispersion curves in an anisotropic plate.
(a) SH_1 and S_o cross over; (b) SH_1 and S_o split.

$$\Delta K = \frac{K\pi^2 C}{35\ T} (25\sqrt{2}\ W_{400} - 4\sqrt{5}\ W_{420} \cos 2\alpha + 6\sqrt{35}\ W_{440} \cos 4\alpha) \qquad (2)$$

where $K = \frac{\pi}{b}$, and α is the wave propagation direction with respect to rolling direction. Therefore, explicit expressions for W's can be easily derived and expressed in terms of velocity differences at 0°, 45°, and 90° propagation directions. This technique has been evaluated using exact solutions for wave propagation in orthotropic free plates of polycrystalline Cu material [11] and found to work satisfactorily [12]. Table I shows the comparison of true W's and W's obtained from Eq. (2).

In contrast to the techniques mentioned earlier, this technique eliminates the need for absolute measurement of velocities and reduces the sensitivity to the average isotropic moduli.

Table I. Comparisons of exact W's and their estimations from Eq. (2)
$(\times 10^{-3})$

	W_{400}	W_{420}	W_{440}
Exact	1.00	1.00	1.00
Voigt	1.12	1.33	1.13
Hill	1.02	1.21	1.03
Reuss	0.91	1.08	0.91

CHARACTERIZING TEXTURE OF HEXAGONAL POLYCRYSTALS

For materials with hexagonal crystallites, there are five nonzero and independent ODC's. W_{200}, W_{220}, W_{400}, W_{420}, and W_{440}. As for cubic crystallites, one again can express the elastic constants and wave speeds in terms of the ODC's and then solve for the latter. The result is

$$SH_o: \quad B\, W_{400} - \sqrt{5}\, A_3\, W_{200} = \frac{105\sqrt{2}\,\rho}{16\pi^2} \left[V_{SH_o}^2(45) + V_{SH_o}^2(0) - \frac{2T}{\rho} \right];$$

$$W_{440} = \frac{3\sqrt{35}\,\rho}{16\pi^2 B} \left[V_{SH_o}^2(45) - V_{SH_o}^2(0) \right];$$

$$S_o: \quad [3 + 8\, P/L + 8\, P^2/L^2]\, B\, W_{400} + 2\sqrt{5}\; \{[1 - 2\, P^2/L^2]\, A_1 - P/L\, A_2\} W_{200}$$

$$= \frac{105\sqrt{2}\,\rho}{32\pi^2} \left[V_{S_o}^2(0) + V_{S_o}^2(90) + 2V_{S_o}^2(45) - \frac{4(L - P^2/L)}{\rho} \right]; \qquad (3)$$

$$(1 + 2\, P/L)\, B\, W_{420} + \sqrt{3}\, [A_1 + (P/L)A_2] W_{220}$$

$$= \frac{21\sqrt{5}\,\rho}{32\pi^2} \left[V_{S_o}^2(90) - V_{S_o}^2(0) \right];$$

$$W_{440} = \frac{3\sqrt{35}\,\rho}{32\pi^2 B} \left[V_{S_o}^2(0) + V_{S_o}^2(90) - 2V_{S_o}^2(45) \right],$$

where L, P, T are elastic constants for the corresponding isotropic material. A_1, A_2, A_3, and B are elastic anisotropy constants. These constants are related to the elastic moduli of single crystals and, depending on the averaging scheme, the relations vary [7,14]. Regardless of the averaging scheme, there always exists the relation $A_1 + A_2 + A_3 = 0$.

One can see that the situation is considerably more complex than for the cubic case. The determination of W_{440} is very similar, being possible from the angular dependence of either the S_o or SH_o mode velocities. Absolute velocity measurements are necessary to determine W_{200} and W_{400}. The required algorithms are sensitive to the values of the isotropic moduli, and one must further solve a pair of linear equations involving SH_o and S_o mode data to separately determine W_{200} and W_{400}. The stability of this further inversion has not yet been investigated, but the erratic predications from cubic case imply questionable reliability for doing so. . Furthermore, there is insufficient information to separately determine W_{220} and W_{420}, although the indicated linear combination depends only on a relative measurement and should be determined with high precision.

Even more than in the cubic case, important additional information appears to be in the behavior of the Lamé modes. At the first point of tangency, one finds that

$$\Delta K(SH_1) = \frac{K\pi^2}{105T} \left[(\sqrt{10}\ A_3 W_{200} + 6\sqrt{2}\ B W_{400}) + \right.$$
$$\left. + (2\sqrt{15}\ A_3 W_{220} + 8\sqrt{5}\ B W_{420}) \cos 2\alpha + 4\sqrt{35}\ B W_{440} \cos 4\alpha \right];$$

$$\Delta K(S_o) = - \frac{K\pi^2}{105T} \left[(2\sqrt{10}\ A_3 W_{200} + 19\sqrt{2}\ B W_{400}) + \right.$$
$$\left. + (4\sqrt{15}\ A_3 W_{220} - 12\sqrt{5}\ B W_{420}) \cos 2\alpha + 2\sqrt{35}\ B W_{440} \cos 4\alpha \right]; \tag{4}$$

$$\Delta K = \Delta K(SH_1) - \Delta K(S_o) = \frac{K\pi^2}{105T} \left[(3\sqrt{10}\ A_3 W_{200} + 25\sqrt{2}\ B W_{400}) + \right.$$
$$\left. + (6\sqrt{15}\ A_3 W_{220} - 4\sqrt{5}\ B W_{420}) \cos 2\alpha + 6\sqrt{35}\ B W_{440} \cos 4\alpha \right].$$

It is clear that there is now ample basis to separately determine various linear combinations of W_{220} and W_{420} from angular dependences of the velocities. Hence these constants can be uniquely determined. Furthermore, one can obtain one combination of W_{200} and W_{400} from relative measurements of the shifts of the SH_1 and S_o modes. Further measurements must be sought to obtain a second linear combination from relative rather than absolute data since the shifts $\Delta K(SH_1)$ and $\Delta K(S_o)$ are defined with respect to the Lamé mode point of an isotropic polycrystal material. Thus it can not be expected that the angle independent part of these quantities will be a basis for accurate predictions for W_{200} amd W_{400}.

CONCLUSIONS

Use of higher order mode information to improve the accuracy of determination of W_{400} in cubic materials is first discussed. In particular, it is shown how this ODC can be obtained from the relative shift of the lowest order Lamé modes. In addition, new equations are then presented for the characterization of texture in hexagonal materials. Based on measurements of the velocities of the SH_o and S_o plate modes, it is shown that W_{440} can be determined from relative angular variations, W_{400} and W_{200} can be determined from absolute measurements, and only a linear combination of W_{220} and W_{420} can be determined from relative measurements. The use of higher order Lamé mode behavior is proposed to gain additional information. It is shown that W_{220} and W_{420} can now be separated from angular variations and one linear combination of W_{200} and W_{400} can be determined from relative measurements.

88

ACKNOWLEDGMENTS

Ames Laboratory is operated for the U. S. Department of Energy by Iowa State University under contract no. W7405-ENG-82. This work was supported by the Office of Basic Energy Sciences, Division of Materials Sciences.

REFERENCES

1. R. B. Thompson, S. S. Lee and J. F. Smith, J. Acoust. Soc. Am. 80, 921 (1986).
2. R. B. Thompson, S. S. Lee and J. F. Smith, Ultrasonics 25, 133 (1987).
3. R.-J. Roe, J. Appl. Phys. 36 (6), 2024 (1965).
4. C. M. Sayers, J. Phys. D: Appl. Phys., 15, 2157 (1982).
5. H.-J. Bunge, Texture Analysis in Materials Science, Translated by P. R. Morris (Butterworths, Berlin, 1982).
6. R.-J. Roe, J. Appl. Phys. 31 (5), 2069 (1966).
7. C. M. Sayers, Ultrasonics 24 (5) 289, (1986).
8. S. S. Lee, J. F. Smith and R. B. Thompson, in Nondestructive Characterization of Materials, edited by J. F. Bussiere (Plenum Press, New York, 1987), p.155.
9. S. J. Wormley, R. B. Thompson and Y. Li in Review of Progress in Quantitative Nondestructive Evaluation 7B, edited by D. O. Thompson and D. E. Chimenti (Plenum Press, New York, 1988), p. 1639.
10. M. Hirao, K. Aoki, and H. Fukuoka, J. Accoust. Soc. Am., 81 1434 (1987).
11. Y. Li and R. B. Thompson, in Review of Progress in Quantitative Nondestructive Evaluation 8, edited by D. O. Thompson and D. E. Chimenti (Plenum Press, New York, in press).
12. Y. Li and R. B. Thompson, in Review of Progress in Quantitative Nondestructive Evaluation 8, edited by D. O. Thompson and D. E. Chimenti (Plenum Press, New York, in press).
13. B. A. Auld, Accoustic Fields and Waves in Solids, Vol. II (John Wiley and Sons, 1973).
14. Y. Li, Notes on relationship between C_{ij} for hexagonal polycrystallites and W_{LMN}. Unpublished, (1987).

PART III

Degradation (Creep, Fatigue, Etc.)

MATERIALS PROPERTIES CHARACTERIZATION
IN FOSSIL-FUEL POWER PLANTS

S. M. Gehl, R. Viswanathan, and R. D. Townsend
Electric Power Research Institute
3412 Hillview Avenue
Palo Alto, California 94303

ABSTRACT

Significant changes in microstructure occur in many fossil plant components that operate for long times at elevated temperatures. Some examples include hydrogen damage in steam-generating boiler tubes, carbide coarsening in superheater tubes, creep cavitation in the heat-affected zones of thick-section weldments, and temper embrittlement in turbine rotors and disks. Quantitative estimates of the remaining useful lives of components affected by these damage mechanisms require that the degree of damage be determined and related to the life-limiting mechanisms.
The paper discusses recent progress in the development and field demonstration of nondestructive techniques for characterizing the damage types listed above. This research has produced several products that have achieved or are close to commercial status. The research has also helped to define some of the limitations related to access, sensitivity, and signal interpretation of the nondestructive techniques. The paper discusses how these limitations can often be mitigated by using several nondestructive techniques in concert to obtain a more complete picture of material condition or by removing small specimens for detailed laboratory analysis.

INTRODUCTION

Utilities employ a variety of approaches to reduce the cost of fossil-fired electricity generation and improve ability to respond to changing and uncertain electricity demand. Programs to extend the life of current plants, convert base-load plants to cycling operation, and develop predictive maintenance strategies are examples of this trend. Techniques for assessing the useful life of key components are needed in such programs to determine inspection and maintenance intervals and estimate the true lifetime cost of a planned change in operating conditions. The use of life assessment procedures in utilities is in its infancy, but is expected to grow as the reliability of the techniques improves.

The Electric Power Research Institute (EPRI) has introduced a three level approach to life assessment [1]. As illustrated in Figure 1, the three levels are characterized by progressive increases in the information required, analytical detail, and cost of performing the evaluation. To minimize the level of effort expended, higher evaluation levels are undertaken only if the lower levels fail to yield the desired lifetime.

For highly stressed components, remaining life is determined by the time to initiate a crack and the time for the crack to propagate to a critical size. The difficulty in life assessment is that crack initiation times and crack propagation rates differ widely among nominally identical components because of minor variations in alloy chemistry, thermal mechanical treatment, and in-service aging. Figure 2 shows that aging effects can simultaneously accelerate crack growth and reduce the critical size for unstable crack propagation, thus drastically shortening life. This illustration points out the need for the development of reliable nondestructive techniques to characterize the metallurgical condition and mechanical properties of aged components. Materials condition monitoring

EPRI Three Level Assessment

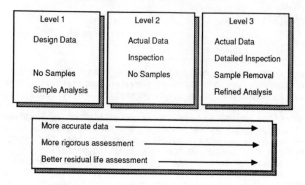

Figure 1. The EPRI three-level approach to life assessment. Level 1 relies on design data to estimate component dimensions, stresses and temperatures and makes use of a simplified analytical treatment. The extent of inspection and the detail of analysis increases progressively with the other levels of assessment.

Effect of Aging on Flaw Growth

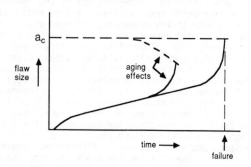

Figure 2. Aging effects on residual life. Creep damage can accelerate flaw growth, while temper embrittlement reduces the critical flaw size for unstable crack propagation.

techniques also have the potential for detecting damaging microstructural changes before cracks form, thus giving the utility advance warning of the need to repair or replace the component in question. Materials characterization methods are applicable to the second and third levels of life assessment. At either level, reliable materials characterization techniques can improve the accuracy and reduce the cost of remaining life estimates.

PARTIALLY DESTRUCTIVE CHARACTERIZATION METHODS

It's convenient to divide "nondestructive" methods for characterizing material condition into methods that are in reality partially destructive and those that are completely nonintrusive. Until recently, partially destructive techniques required the removal of boat or plug samples from the component. The advantage of these methods is that once a specimen is removed, it can be subjected to destructive examinations and mechanical property tests that would not be possible in the field. The obvious disadvantage is that the areas from which boat or plug samples are removed must be weld repaired. In addition to the time and expense of making the repair, a post weld heat treatment is frequently required, which may itself be the cause of subsequent cracking.

This problem has been addressed by the recent development of techniques for the field removal of small samples from large components in geometries that provide material for mechanical property tests and metallurgical examinations. If a sufficiently small piece is removed from the component, there is no need for weld repair. Grinding or buffing to smooth the edges of the cut surface is usually the only repair required after removal of the miniature specimens.

One example of small sample removal is the minispecimen technique for stress rupture testing developed for EPRI by the Central Electricity Generating Board [2]. In this method, a small wedge of material is removed from a pipe or header using a drilling jig. A cylindrical sample approximately 15mm long and 3mm in diameter is machined from the wedge. Grip ends are electron-beam welded to the cylinder to form a tensile specimen. The specimen is usually loaded in creep to the service stress at a temperature above the service temperature. Extrapolation techniques have been developed to allow the estimation of remaining life at the service temperature from a single elevated temperature isostress test. The ability to obtain data from a single specimen minimizes the need for sample removal. By conducting the test at an elevated temperature, remaining life estimates can usually be obtained in about 1,000h, as opposed to the 10,000h or more required for tests performed at the service temperature.

A second approach to removing small specimens is a scoop-type sampling device developed for removing small disk samples from turbine rotors, piping, and headers [3]. This device is operated remotely and can remove samples from rotor bores and other restricted access locations. The sample diameter ranges from 13 to 25mm and the thickness from 0.8 to 2.5mm. The time required for specimen removal is approximately 30 to 60 minutes. Initially, these samples have been used for metallographic examination and alloy chemistry analysis. However, small specimens for punch tests and other ultraminiature mechanical property measurements can be prepared from the disk samples. The scoop sampling technique appears to be applicable to a variety of component geometries and can remove specimens in a relatively short time. Therefore, the use of this or similar devices is expected to grow along with ultraminiature test methods.

A disadvantage shared by all sampling techniques, including boat and plug sampling, is the small size of the sample relative to the spatial variations in the properties of interest. Because of heat-specific variations in creep resistance, for example, a sample removed from a single spool piece cannot be presumed to be representative of an entire steam line. Even within a single spool piece or fitting, local variations in operating temperature or stress can produce significant local differences in mechanical behavior. Ultraminiature specimens are a particular concern in this regard because they are obtained from the near-surface region of the component, and therefore are not sensitive to through-thickness variations in properties, if present. Care should always be exercised in planning sample removal to verify that the specimens obtained are representative of the component. In some cases, removal of multiple specimens may be necessary to provide an adequate characterization of condition.

NONINTRUSIVE TECHNIQUES

The problem of local variations in material properties disappears if truly nondestructive techniques that provide rapid, low-cost, and accurate assessments of component condition are available. Although this ideal situation has yet to be achieved, monitoring techniques incorporating significant advances in component characterization are now available and several others are nearing commercial status.

Microstructural examination, mechanical testing, acoustic properties, electrical and magnetic measurements, chemical and electrochemical response, and other principles have formed the basis for materials characterization techniques used on power plant components. A complete review of the available technology is beyond the scope of this paper. Instead, we will focus on several emerging technologies that have the potential for accurate life assessments and reduced inspection cost.

HARDNESS MEASUREMENTS

Hardness measurements have been used for many years as a quality control measure and to assess changes in the room-temperature mechanical properties of service-exposed components. More recently, attempts have been made to correlate hardness directly with creep and fatigue life expenditure. An example is the use of microhardness to estimate the extent of creep damage in Cr-Mo-V turbine rotors and blade attachments [4, 5]. In this research, the combined effects of thermal softening and creep cavitation were shown by plotting the ratio of posttest to initial hardness against a stress-corrected Larson-Miller parameter, as illustrated in Figure 3. Hardness measurements have also been employed in other recent studies [6, 7], and are being investigated in EPRI project RP2481 on advanced NDE methods for turbine rotor life assessment.

Although the initial results using hardness ratios are promising, there appear to be circumstances in which the technique will not work. In one recent case, a highly stressed turbine blade attachment failed by a creep cracking mechanism with no detectable hardness change [8]. Additional studies are needed to establish the range of applicability of the hardness ratio technique.

REPLICA METALLOGRAPHY

Microstructure examinations of large components are usually performed by metallographic preparation in the field followed by plastic replication of the polished and etched surface. A standard practice for replica field metallography is now available [9]. Recent interest in replica metallography has focused on the use of quantitative stereology techniques to assess microstructural condition.

Hardness Ratio in Creep-Tested Rotor Steel

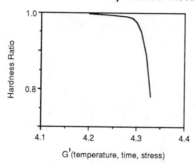

Figure 3. Ratio of post exposure hardness to initial hardness for crept Cr-Mo-V rotor specimens plotted against a stress-corrected Larson-Miller parameter (after Ref. 5).

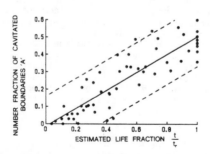

Figure 4. Relationship of the "A" parameter to creep life expenditure for 1Cr-0.5Mo steel [10].

An example of quantitative stereology applications is the use of the number fraction of grain boundaries containing creep voids, called the "A" parameter, as a means of estimating creep life expenditure [10]. This parameter was selected in part because it can be measured easily in either optical or scanning electron microscopes, and because it is relatively insensitive to the magnification at which the replica is viewed. In contrast, parameters such as the cavity volume fraction or number density of cavities are much more sensitive to magnification as well as the details of metallographic preparation.

Figure 4 shows the relationship between creep life expenditure and the "A" parameter for heat-affected zone material from two heats of 1Cr 0.5Mo steel. Some of the scatter evident in the graph is due to differences in cavitation for the two heats of material and differences in test temperature; however, much of the scatter is unexplained. The minimum bounding curve of the scatter band can be used to give a conservative estimate of remaining life. These and other similar data on Cr-Mo and Cr-Mo-V steels indicate that standard techniques for specimen preparation and image analysis should be used to obtain consistent results [10]; and that different correlations between remaining life and the "A" parameter must be used for Cr-Mo and Cr-Mo-V steels [5, 10, 11]. Therefore, each laboratory making "A" parameter measurements should calibrate the metallographic procedure on the same alloy as the component under study.

Replica techniques have also been used to characterize carbide particle morphology and spacing in Cr-Mo steels [12]. In this case, an extraction replica technique is used to remove carbide particles for subsequent examination in a transmission electron microscope. The electrolytic polishing, etching, and matrix dissolution steps would require special modification for field use. In addition, work to date has employed vacuum equipment for deposition of the thin carbon replica. To get around the need for vacuum equipment, a field technique would require a two-stage plastic-carbon replica. Carbide particles on a plastic extraction replica would be transferred to a carbon film in the laboratory for transmission electron microscopy.

Because of the complexity of this method, the requirement for special equipment, and the development effort required to make it suitable for field use, it is unlikely that carbide particle spacing measurements will be widely used in the near future. One possible exception is through the use of miniature specimens, which would eliminate the need for adapting the extraction replication procedures for field use.

The difficulties in preparing specimens and the fact that measurements are confined to surface or near-surface locations frequently offset the advantage of direct microstructure characterization inherent in metallographic methods. Therefore, the attention of researchers is increasingly turning toward indirect methods of assessing microstructural condition. Some of these methods are discussed in the following sections.

ULTRASONIC METHODS

The velocity of sound in steels is a function of several microstructural characteristics, such as grain size and carbide morphology and distribution [13, 14]. The presence of microvoids caused by hydrogen damage and creep also affects sound velocity [15, 16]. The development of velocity measurement into a reliable, field-usable NDE technique has application in remaining life estimation of boiler superheater tubes, headers, main and reheat steam lines, and carbon steel water wall tubes. Currently, the most common methods of inspecting these components are specimen removal and replication for creep-damaged components, and attenuation of a back-wall-reflected ultrasonic pulse in the case of hydrogen damage. In comparison, ultrasonic velocity measurement is rapid, completely nondestructive, and can inspect the through-wall thickness of most components.

A schematic drawing of the EPRI-developed hydrogen damage detection technique is shown in Figure 5. Velocity measurements are capable of locating even small amounts of hydrogen damage in water wall tubes. In addition, the technique is highly specific for hydrogen damage [16, 17]. In two field demonstrations, comparisons of the ultrasonic velocity and ultrasonic attenuation methods were performed. The results indicated that the velocity measurement reliably detects all areas of hydrogen damage and does not record indications in undamaged areas. Independently performed attenuation measurements also located the hydrogen damage but had a relatively high false call rate [17]. However, since attenuation measurements can usually be made more rapidly than velocity measurements, EPRI is evaluating the combined use of the two methods. In this approach, a screening examination would be made with the attenuation method to identify suspect areas, followed by a velocity measurement to confirm the presence of hydrogen damage.

Laboratory studies indicate that creep cavitation also reduces ultrasonic velocity if the cavitation is distributed over a large enough volume. However, creep damage confined to a small volume, as in the heat affected zone of weldments, is somewhat more difficult to detect. Recent studies indicate that measurement of the energy backscatter of an ultrasonic pulse may be a suitable means of detecting creep cavitation in plant components [17].

BARKHAUSEN NOISE ANALYSIS

Magnetic hysteresis loops contain a fine structure, shown in Figure 6, which results from the nonuniform motion of domain walls [18]. Domain wall motion is impeded by obstacles, such as voids, carbide particles, and local stresses, until a sufficiently high applied magnetic field overcomes the pinning effect. Once this occurs, the domain wall travels rapidly through the structure until another obstacle is encountered. The resulting discontinuities in the hysteresis loop are called "Barkhausen noise." The noise signal gives a measure of the distribution of obstacles in a component. This technique is called Barkhausen noise analysis (BNA). The sensitivity of BNA to stress and microstructure makes the technique useful for characterizing the changes that occur during high-temperature service.

The advantages of BNA are the measurement speed and absence of requirements for surface preparation (other than brushing to remove loose scale). In many cases, the microstructural effects are negligible and the Barkhausen signal can be used as a stress indicator. That is, in a microstructurally homogeneous part, increasing BNA values indicate an increasing tensile stress. The BNA method gives the direction of stress in addition to magnitude. By applying the magnetic field in different directions and measuring the Barkhausen signal, the principal stress directions and the magnitude and sign of the principal stresses can be determined.

In one application, BNA was used to inspect a carbon steel strap that supported one end of a reheat outlet header [19]. The strap contained a splice weld, as shown in Figure 7, and experienced temperatures in excess of 510oC. Figure 8 is a map of the longitudinal BNA values measured on the strap, showing ridges of high BNA values on either side of the splice weld. The ridges suggested the presence of high tensile stresses in those regions. Follow-up surface replication showed creep voids and microcracks corresponding with the location of the ridges.

The stress state in this demonstration was uniaxial tension. Other applications of the BNA technique have explored more complex geometries and stress states [20]. The results of these studies indicate that BNA is an excellent means of performing a rapid screening inspection to identify possible problem areas. However, BNA by itself cannot reliably distinguish

98

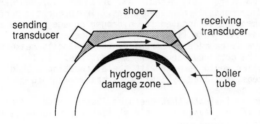

Figure 5. Hydrogen damage technique based on ultrasonic velocity measurement. An ultrasonic pulse travels a constant known distance between two transducers. The velocity is determined from the time of flight of the pulse. Regions of hydrogen damage retard the time of arrival at the receiving transducer.

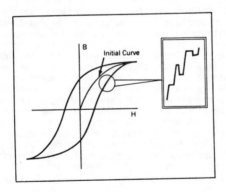

Figure 6. Illustration of the Barkhausen effect, which is caused by nonuniform domain wall motion.

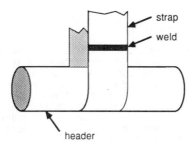

Figure 7. Schematic of splice weld in carbon steel hanger strap.

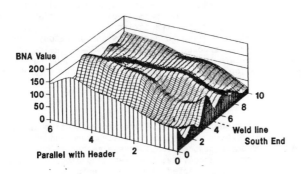

Figure 8. BNA Inspection grid and resulting 3D plot of longitudinal BNA values (arbitrary units) on a support strap containing a splice weld. The weld lies parallel with the header at 5 in. from the reference line.

between local stress gradients and local microstructure gradients. Therefore, BNA results need to be confirmed and further elucidated with other NDE techniques that give a more detailed and quantitative picture of component condition.

CHEMICAL ETCHING FOR TEMPER EMBRITTLEMENT

Temper embrittlement is the progressive reduction in toughness of alloy steels, brought about by exposure at temperatures in the range of 300-600oC. It is caused by impurity segregation to grain boundaries and appears as a change in the low-ductility fracture mode from transgranular cleavage to intergranular fracture. Temper embrittlement is a particular concern for low-pressure turbine rotors, which may be operated at embrittling temperatures for periods in excess of 150,000 hours.

Currently, advanced surface analysis methods such as Auger electron spectroscopy (AES) are available to characterize grain boundary composition. This technique is expensive and requires the removal of a specimen from the component.

EPRI is developing a low-cost nondestructive chemical etching technique to quantify temper embrittlement [21, 22]. The technique is based on the relationship between grain-boundary phosphorus segregation and susceptibility to etching by picric acid. Etching susceptibility is measured by pressing a softened plastic replicating tape into the polished and etched surface of the component and measuring the depth of etching along the grain boundaries on the replica.

As shown in Figure 9, the etching depth is strongly correlated with the extent of phosphorus segregation in Ni-Cr-Mo-V steels. The contribution of grain boundary segregation to loss of fracture toughness in these specimens was estimated by measuring the increase in ductile-to-brittle transition temperature produced by thermal aging. The relationship between etch depth and ductile-to-brittle transition is shown in Figure 10.

The picric acid etch test has given good results in laboratory tests on artificially aged and service-exposed heats of Ni-Cr-Mo-V steels, the latter obtained from low-pressure turbine rotors. The test is being qualified for field application and will be extended to Cr-Mo-V steels for high-pressure rotor applications.

DISCUSSION

The nondestructive methods described above are representative of the technologies being applied to characterize the microstructure and properties of service-aged components. Although these methods provide a greatly improved life assessment capability, the preceding discussion also shows that there are limitations associated with the techniques.

For example, Figure 4 shows a relatively large amount of scatter in the relationship between life expenditure and the "A" parameter. Therefore, assessments based solely on a measurement of "A" would have to use the conservative lower bounding curve, which in most cases constitutes a gross underestimate of remaining life. A second problem with all indirect methods is the so-called "inverse problem" of NDE. Because most of the NDE techniques are not uniquely related to a particular aspect of microstructure, multiple interpretations of the NDE result are possible. An example of this problem is BNA, which is sensitive to both microstructural features (cavities, second-phase particles, etc.) and to local stress variations. A BNA measurement by itself usually cannot distinguish between these alternate interpretations.

Faced with the limitations of the nondestructive techniques, users frequently employ several methods in parallel to obtain a more complete picture of component condition. In situations where creep is expected to

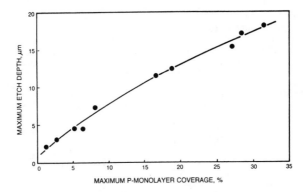

Figure 9. Correlation between depth of penetration in picric acid etch test and phosphorus segregation measured by Auger electron spectroscopy.

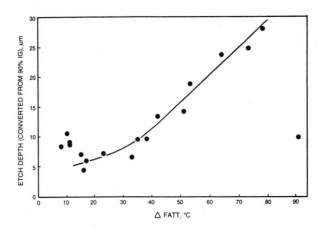

Figure 10. Relationship between etch depth and the change in ductile-to-brittle transition temperature produced by long-term annealing. The ductile-to-brittle transition is measured by the fracture appearance transition temperature (FATT).

be the life-limiting mechanism, the removal of miniature specimens for accelerated life mechanical testing is a useful means of supplementing microstructural techniques and directly assessing the effects of alloy chemistry and thermomechanical treatment variations.

Ultrasonic velocity measurements for porosity detection present an interesting case study in that, while carbide particle distribution and grain size also affect velocity, their effect is small compared with porosity effects. For practical purposes, we can ignore carbide particle and grain size effects and assume that any significant changes in velocity are due to porosity. The relative absence of interfering effects is responsible for the high accuracy of velocity measurements in detecting hydrogen damage and creep cavitation.

Electromagnetic methods offer a means of rapidly measuring several parameters that can be related to the properties of interest. For ferritic steels, magnetic hysteresis loop and eddy current evaluations using simple coil probes provide simultaneous measurements of parameters such as conductivity, permeability, coercive force, remanent magnetization, and Barkhausen effect. With a suitable set of calibration specimens, magnetic measurements should be able to provide a relatively unambiguous picture of component condition. Considerable effort is required, however, to develop the necessary correlations for the alloys of interest in fossil power plants.

An additional issue is the cost of component condition assessments relative to the benefit of knowing the remaining life. As Figure 1 indicates, the cost of assessing remaining life increases with the accuracy of the estimate. In general, the benefit of knowing the remaining life also increases with increasing accuracy until a point of diminishing returns is reached. EPRI is developing guidelines to allow utilities to assess the plant- and system-specific factors that determine the appropriate level of life assessment for a given application. A flexible approach is being adopted in these guidelines to allow for the expected improvements in the technology of condition characterization and life assessment.

CONCLUSIONS

Life assessment activities are anticipated to become increasingly important as utilities emphasize the need for maintaining operational flexibility and maximizing the economic return from existing generating capacity. Nondestructive techniques for characterizing component condition will be an important part of the life assessment effort. Recently developed techniques have greatly improved the accuracy and speed of material condition assessments and resulted in several quantitative correlations between condition and remaining life of service aged components. Further advances in this technology are expected to consist of the application of multiple inspection techniques to gain additional accuracy and reduce the ambiguity of nondestructive evaluations. The result will be a suite of NDE methods that can be integrated into utility predictive maintenance programs with the goal of optimizing inspection and repair intervals and reducing overall maintenance costs.

ACKNOWLEDGEMENTS

The authors would like to acknowledge the efforts of the individuals and organizations who performed the EPRI-funded work described in this paper. The following individuals have made notable contributions: A. S. Birring (Southwest Research Institute), S. M. Bruemmer (Battelle Northwest Laboratories), R. H. Richman (Daedalus Associates), M. C. Askins (Central Electricity Generating Board), and M. S. Shammas (Central Electricity Generating Board).

REFERENCES

[1] R. B. Dooley and J. Byron, "The EPRI Approach to Fossil Plant Life
     Extension," in Proc. International Conference on Life Assessment and
     Extension, vol. 1, pp. 101-105, The Hague, Netherlands, June 13-15,
     1988.

[2] "Remaining Life Estimation of Boiler Pressure Parts, Volume 2:
     Miniature Specimen Creep Testing," EPRI Report CS-5588, Vol. 2,
     January, 1988.

[3] A. McMinn, D. Mercaldi, and D. Mauney, "Evaluation of Rotor Materials
     by Remote Sampling," in Proc. of the Second EPRI Fossil Plant
     Inspections Conference, San Antonio, Texas, November 29-December 1,
     1988.

[4] K. Akiyama, T. Shiota, H. Ikawa, K. Kwamoto, T. Goto, and H. Karato,
     "Life Assessment of Turbine Components by Integrated Techniques,"
     Proc. of the American Power Conference, Chicago, IL, April 27-29,
     1987.

[5] Y. Kadoya, H. Karato, T. Goto, Y. Kadoya, and K. Kawamoto, "State of
     the Art NDE Techniques for Crack Initiation Life Assessment of High
     Temperature Rotors," in Proc. International Conference on Life
     Assessment and Extension, vol. 2, pp. 9-19, The Hague, Netherlands,
     June 13-15, 1988.

[6] S. Ishizaki, et al., "Nondestructive Residual Life Evaluation
     Techniques of Boiler Materials," in Proc. International Conference on
     Life Assessment and Extension, vol. 1, pp. 54-64, The Hague,
     Netherlands, June 13-15, 1988.

[7] A.M. Bissel, B.J. Cane, and J.F. Delong, "Remanent Life Assessment of
     Seam Welded Pipework," in Proc. International Conference on Life
     Assessment and Extension, vol. 2, pp. 203-211, The Hague,
     Netherlands, June 13-15, 1988.

[8] "Metallography and Stress Analysis of a Cracked First Stage Rim
     Attachment," EPRI RP2481-6 Interim Report, May, 1988.

[9] "Emergency Standard Practice for Production and Evaluation of Field
     Metallographic Replicas," ASTM Standard ES 12-87, American Society
     for Testing and Materials, Philadelphia, PA.

[10] M.S. Shammas, "Metallographic Methods for Predicting the Remanent
     Life of Ferritic Coarse-Grained Weld Heat Affected Zones Subject to
     Creep Cavitation," in Proc. International Conference on Life
     Assessment and Extension, vol. 3, pp. 238-244, The Hague,
     Netherlands, June 13-15, 1988.

[11] J. Maguire and D. J. Gooch, "Metallographic Techniques for Residual
     Life Assessment of 1CrMoV Rotor Forgings," in Proc. International
     Conference on Life Assessment and Extension, vol. 2, pp. 116-124, The
     Hague, Netherlands, June 13-15, 1988.

[12] "Remaining Life Estimation of Boiler Pressure Parts, Volume 3: A
     Metallographic Technique for Life Assessment of Parent 1CrO.5Mo
     Steel," EPRI Report CS-5588, vol. 3, April, 1988.

[13] E.P. Papaclakis in Physical Acoustics, Vol. 12, W.P. Mason and R.N.
      Thurston, eds, pp. 228-374, Academic Press, New York, 1976.

[14] R. Klinman, et al., Materials Evaluation, 39, pp. 1116-1120 (1981).

[15] H. Willems, "Characterization of Creep Damage by Means of Ultrasonic
      Techniques," in Proc. International Conference on Life Assessment and
      Extension, vol. 2, pp. 86-91, The Hague, Netherlands, June 13-15,
      1988.

[16] A.S. Birring, D.G. Alcazar, J.J. Hanley, G.J. Hendrix, and S.M. Gehl,
      "Detection of Hydrogen Damage by Ultrasonics," in Boiler Tube
      Failures in Fossil Power Plants, EPRI CS-5500-SR, November, 1987.

[17] A.S. Birring, D.G. Alcazar, J.J. Hanley, and S.M. Gehl, "Ultrasonic
      Assessment of Creep and Hydrogen Damage in Fossil Plant Components,"
      in Proc. of the Second EPRI Fossil Plant Inspections Conference, San
      Antonio, Texas, November 29 - December 1, 1988.

[18] G.A. Matzkanin, et al., "The Barkhausen Effect and Its Application to
      Nondestructive Evaluation," Southwest Research Institute Report
      NTIAC-79-2, October, 1979.

[19] S.M. Gehl, J. Scheibel, J.J. Yavelak, and G. Lamping, "Advanced
      Inspection Technology for Plant Life Extension," Jt. ASME/IEEE Power
      Generation Conference, Miami, FL, October 4-8, 1987, Paper No. 87-
      JPGC-Pwr-D.

[20] J.J. Yavelak and J. Hickson, "Using BNA to Guide Boiler Condition
      Assessments," in Proc. of the Second EPRI Fossil Plant Inspections
      Conference, San Antonio, Texas, November 29-December 1, 1988.

[21] "Grain Boundary Composition and Intergranular Fracture of Steels,"
      EPRI Report RD-3859. Vol. 1, January, 1985, Vol. 2, November, 1986.

[22] R.H. Richman, R.L. Cargill, S.M. Bruemmer, and S.M. Gehl,
      "Nondestructive Methods for Assessing Temper Embrittlement of In-
      Service Components," in Proc. of the Second EPRI Fossil Plant
      Inspections Conference, San Antonio, Texas, November 29-December 1,
      1988.

NDT-TECHNIQUES FOR MONITORING MATERIAL DEGRADATION

PAUL HÖLLER AND GERD DOBMANN
Fraunhofer-Institut für zerstörungsfreie Prüfverfahren, Universität Geb.37,
D-6600 Saarbrücken 11, Federal Republic of Germany

ABSTRACT

The early detection of material degradation by NDT is of basic importance for safe and economic operation of many industrial components. Depending on the type of component as well as on service exposure conditions,
NDT-techniques, which are sensitive to deformation, erosion, wear, porosity, microcracks and so on, are required. At present, only very few techniques are available for monitoring early stages of degradation. An overview of NDT-techniques already in use as well as under development in Germany for this problem is given. Special emphasis will be laid on the presentation of recent advances towards better detectability and objectivity.

INTRODUCTION

There is an inseparable connection between material application and
material degradation (degradation of material mechanical properties) during the designed lifetime. Moreover material degradation is an inherent
problem in every kind of technology, but also in every new one. Material
degradation may cause damage which can endanger human safety and the environment. High-grade materials can be unnecessarily destroyed and a lot of
energy can be wasted. It is evident that the economy is influenced and that
introduction of new technologies is hindered.
This point is illustrated in the following two examples. A British
(1969) and an US-study (1978) [1] show that damage caused by corrosion produce costs of 3.5% and 4.2% respectively of the gross national income per
year in these western industrial societies. The costs for maintenance and
repair caused by the degradation of materials in civil engineering in the
FRG amounts to 40-60 billion DM per year [2].
Therefore, the investigation of the reasons for material degradation
and the development of possibilities for suitable protection is an important task for each industrial society. All the available knowledge should
be applied to develop techniques to assure maximum lifetime of materials
and to secure the safe application of technical products.
Now NDT-experts are asked to develop techniques to detect, to characterize and to quantify material degradation. NDT as a tool in quality assurance is mainly asked to detect the early stages of degradation, in order
- to avoid further degradation
- to develop protection strategies or
- to characterize the essential parameters
which are the input values for an algorithmic lifetime prediction. Table 1
gives an overview of NDT-techniques already in use as well as under development.

Types of degradation		NDT-technique
<u>Corrosion</u>	wastage	optical, borescope
		UT - wall thickness
		UT - SH plate waves
		EC - heat exchanger tubes
		X-ray
	cracking	UT conventional + UT phased array
		EC - mainly austenitic components
		X-ray
		Liquid penetration
		Magnetic particle inspection

Techniques for early detection are not available for field applications.

<u>Fatigue</u>	conventional techniques after
	crack formation
	problem: crack-closure
	(LCF,compressive stress)

Techniques for early detection are not available for field applications, laboratory investigation.

<u>Creep</u>	replication (microscopy)
	hardness

UT-velocity, 3MA, electric resistance, capacitive strain gages under development for field applications

<u>Thermal aging</u>	hardness, replication

Techniques for early detection are not available, 3MA under development

<u>Wear</u>	cutting tool condition monitoring by AE under
	development in production technology

<u>Table 1</u>: State of the Art of NDT for different Types of Material Degradation

From Table 1, we can state that in most of the cases the detectability of all conventional techniques starts with the crack formation or macroscopic damage formation. Most of the efforts for the development of the NDT-techniques have been related to automation, to transducer manipulation, to data acquisition, data reduction and data evaluation, in order to reliably detect, classify and size the defect [3,4].

A new field in NDT in the FRG is starting under the headline "micro-NDT", with the aim of developing and optimizing techniques with spatial resolution in the μm-range. For micro defect detection, for example a micro X-ray tomography is under development [5], but only small components can be inspected at present.

The characterization of microstructures as part of micro-NDT was an important aim of research and development activities in the last decade [6,7,8]. Some new electromagnetic and UT-techniques are under development which have great potential for microstructure analysis and, therefore, also for the early detection of material degradation. Some of these techniques are reported in detail in this paper and other papers at this conference. The contribution of H. Willems in this proceedings gives the state of the art of creep damage characterization by NDT.

MATERIAL DEGRADATION BY CORROSION

Since 1974 the German ministry for research and technology sponsors the R+D-programme "corrosion and corrosion protection". Up to 1983, 409 projects with a total amount of 108.9 million DM in the first stage have been performed (51.8% from industry). Up to 1986, in the second stage of the programme 146 projects with a total amount of 52.5 million DM have been sponsored (28.8 million DM from the ministry, 23.7 million DM from industry [1]).

In this programme IzfP has developed an eddy-current-technique for the detection and quantification of intergranular stress corrosion cracking (IGSCC) in austenitic steels. The material degradation begins with microcracks along the grain boundaries and crack-fields with randomly oriented cracks occur. If the degradation is restricted to the near surface zone and can be detected, in practice the damage can be removed by grinding, provided the remaining wall thickness remains large enough. Normally eddy current techniques are designed for single crack detection and sizing in relation to artifical slots of known depth and length.

It is obvious that IGSC-crack-fields cannot be detected in the frequency range normally used for single crack detection (up to 200 kHz). The reason is that in this case eddy-currents cannot exist in the degraded volume. The damaged zone influences the diffusion of eddy-currents in the same way as the lift-off effect. Fig. 1 documents this fact in the lower part. With 30 kHz frequency, both lift-off effect and IGSC-crack-fields show the same phase direction in the complex impedance plane. The effects can be separated at higher frequencies ($\geq$ 1 MHz). For inspection the used frequency range depends on the number of cracks in the sensed volume and the damaged sheet thickness. In the upper part of Fig. 1 the facts are demonstrated for a frequency of 2 MHz. Now both effects can be separated clearly.

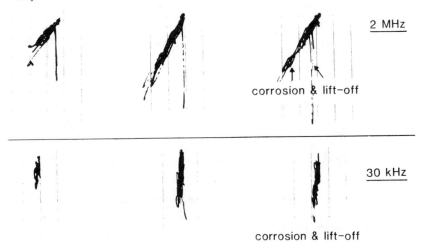

2 MHz

corrosion & lift-off

30 kHz

corrosion & lift-off

Fig. 1: Crack-Fields, Low- and High- Frequency Results for Lift-Off and Corrosion

Though the lift-off now can be suppressed as a disturbing parameter by the well-known phase selection the result of the investigations is, that the remaining signal (after phase selection) could not be correlated with the degradation-depth. The reason is, that this signal is not uniquely related only to the degradation-depth, it is also influenced by the degradation-density.

A solution has been found by a multifrequency approach. R. Becker [9] has demonstrated, that for low frequencies the lift-off signal is a superposition of the real lift-off of the coil and the apparent lift-off produced by the IGSC-crack-fields. Therefore as a first step the inspection task is to measure the real lift-off of the probe. This is performed with a high frequency, where now the damage effect is suppressed by phase selection. For frequencies larger than 1 MHz the phase angle between both effects is again larger than 30°. Then the low-frequency signal is corrected by the real lift-off. The remaining signal after subtraction is only a function of the degradation-depth.

Nevertheless it is obvious, that this approach cannot be used in situations where the IGSCC occur in austenitic weldments and where a single deep crack is dominating in the crack field. In the first case magnetic residual-δ-ferrite-phases disturb the evaluation (permeability changes from 1 to 8), in the second case the resulting signal is dominated by the single crack but with an uncertain amount.

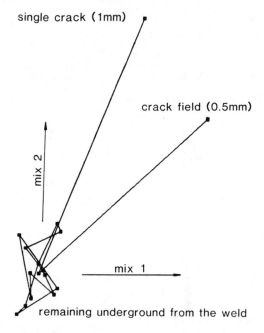

Fig. 2: Defect Characterization

Therefore the two-frequency-approach has been changed to a four-frequency approach using the multi-mixing-concept [10]:
- one frequency is selected lower than 30 kHz

- δ-ferrite- and lift-off-effects are suppressed in the signal components of all frequencies
- in order to obtain a unique result, the signal components of all frequencies ≥ 30 kHz, related to the IGSC-crack-fields are also suppressed
- two multifrequency approaches are selected (Mix1, Mix2), both in a similar way related to the suppression of the disturbing parameters
- in the classification plane (Mix1-Mix2-plane) the indications of crack-fields and large single cracks can be separated in well-defined angular regions.

This complicated optimization process can be performed at IzfP with existing software. The result is documented in Fig. 2 and Fig. 3. Fig. 2 shows the Mix1, Mix2 classification plane; δ-ferrite- and lift-off-signals are suppressed. The remaining noise-signals are indicated around the zero point. A single crack (1 mm depth, artifical slot) and an IGSC-crack-field (0.5 mm degradation-depth) are separated. Fig. 3 documents the results in another representation (pen-recorder). The detectability for single cracks as well as for IGSC-crack-fields is in the range 0.8 mm. The emphasis of further optimizations will be the probe design for temperatures of ~ 250°C (in-service inspection temperature).

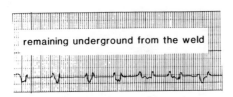

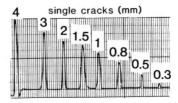

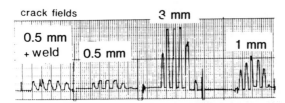

Fig. 3: Defect Signals cleared of Disturbances

The second material degradation example is corrosion related to cathodic stress corrosion (hydrogen induced stress corrosion cracking). This type of corrosion contributes only 2% to the total amount of corrosion damage and is restricted primarily to high-strength steels. But related to the risk of the damages, it is at the top of the list.

Particularly endangered are weldments in vessels and tubings where service loads superimposed on residual stresses favour the stress corrosion cracking. Up to now there exists no NDT-technique which can detect this type of corrosion before formation of the crack. Further there is no approach to quantify the degradation at early stages. The primary reason for the damages caused by cathodic stress corrosion cracking is hydrogen on the atomic level which can diffuse into the material. Lattice defects like pores, grain boundaries, dislocations and inclusions are locations where a recombination can occur which results in locally concentrated high compressive stresses. Tough steels respond with blistering, brittle steels with cracking. The critical hydrogen concentration necessary for crack initiation depends on the material and the nature of lattice defects [11].

Micromagnetic NDT-techniques react with sensitivity to microstructure and residual stress changes [12]. Together with the BASF company (chemical industry) IzfP has recently investigated the sensitivity of the Magnetic Barkhausen noise as a micromagnetic technique to detect this type of corrosion. This is of special interest in oil field applications, for example in the eastern mediterranean sea (off-shore) where the oil has high contents of seawater and hydrogen sulphide (H_2S).

Martensitic and martensitic annealed microstructures of the steel grade X 20Cr 13 under bending stresses in a H_2S saturated NaCl-solution were investigated by I.Altpeter of IzfP [13] in cooperation with BASF. During the exposure time, magnetic Barkhausen noise was measured continuously. Fig. 4 shows the experimental set-up; the 3MA-device, described elsewhere [14], was used for evaluation.

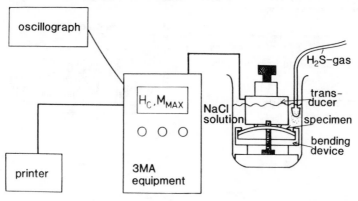

Fig. 4: Experimental Set-Up

Fig. 5 shows this equipment and the handy U-shaped magnetic yoke transducer. Two micromagnetic quantities are derived, the maximum amplitude of the Barkhausen noise M_{max} and H_{cM}, the magnetic field strength belonging to this maximum. The yoke transducer magnetizes the material under inspection dynamically in a hysteresis with a fixed magnetizing frequency (sinusoidal) and a fixed field amplitude. Both of these parameters can be adjusted.

Fig. 6 documents the results of an experiment with the martensitic microstructure. The specimen was under a bending load of $\sigma_+ = 400$ MPa during an exposure time of 130 minutes. At first M_{max} decreases slowly, then increases to a maximum (at 110 minutes) before it decreases rapidly up to the rupture of the specimen. By metallographic investigations it has been verified that the maximum in the curve occurs before macro-cracking. Though the signal increase is a real microstructural effect and can be used for early warning before damage. The slow decrease in the amplitude can be interpreted with the increasing inner compressive stress, produced by the hydrogen. The rapid increasing cannot be understood in detail up to now. Undoubtedly, the fact documents an increasing of irreversible magnetic processes, i.e. the density of magnetic sensitive lattice defects is increasing.

As a second step a martensitic annealed microstructure was investigated experimentally during 180 minutes of exposure time (bending stress σ_+ = 276 MPa). Here we can observe a 20% decrease in M_{max} related to the value at the beginning (and comparable to the behaviour in Fig. 6). Furthermore we can observe changes in the Barkhausen-noise-profile curves, M as function

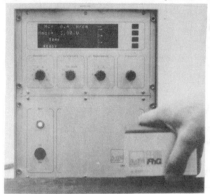

Fig. 5: 3MA-Device, the Barkhausen Noise Instrument

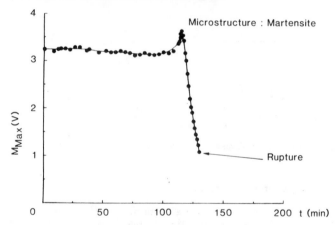

Fig. 6: Amplitude of Magnetic Barkhausen Noise as a Function of Exposure Time

of the magnetic field strength, The first peak in the double peak curves decreases, whereas the second is nearly constant (Fig. 7).

The specimen was exposed in the autoclave bath (pH-3) for 17 hours but without load. At the beginning of a new experiment under load we state that the double peak curve has vanished totally (Fig. 8). Only a single peak occurs (H_{cM} is increased from 30 A/cm upto 39 A/cm, 30%). After 2 hours we have observed the rupture. During this exposure time, M_{max} decreases continuously.

The last measured M_{max}-value (specimen state 5 in Fig. 8) has been verified at each half of the ruptured specimen. From the results it is obvious that the effect is caused by microstructural degradation and not cau-

112

sed by the macro-crack.

These initial investigations show that the 3MA-technique has a high potential for cathodic stress corrosion detection and characterization. Further research should be continued.

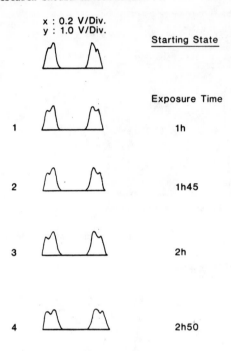

x : 0.2 V/Div.
y : 1.0 V/Div.

Starting State

Exposure Time

1   1h

2   1h45

3   2h

4   2h50

Fig. 7: Magnetic Barkhausen Noise Profile Curves as Function of Exposure Time

MATERIAL DEGRADATION BY FATIGUE

Material fatigue is correlated with micro- or macro-plastic-deformation. The dislocation network and dislocation density are influenced. The 3MA technique [12] has strong potential for detecting plastic deformation in magnetic materials. It is a well known fact for instance that the coercivity of a magnetic material is growing with dislocation density [15]. The dislocations hinder the magnetization process i.e. the lattice defects are pinning points for the Bloch walls. Therefore all 3MA-quantities are sensitive to this kind of material degradation.

In the nuclear safety programme of the German ministry for research and development, IzfP has investigated these effects. For this purpose a special set of specimen of fine grained bainitic tough structural steel (22NiMoCr 37) has been designed. This steel is used in FRG for pressure vessels and pipes in the primary circuit of nuclear power plants.

These specimen have been deformed plastically with well defined plastic strain (0.5%, 1%, 2%, 5%, 10%). The strain hardening has been characterized by Vickers-hardness (HV10); but a sensitivity has been obseved only for residual strain $\geq$ 1% (205 HV10-1%, 224 HV10-10%). On the other hand 3MA-quantities are sensitive especially in the lower range of plastic deformation. Fig. 9 documents this fact with the Barkhausen-noise profile curves (magnetizing frequency 0.05 Hz, magnetic field amplitude 100 A/cm, tape-recorder-head as transducer with bandwidth 0.5-20 kHz and centre frequency 10 kHz). We can observe a large increase of the H_{cM}-value (magnetic field value which corresponds to the peak-maximum) especially between 0.5%

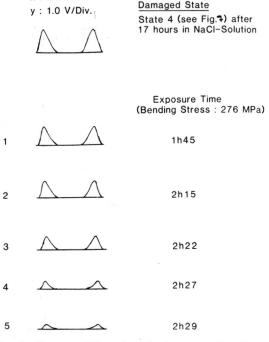

x : 0.2 V/Div.
y : 1.0 V/Div.

Damaged State
State 4 (see Fig.7) after
17 hours in NaCl-Solution

Exposure Time
(Bending Stress : 276 MPa)

1	1h45
2	2h15
3	2h22
4	2h27
5	2h29

Fig. 8: Magnetic Barkhausen Noise Profile Curves as Function of Exposure Time

and 1% residual strain. The sensitivity decreases for larger values of plastic deformation. Fig. 10 gives an overview of these results.

The Barkhausen-noise amplitude M_{max}, the field value H_{cM} and the half-width at 75% of maximum value of the incremental permeability $\Delta\mu_{\Delta 75}$ [14] are discussed as a function of the plastic strain ε_{pl}. Of a special interest is the relatively high sensitivity of $\Delta\mu_{\Delta}$ which increases up to 500% (10% residual strain) compared to the "as delivered" state.

Fig. 11 shows the evaluation of upper harmonics in the tangential magnetic field strength as a function of time during a hysteresis cycle, a technique which has been developed recently [16]. In addition to the coercivity H_{c0}, a distortion factor K can be derived. Fig. 12 shows the results with these two quantities, bu now the magnetizing frequency was increased to 50 Hz. It should be mentioned that technical components normally are designed for plastic deformations below 2%.

Therefore the 3MA-technique has special potential for practical applications in industry.

22 NiMoCr3 7

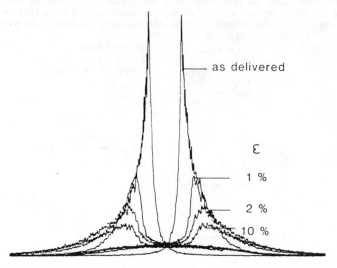

as delivered

ε

1 %

2 %

10 %

Fig. 9: **Magnetic Barkhausen Noise Profile Curves as Function of Plastic Deformation**

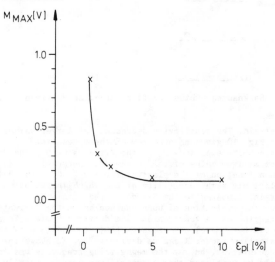

Fig. 10a: M_{max} as a Function of Plastic Deformation

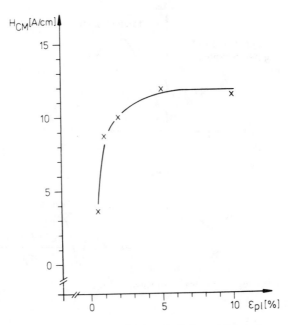

Fig. 10b: H_{cM} as a Function of Plastic Deformation

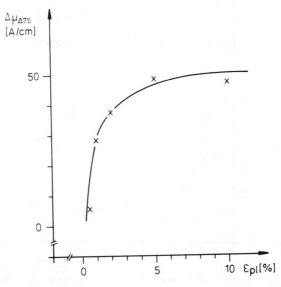

Fig. 10c: $\Delta\mu_{\Delta75}$ as a Function of Plastic Deformation

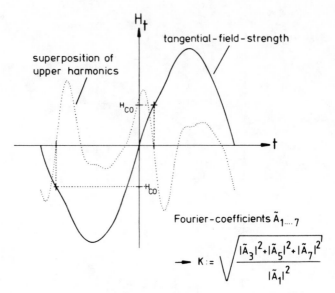

Fig. 11: Deduction of Measured Quantities

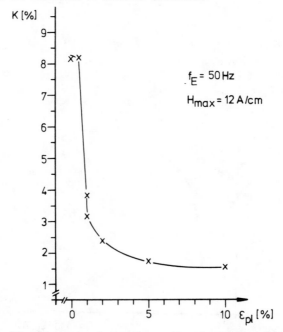

Fig. 12a: Distortion Factor K as a Function of Plastic Deformation

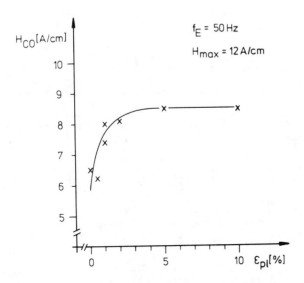

Fig. 12b: Coercivity H_{co} as a Function of Plastic Deformation

CONCLUSIONS

The early detection and quantification of material degradation by NDT is a strong demand in all kind of industrial applications. NDT-techniques sensitive to microstructure characterization should be optimized for the solution of these special tasks. Three examples have been discussed which show that new approaches are under development. Further intensive research should be carried out.

REFERENCES

[1] Dechema: Korrosion und Korrosionsschutz, ein Orientierungsrahmen für anwendungsbezogene Verbundforschung, Frankfurt, 1988
[2] E. Lübbert, in Zerstörungsfreie Prüfung im Bauwesen, edited by G.Schickert and D. Schnitger (Bundesanstalt für Materialprüfung, BAM, Berlin, 1986), introduction.
[3] W. Kappes, R.K. Neumann, K.K. Stanger, F. Höh: Datenaufnahme und -verarbeitung bei der automatisierten Ultraschallprüfung mit dem ALOK-Verfahren. Tagungsband zum Seminar Automatisierung in der Ultraschallprüfung, Stand der Technik und Entwicklungstendenzen bei mobiler Prüftechnik, Nov. 7-8, 1988, DGZfP/BAM, Berlin.
[4] Müller, Schmitz, Schäfer: Recent Experiences in NDE of Reactor Pressure Vessels with LSAFT. 8th Intern. Conference on NDE in the Nuclear Industry, Nov. 17-20, 1986, Kissimmee, Florida.
[5] M. Maisl, H. Reiter, P. Höller: Micro Radiography and Tomography for High Resolution NDT of Advanced Materials and Microstructural Components, to be published in Proceedings of ASME Winter Annual Meeting November 27-December 2, Chicago,IL.

[6] R.E. Green, Jr., C.O. Ruud: Proceedings of Nondestructive Methods for Material Property Determination. April 6-8,1983, Hershey, Plenum Press, New York, 1984.

[7] J.F. Bussière, J.P. Monchalin, C.O. Ruud, R.E. Green, Jr.: Proceedings of Nondestructive Characterization of Materials II. July 21-23, 1986, Montreal, Plenum Press, New York, 1987.

[8] P. Höller, V. Hauk: Proceedings of Nondestructive Characterization of Materials III. October 3-6, 1986, Saarbrücken, Springer Verlag, Heidelberg, 1988/89 (to be published)

[9] R. Becker and K. Betzold: Computer aided design of eddy current weld inspection, optimization of a multifrequency approach. Proceedings of the International Conference on Improved Weldment control with Special Reference to Computer Technology, Vienna, 1988.

[10] R. Becker, K. Betzold, G. Dobmann and P. Höller: Progress in Defect Detection and Characterization with Multi-Multi-Frequency-Eddy-Current and Current Perturbation Testing. Proceedings of the 9th International Conference on NDE in the Nuclear Industry, Tokyo, 1988.

[11] H. Spähn, in Werkstoffkunde Eisen und Stahl, Teil I, Bd. 2 (Verlag Stahleisen, Düsseldorf, 1983), p. 648.

[12] G. Dobmann, W.A. Theiner and R. Becker in Proceedings of Nondestructive Characterization of Materials III, Oct. 3-6, 1986, Springer Verlag, Heidelberg, 1988/89 (to be published).

[13] I. Altpeter, internal IzfP-report EM-IV-6.7, 1984.

[14] W.A. Theiner, B. Reimringer, H. Kopp in Proceedings of Nondestructive Characterization of Materials III, Oct. 3-6, 1988, Saarbrücken, Springer Verlag, Heidelberg, 1988/89 (to be published).

[15] H. Kronmüller in A. Seeger: Moderne Probleme der Metallphysik, Springer Verlag, Heidelberg, 1966.

[16] G. Dobmann, H. Pitsch in Proceedings of Nondestructive Characterization of Materials III, Oct. 3-6, 1988, Saarbrücken, Springer Verlag, Heidelberg, 1988/89 (to be published).

NONDESTRUCTIVE DETECTION OF MATERIAL DEGRADATION
CAUSED BY CREEP AND HYDROGEN ATTACK

A. S. BIRRING
Southwest Research Institute, 6220 Culebra Road, San Antonio, Texas

ABSTRACT

Hydrogen attack and creep damage may reduce the fracture toughness as well as the strength of steels. This reduction is caused partially by the presence of cavities and microcracks at the grain boundaries. A large reduction in fracture toughness can make some components unsafe for operation. Ultrasonic tests performed on damaged samples showed a decrease in wave velocity and an increase in backscatter. The longitudinal-wave velocity decreased more than shear-wave velocity. Where the fracture toughness was reduced by 43 percent, samples with hydrogen attack showed a tenfold increase in backscatter. Such results demonstrate the potential for ultrasonic nondestructive testing to quantify damage and assess the component's remaining life.

INTRODUCTION

Degradation of materials in service has become a major concern for the petrochemical and electric utilities. Degradation can occur from adverse combinations of such factors as temperature, stress, chemical reaction, and time. In particular, hydrogen attack (see Figure 1) is produced by the effect of hydrogen pressure and temperature [1,2], while creep is produced by the combined effect of stress and temperature [3,4]. Material degradation gradually decreases the fracture toughness and creep strength of the material and hence reduces its remaining life. An assessment of such damage may be required during service because such degradation may not have been accounted for in the original design, or the performance criteria may have changed after the component was commissioned.

One such example is the safe operating limit for steels in hydrogen environments. The American Petroleum Institute (API) recommended maximum operating temperatures and hydrogen partial pressure for steels through its publication API 941 [5]. Written in 1977, the publication was revised in 1983 to more conservative temperatures after hydrogen damage was reported in steels operating at allowable temperatures in accordance with the 1977 guidelines.

Assessment of material damage is also required to determine the remaining life of components and to assess creep damage in fossil-plant components operating at high temperatures and stresses. Accurate measurement of creep damage in these latter components can provide the basis for defining new operating conditions and limits which will serve to extend their life and, thus, delay or avoid building new and expensive plants.

Hydrogen Damage - Background

The importance of measuring hydrogen damage has instigated worldwide interest in the development of suitable nondestructive test (NDT) methods. Researchers in both the petrochemical and power industries are aggressively developing NDT methods that can measure the amount of degradation and relate it to the fracture toughness of the material.

Nondestructive detection of hydrogen damage has been performed by measuring the loss of ultrasonic wave velocity. Microcracks and cavities produced by hydrogen attack lower the elastic modulus, thereby reducing the ultrasonic velocity. A small change in ultrasonic velocity is difficult to

Microcrack

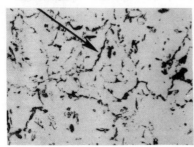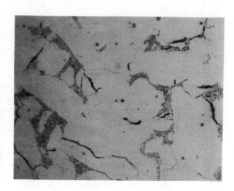

Figure 1. Microcracks produced by hydrogen attack. (a) Early
stages (mag. = 400X). (b) Severe damage (SwRI micrograph No.
28613, mag. = 500X).

measure in components when the thickness is not accurately known.
Researchers have found that the ratio of longitudinal-to-shear wave transit
time can be related to hydrogen attack. This ratio increases with damage;
and results have been verified by Watanabe et al. [6], Birring et al. [7],
and Senior [8]. The velocity-ratio technique successfully detects the last
stages of hydrogen attack when hydrogen attack has produced uniform porosity
(microcracks) in the material, which reduces the velocity. However, the
technique may not successfully detect isolated porosity.

Another technique to detect hydrogen attack is measurement of attenuation. Loper et al. [9] and Latimer et al. [10] applied ultrasonic attenuation to detect hydrogen attack in fossil-fired boiler tubes. An increase in
attenuation greater than 3 dB across the thickness of a boiler tube [approxi-
mately 6.3 mm thick (0.25 inch)] was considered to be indicative of hydrogen
attack damage [9]. The attenuation technique worked very well when the
surface of the component on the side opposite the transducer was smooth
[e.g., a smooth inside surface (ID) when the transducer was placed on a
smooth area of the outside surface (OD)]. Reliability of the attenuation
technique deteriorated sharply on rough surfaces that scattered ultrasound.

Yajima [11] applied an ultrasonic pulse-echo detection technique to
detect cracking caused by hydrogen attack. He reports the method could
detect hydrogen attack in a 140-mm thick vessel if there were 150 to 200
intercrystalline cracks 30 to 40 micrometers in length per square millimeter.
He also recommends selecting test frequency according to the vessel wall
thickness.

Creep Damage - Background

While NDT techniques to detect hydrogen attack generally have been
limited to ultrasonics, a variety of techniques have been investigated for
creep damage. Neubauer and Wendel [12] suggested replication to detect creep
cavities on the grain boundaries (see Figure 2). This technique uses a thin
plastic tape applied to the polished surface of the metal. The tape is
removed, and the cavities are observed under an optical microscope. Shammas

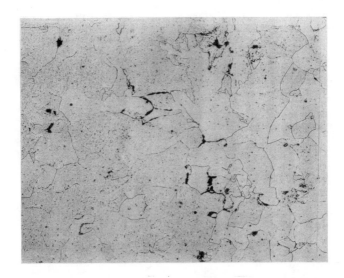

Figure 2. Creep cavities (black spots on grain boundaries) detected by replication in a steam lead removed from service in a fossil plant. The replication technique is slow for field inspection and is limited to surface inspection (SwRI micrograph No. 34097, Mag = 200X).

[13] extended this technique by correlating the number of grain boundaries undergoing cavitation to the remaining life.

Akiyama et al. [14] evaluated both the hardness test and replication techniques to determine creep damage in Cr-Mo rotor steels. Using these NDT techniques with stress analysis, they developed rules to estimate remaining life of blade-attachment areas in turbine disks. Likewise, Goto [15] and Kimura et al. [16] successfully used hardness measurements to assess creep damage. Goto was able to correlate eddy current measurements to the Larson Miller parameter in a 304L stainless steel. He used the X-ray diffraction measurements to estimate creep strain. Using these same approaches, Sakurai et al. [17] developed a creep evaluation system for turbine rotors. This system consisted of three different apparatuses: hardness testing, electrical resistivity, and a visual microcrack detection unit. Based on the principle that creep reduces hardness and increases electrical resistivity of steels, the system can detect creep damage in rotor bores. Another system to take replicas in rotor bores has been developed by Mitsubishi Heavy Industries [18].

While all these previously discussed techniques (including replication, hardness, resistivity, x-ray diffraction, and visual) can detect creep, each has the same major shortcoming: it is limited to surface inspection. Limited surface inspection can be deceiving when inspecting piping systems from the OD because creep may initiate from the weld cusp at the mid-thickness or from the ID. Conversely, inspection from the ID can be very expensive, as it requires special tools to inspect long piping systems such as steam lines, reheat lines, and headers in power plants. The petrochemical industry does not, generally, consider surface inspection as a reliable approach. Interestingly, while both hydrogen attack and creep produce similar damage such as microcracks, NDE techniques for hydrogen attack are not surface limited. The author expects that future NDE development efforts in creep assessment also will be directed toward volumetric damage assessment.

Recently, some work in detection of volumetric creep damage has been reported. Willems et al. [19] and Birring [20] have shown that creep reduces ultrasonic velocity of longitudinal (L), shear (S), surface, and creeping waves. Another technique applied for creep damage is backscattering. Nakashiro et al. [21] and Birring [20] have both found that backscattering is increased by creep cavitation.

These past efforts illustrate the vast amount of work being done in search of a satisfactory NDT approach for detection of material degradation. The goal is to quantify the damage nondestructively and to correlate it with the remaining service life fracture toughness. This paper discusses some recent results in which correlation has been partially obtained.

DETECTION OF HYDROGEN ATTACK

Two approaches, namely velocity and backscatter, have been investigated to detect hydrogen attack. These approaches are discussed in the following two subsections.

Velocity Technique

Ultrasonic velocity measurements were made on boiler tubes taken from fossil plants, as shown in Figure 3. Refracted L-waves (5 MHz) were transmitted along a chordal path, with the centerline of the beam tangential to the mid-thickness. Velocity was calculated from the transit time of the earliest arrival signal. Ultrasonic refracted L-wave velocities $v\ell$ in the hydrogen-attacked regions dropped between 1.4 and 9.3 percent. A drop of less than 0.8 percent was observed on tubes with no damage. Refracted S-waves with a drop in velocity vs of 1.2 to 4.3 percent were found to be less sensitive than L-waves.

A quantitative correlation between the depth of damage (h) and ultrasonic velocity drop was also obtained. Figure 4 shows a poor initial correlation due to the wide scatter in the data. This scatter was caused by the velocity drop being dependent on both the depth and the volume of metal through which the wave travels. To compensate for this effect, a variable called effective depth (h_{eff}) was used. The effective depth is equal

$$h_{eff} = \frac{(S_1 + S_2)}{2 \times 14} \times h \text{ (depth of damage)} \tag{1}$$

to where S_1 and S_2 (shown in Figure 3) are in mm, with a maximum value of 14.0 mm each. A value greater than 14 mm does not influence the wave velocity, as it is not in the wave path. Using the effective depth, the velocity drop correlated quite well with the damage (see Figure 5). This plot clearly shows the potential of ultrasonic tests to quantify damage.

Field tests were conducted at three utilities--Public Service of Colorado (PSC), Los Angeles Department of Water and Power (LADWP), and Saskatchewan Power (SKP)--to verify the velocity drop technique shown in Figure 3. At all three sites, the inspection was conducted in a two-step procedure. First, an inspection was performed to detect tubes with ID corrosion. This step was taken because hydrogen damage in boiler tubes is always associated with corrosion. The conventional inspection used for corrosion is a normal-incidence transducer applied to measure the loss of the fourth backwall [10]. A significant loss of fourth backwall amplitude indicates the possibility of ID surface corrosion. (Loss of fourth backwall also can be caused by other factors such as ID scale, OD corrosion, poor coupling, and delaminations in the steel.) After suspect tubes were

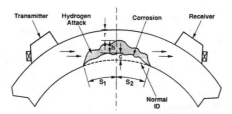

Figure 3. Configuration to measure reduction of ultrasonic-wave velocities in boiler tubes for detection of hydrogen damage.

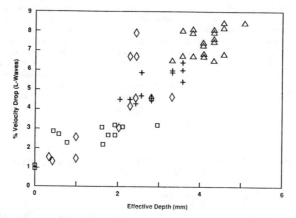

Figure 4. Correlation between depth of hydrogen damage (h) and loss of ultrasonic L-wave velocity on four tubes represented by the various symbols. Correlation coefficient = 0.62.

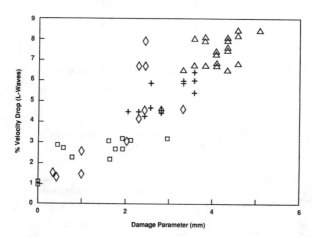

Figure 5. Correlation between effective depth of the hydrogen damage parameters (DP) and loss of ultrasonic L-wave velocity on four tubes represented by the various symbols. Correlation coefficient = 0.81.

identified, the velocity drop technique was applied to identify tubes with hydrogen damage. Results of the three field tests showed complete success with the velocity method.

In particular, at one of the sites, 72 tube locations suspected to contain corrosion after applying the fourth backwall amplitude technique were reinspected with the velocity technique. Tests indicated that of the 72 tubes suspected, only one tube was attacked by hydrogen. The damage, localized along a 1-foot, 4-inch length, was verified by metallurgical tests. The field test results, summarized in Table I, show the benefit of using the velocity drop technique. If only the amplitude technique were applied, unnecessary tube replacements would have occurred.

When applying the velocity technique, an additional concern is the influence on time measurement by other parameters including time measurement accuracy, influence of coupling, and microstructure. Time measurement accuracy was determined by taking measurements off the CRT screen. The time sensitivity used during the measurements was 50 ns/div. The influence of coupling was measured by taking time measurements at the same location in a tube. The transducer was removed and reapplied on the tube OD after each measurement. The influence of microstructure was measured by taking time measurements at several locations of four tube samples that had no hydrogen damage. The influence of such parameters was determined during the field tests and is given in Table II. These results show that the velocity

Table I

SUMMARY OF FIELD TESTS TO DETECT HYDROGEN-DAMAGED TUBES IN
FOSSIL-FIRED BOILERS

Utility	No. of Tubes Suspected by by the Amplitude Technique	No. of Tubes Identified by the Velocity Method	Tubes Cut for Metallurgical Tests	Tubes with Hydrogen Damage Verified by Metallurgical Tests
LADWP	19	0	3	0
PSC	72	1	4*	1
SKP	3	3	3*	3

*Includes the tube(s) identified by the velocity technique.

Table II

VELOCITY TECHNIQUE: INFLUENCE OF PARAMETERS
ON TIME MEASUREMENTS

Parameter	Effect on L-Wave Velocity Measurements (%)
Microstructure	0.69
Coupling	0.23
Time Measurement Accuracy	0.11
Total Effect of the Above Parameters	0.73

technique is not significantly influenced by other parameters because hydrogen damage produces L-wave velocity changes in the range of 1.4 to 9.3 percent.

Backscatter Technique

The ultrasonic backscatter technique [22] was applied on samples obtained from a 98-mm thick ASTM-A204-Grade-A steel vessel operating in a petrochemical plant. Figure 6a shows the configuration to take backscatter measurements. Three 25-mm thick samples were cut from the ID, mid-thickness, and OD. Metallographic analysis showed very few microcracks on the sample removed from the ID (see Figure 1), so the damage was estimated to be in the early stages. The OD sample had no damage.

Results of the ultrasonic and material parameter measurements, given in Table III, show that the velocity measurements displayed a very small decrease (1 percent). Based only on the velocity measurement, a conclusion could be reached that the samples had no damage. The ratio of $V_S/v\ell$ at the ID was also only 0.553, which is close to the nominal value even though the sample had damage. Attenuation measurements showed a definite increase from 1.7 dB at the OD to 3.2 dB at the ID, which could indicate damage. Attenuation measurements, however, can result in false indications if the ID surface is pitted and scatters ultrasound. The most prominent results were determined from the backscatter measurements, which displayed a significant increase in amplitude at the ID--about 10 times over the ID.

The backscattering results (see Figure 7) were very important, as backscattering was found to be more sensitive for detecting hydrogen attack on the samples than velocity measurements. Taking only velocity measurements would have led to an incorrect conclusion that the ID sample had no hydrogen attack.

The major advantage of using backscatter measurements to determine hydrogen attack is that the technique analyzes only the sound scattered within the material and is not affected by the condition of the ID surface. In fact, backscatter measurements can be taken on geometries that do not produce a backsurface reflection, as required by velocity and attenuation measurements.

Table III

ULTRASONIC AND MATERIAL TOUGHNESS MEASUREMENTS

Sample	Attenuation (10 MHz) (db/cm)	L-Wave Velocity Change*	S-Wave Velocity Change*	Backscatter Amplitude (10 MHz) (mV)	Charpy Absorbed Energy (Joules)	Fracture** Toughness MPa√m
OD	1.7	0	0	2	40	134
Mid-Thickness	2.4	-0.7	-0.5	15	25	Test Malfunction
ID	3.2	-1.2	-0.7	20	24	76.1

*Reference to OD sample.
**Measured using ASTM E813 and E399.

126

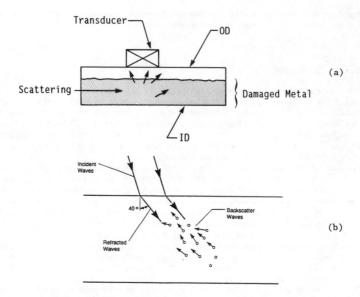

Figure 6.   Configuration to measure ultrasonic backscatter using (a) normal incidence L-waves and (b) refracted S-waves.

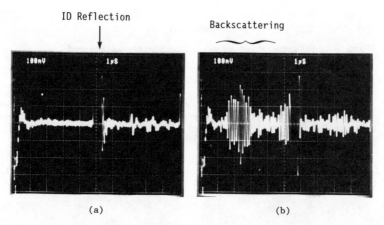

Figure 7. Backscattering from hydrogen-attack measurements using a 10-MHz transducer. (a) Backscattering from a sample cut from the outside surface, which had no hydrogen attack. (b) Backscattering from a sample cut from the inside surface, which had hydrogen attack.

DETECTION OF CREEP DAMAGE

The results of velocity and backscatter techniques applied on creep samples are described.

Velocity Technique

Both velocity and backscatter techniques were applied to detect creep damage. Creep test samples were made from alpha-iron samples to avoid any effects of carbide inclusions on ultrasonic scattering. Ultrasonic tests included velocity measurements of L, S, surface, and creeping waves. Results showed that the L-wave velocity decreased by as much as 4.7 percent; the S-wave velocity decreased up to 2.5 percent; the creeping wave, by 5.8 percent; and the surface wave, by 3.8 percent [8].

In general, the ultrasonic velocity measurements were very encouraging. A decrease in ultrasonic velocity was observed for all four types of wave modes. The application of a specific wave mode for creep detection depended on the geometry of the part and the expected location of creep damage. Surface and creeping waves can be applied when damage is localized close to the surface. Surface- and creeping-wave velocities are generally easier to measure, since calculation for velocity does not require the thickness of the material under test. The thickness of the material must be known for L- and S-wave velocity measurement. L- and S-waves would be more applicable for bulk damage across the specimen thickness.

A qualitative relationship between creep damage and a decrease in ultrasonic velocities was also obtained. Figure 8 shows a plot of the decrease in the L- and S-wave velocities versus creep strain in each sample. The figure shows a general decrease in velocity with increased creep strain, with the L-wave decreasing more than the S-wave velocity. Such a relationship is important when assessing the amount of creep damage in the material and can be used to predict the amount of damage. Figure 8 shows a general decrease in velocity with increased scatter in the data. The increased scatter occurs because the minimum velocity at a single point in the specimen is being correlated to the average strain in the specimen. A better correlation would be expected if the velocity measured at a specific point were plotted against the grain boundary or cavitation at that point.

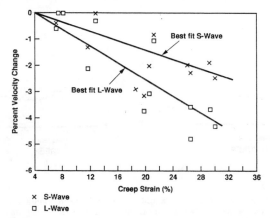

Figure 8. Decrease in L- and S-wave velocities with creep strain. The least squares best fit shows a greater decrease in L-wave velocity compared to S-wave velocity.

Backscatter Technique

Ultrasonic backscattering measurements were taken on the alpha-iron creep samples using 25-MHz S-waves refracted at 40-degrees. Backscatter increased in the samples with creep damage (see Figure 9). Tests were also performed with a normal incidence using 25 MHz L-waves, but no increase in backscatter was noticed. Backscatter increased with S-waves because: (1) orientation of the cavities and microcracks was such that it produced greater backscattering with the wave incident at an angle (S-waves) rather than with a normal incidence (L-waves) and (2) backscattering increased with the shorter wavelengths, as theory predicts that it should. (Wavelengths of S-waves are almost half those of L-waves.)

CALIBRATION TO ASSESS MATERIAL DEGRADATION

Calibration procedures are required in ultrasonic testing to obtain a reference measurement for attenuation, velocity, and backscatter. A realistic reference is important, since ultrasonic waves are also influenced by microstructure (carbides and grain size), stress, temperature, and anistropy. Thus, the reference material should be of the same material as the test specimen and in the same condition (stress, temperature, etc.). To account for such parameters, the reference should be selected from the same test piece in an undamaged portion as close as possible to the damaged section. For example, during the tests for detecting hydrogen damage in boiler tubes, the reference was taken on the same tube in close proximity to the damaged area.

CONCLUSIONS

Interest in nondestructive testing for material degradation has increased with the advent of life extension and the greater awareness of threats to safety. Two ultrasonic techniques, velocity and backscatter, have been shown to have advantages over conventional surface inspection techniques such as replication, hardness, and resistivity. The major advantage of ultrasonics

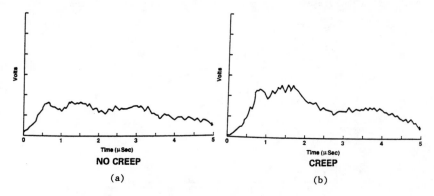

Figure 9. Backscatter results with 25-MHz S-waves at a 40-degree refracted angle in an iron sample. (a) No creep. (b) Creep damage. The number of waveforms averaged 20.

its ability to assess volumetric damage. Both velocity and backscatter were able to detect damage. The backscatter technique, however, was also successful in detecting damage in its early stages. Field tests using the velocity technique were successful for detection of hydrogen damage in the fossil-fired boiler tubes.

ACKNOWLEDGEMENTS

The author thanks John J. Hanley, David G. Alcazar, and Gary J. Hendrix for their help in conducting the ultrasonic tests. Dr. Stephen M. Gehl of the Electric Power Research Institute is acknowledged for financial support to do this work. Mr. Koji Kawano of Idemitsu is thanked for his support in providing hydrogen-damaged samples and in planning the experiments.

REFERENCES

1. B. J. Cooke, C. D. Buscerni, L. J. Clien, and W. E. Erwin, Hydrogen Attack on Cr-Mo Reactor Steels, presented at the Mid-Year Meeting of the API Division of Refining, Houston, Texas, May 1988.

2. R. B. Dooley and H. J. Westwood, in Analysis and Prevention of Boiler Tube Failures (Canadian Electrical Association, Montreal, Canada, Research Report 83/237 G-31, November 1983).

3. D. J. Gooch and R. D. Townsend, in CEGB Remanent Life Assessment Procedures [Electric Power Research Institute (EPRI) Life Extension Conf., CS-5208, EPRI, Palo Alto, CA, 1987].

4. R. Viswanathan and R. B. Dooley, Creep Life Assessment Techniques for Fossil Power Plant Boiler Pressure Parts (International Conf. on Creep, Proc., Tokyo, Japan, 1986), p. 349.

5. API, API Standard 941: Steel for Hydrogen Service at Elevated Temperatures and Pressures in Petroleum Refineries and Petrochemical Plants (API, Washington, D.C., June 1977) (Rev. 1983, 3rd Ed.).

6. T. Watanabe, T. Hasegawa, and K. Kato, in Ultrasonic Velocity Ratio Method for Detecting and Evaluating Hydrogen Attack in Steels (Corr. Monitoring in Ind. Plants Using NDT and Electrochem. Methods Proc., ASTM STP 908, (ASTM, Philadelphia, PA, 1986), pp. 153-164.

7. A. S. Birring, D. G. Alcazar, J. J. Hanley, G. A. Lamping, and S. M. Gehl, in Ultrasonic Methods for Detection of Service-Induced Damage in Fossil Plant Components, Life Extension and Assessment: Nuclear and Fossil Power-Plant Components, ed. C. E. Jaske (ASME, PVP, Vol. 138, NDE, Vol. 4, June 1988), p. 71-76.

8. P. Senior and J. Szilad, Ultrasonic Detection of Hydrogen in Pipe Line Steels, Ultrasonics, 42-44 (January 1984).

9. M. H. Loper, R. D. Shoemaker, and J. A. Strump, in Mitigating Forced Outages by Selective Replacement of Boiler Tubes (EPRI Conf. on Failures and Inspection of Fossil Fired Boiler Tubes, Proc., EPRI CS-3272, Palo Alto, CA, April 1983), pp. 6-35 to 6-52.

10. P. J. Latimer, D. M. Stevens, and T. P. Sherlock, in Hydrogen Damage Detection in Fossil Boiler Tubes (EPRI Conf. on Life Extension and Assessment of Fossil Plants, Proc., EPRI CS-5208, Palo Alto, CA, July 1986), pp. 1061-1076.

11. M. Yajima, D. Shozen, and T. Ohe, Detection of High-Temperature Hydrogen Attack in Steels by Ultrasonic Testing (49th API Midyear Mtg., Proc. 63, API, Washington, D. C., May 1984), pp. 44-54.

12. B. Neubauer and V. Wendel, Rest Life Estimation of Creeping Components by Means of Replicas (ASME Inter. Conf. on Advance in Life Prediction Methods, Proc., April 1983).

13. M. S. Shammas, Predicting the Remanent Life of 1Cr1/2Mo Coarse-Grained Heat Affected Zone Material by Quantitative Cavitation Measurements (Central Electricity Generating Bd., Leatherhead, Surrey, Eng., Rep. No. TPRD/L/3199/R87, EPRI RP2253-1, Nov. 1987).

14. K. Akiyama, T. Shiota, T. Maeyama, S. Shimojo, and T. Goto, Non-Destructive Evaluation of Creep Damage for Cr-Mo-V Rotor (Joint ASME/IEEE Power Conf. Proc., Paper 87-JPGC-PWR-30, Miami, FL, October 1987).

15. T. Goto, Study on Residual Creep Life Estimation Using Non-Destructive Material Properties Tests (2nd Inter. Conf. on Creep and Fracture of Eng. Mat. and Structures, Proc., Pineridge Press, 1984), 1355.

16. K. Kimura, K. Fujiyama, and M. Maramatsu, in Creep and Fatigue Life Prediction Based on the Nondestructive Assessment of Material Degradation for Steam Turbine Rotors, High Temperature Creep Fatigue (Elsevier Aplied Science, London 1987), pp. 247-270.

17. S. Sakurai, H. Miyata, K. Iijima, H. Sukekawa, and R. Kaneko, A Life Assessment Method Based on Micro-damage Mechanisms for High Temperature Components (Conf. on Life Extension, Proc., Paper No. 2.4.2, The Hague, The Netherlands, June 1988).

18. T. Goto, Y. Kadoya, T. Takigawa, and K. Kawamoto, An NDE System for the Detection of Early Damage in High Temperature Rotors (Inter. Symposium on Nondestructive Characterization of Mater., Proc., Montreal, Plenum Publishing Corp., NY, 1986).

19. H. Willems, W. Bendick, and H. Weber, Nondestructive Eval. of Creep Damage in Service Exposed 14 MoV 63 Steel (Inter. Sym. on Nondestructive Characterization of Mater., Montreal, 1986), pp. 451-460.

20. A. S. Birring, M. Bartlett, and K. Kawano, Ultrasonic Detection of Hydrogen Attack in Steels (pending public. in Corrosion, 1989).

21. M. Nakashiro, H. Yoneyama, and A. Ohamoton, in Assessment for Creep Damage of Boiler Tubes by Newly Researched Ultrasonic Techniques (Inter. Conf. on Advances in Mater. Tech. for Fossil Power Plants, ASM International, Metals Park, OH, 1987).

22. K. Goebbels, in Research Techniques in Nondestructive Testing, IV, edited by R. S. Sharpe (Academic Press Publishers, 1980), pp. 87-158.

APPLICATIONS OF POSITRON ANNIHILATION TO THE MONITORING OF FATIGUE
DAMAGE AND CREEP IN TECHNOLOGICAL COMPONENTS

A J ALLEN*, C F COLEMAN**, S J CONCHIE** AND F A SMITH*
* Materials Physics & Metallurgy Division, B521.2;
** Nuclear Physics Division, B7.21, Harwell Laboratory,
Didcot, Oxfordshire OX11 ORA, UK

ABSTRACT

This paper reviews the use of positron annihilation methods for
technological applications, particularly the use of positron annihilation
gamma ray lineshape analysis for the non-destructive assessment of static
deformation, machining processes, high cycle fatigue and creep in metal and
alloy components. The paper includes description of a transportable
lineshape analysis system recently developed for field applications.

INTRODUCTION

Positron annihilation science has developed steadily over the past 30
years and can currently be divided into five general types of study. These
comprise: measurements of the angular correlation of the annihilation gamma
rays to probe electronic structure in metals and alloys[1]; analysis of the
annihilation gamma ray lineshape spectrum[2] to probe materials properties in
metals and alloys; positron lifetime measurements[3] to determine and resolve
a variety of microstructural parameters in both metals and non-metals; low
energy positron beam studies[4] which are likely to rival electron microscopy
for electron energy loss spectroscopy (EELS) applications; and finally
positron emission tomography[5] (PET) which has achieved considerable success
as an industrial inspection technique for the study of oil flow in engines.
Of all these methods positron annihilation lineshape and positron lifetime
studies show considerabe potential as future field-applicable NDE
techniques.

The use of positrons for NDE is not perhaps obvious. They are emitted
by certain radioisotopes as positive beta particles and are the
anti-particles of electrons. In condensed matter, free positrons are
thermalised by multiple scattering and then annihilate with electrons within
typically 200ps. Whereas the annihilating electron must obey the Pauli
exclusion principle and typically has kinetic energy of order one
electronvolt, the positron can thermalise until its kinetic energy is of
order of that for thermal vibrations ($\sim 1/_{40}$eV). Thus it is the momentum and
state of the annihilating electron which dominates the annihilation event.
The mutual annihilation of positron and electron results in the emission of
two almost collinear 511 keV gamma rays. The collinearity and exact energies
of the annihilation gamma rays, as well as the mean positron lifetime itself,
depend on the environment of the annihilating electrons. In this way a
material is probed with one kind of radiation (positrons) which give rise to
another form of radiation (annihilation gamma rays) and these are sensitive
to and carry information on the material properties[6].

REVIEW OF TECHNOLOGICAL POSITRON ANNIHILATION APPLICATIONS

Figure 1 shows schematically the positron annihilation process in
condensed matter when a positron emitting radioisotope is brought into close
proximity to a sample surface. The maximum penetration distance is
determined by the choice of source and the atomic number density in the

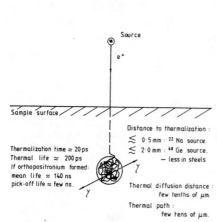

Source

e⁺

Sample surface

Distance to thermalization :
≲ 0·5 mm : ^{22}Na source.
≲ 2·0 mm : ^{68}Ge source.
— less in steels

Thermalization time ≈ 20 ps
Thermal life ≈ 200 ps
If orthopositronium formed:
mean life ≈ 140 ns
pick-off life ≈ few ns.

Thermal diffusion distance :
few tenths of μm.

Thermal path :
few tens of μm.

Fig 1. Schematic of positron
annihilation process

material sampled[7]. It is important
to note that the actual penetration
distances vary continuously from zero
up to a maximum, as defined by the
variable beta emission energies,
typical of any beta decay spectrum.
Clearly in many applications a ^{68}Ge
source is to be preferred to the
other commonly used positron emitting
isotope ^{22}Na since the maximum
emission energy of the former is
1.89 MeV compared to only 0.544 MeV
for the latter. Typical maximum
penetration depths using ^{68}Ge are
therefore less than 0.5mm, ∿1.5mm and
∿2.0mm for iron, aluminium and light
weight resins respectively. These
small but macroscopic depth-sampling
distances render positron annihilation more than simply a surface-
sensitive technique. After thermalization the mean diffusion distance
prior to anihilation is a few tenths
of 1 μm but the mean thermal path is
a few tens of μm. Thus ∿10^5 lattice
sites are sampled on average by
each thermal positron.

    In a metal or alloy positron annihilation can involve valence or
conduction electrons. Whereas valence electron states are bound to
particular atoms, conduction electrons are unlocalised. The Heizenberg
uncertainty princple requires that the former have higher kinetic energy than
the latter which nevertheless have much more kinetic energy than do
thermalised positrons. Thus the energies and momenta of the annihilation
gamma rays are Doppler-shifted by the centre of mass motion of the e⁺e⁻ pair
at annihilation and this effect is greater when the annihilation is with a
valence electron than with a conduction electron. In practice many
annihilation events are recorded using scintillator or intrinsic germanium
detectors, and a statistical distribution is built up of the annihilation
characteristics.

    One of the most successful laboratory applications of positron
annihilation has been to use 2-dimensional ACAR detectors[1] placed some
distance away on opposite sides of a single crystal sample and source. The
statistical distribution of the non-collinearity in emission of the
annihilation gamma rays can be systematically plotted out to determine the
angular momentum distribution of the annihilating electrons in 3 dimensions.
Normally a bell shaped distribution is seen with a parabolic profile
superimposed on a broader Lorentzian component. These distributions can be
associated with the conduction and valence electron bands respectively. The
cut off between the two is usually sharp and gives the momenta and hence
energies at the electron Fermi level. By systematically changing the crystal
orientation it is possible to map out the Fermi level in 3 dimensions. For
impure metals and alloys, this is the most accurate method known for
determining the Fermi surface. The work also has important applications in
the semiconductor industry. However the apparatus is usually at least 6
metres long and mounted in concrete and is therefore not suitable for NDE
applications.

    In positron annihilation gamma ray lineshape[2] measurements the energy
spectrum of the annihilation gammas is measured using a high resolution

intrinsic germanium detector. The 511 keV annihilation peak is Doppler broadened by the centre-of-mass momenta of the annihilation events. As in angular correlation measurements, the Doppler broadening associated with annihilations involving valence electrons is greater than when the annihilations involve conduction electrons. If vacancy defects are introduced into the sample material this has three effects. Firstly the conduction electron band overflows slightly into the vacancies whilst the localised valence electron band cannot. Secondly these vacancies become negatively charged, due to the overflowing conduction electron band. Thus positrons are attracted to vacancy defects and can be trapped there. Finally the annihilation events taking place at these vacancy sites only involve conduction electrons. Since the conduction electron kinetic energies are generally less than those of the valence electrons, the Doppler broadening in the annihilation gamma ray energy lineshape is reduced by the presence of vacancy defects such as vacancies, microvoids and dislocations in the atomic site concentration range $\approx 10^{-7}$ to 10^{-4}. Such vacancy defect concentrations arise in association with plastic deformation, high cycle fatigue, creep or irradiation damage. Positron lineshape analysis is thus particularly suited to NDE assessment of these phenomena[8] in metal or alloy components and is shown schematically in figure 2.

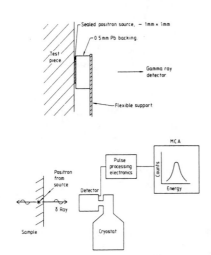

Fig 2. Schematic diagram of lineshape measurements

Positron lifetime[3] distribution 'spectra' provide complementary information to that obtainable from lineshape analysis. Whereas lineshape analysis is essentially a 'one parameter' measurement, lifetime spectra can frequently be time-resolved into components each of which may be associated with a different microstructural effect. In addition, lifetime measurements are particularly well suited to study positronium formation in non-metals[9]. Positronium is formed when the thermalising positron forms a pseudo-stable pair with an electron in the sample material. Parapositronium (with antiparallel e^+ and e^- spins) decays within $\sim$125ps, but ortho positronium (with parallel e^+ and e^- spins) has a natural life of $\sim$100ns. In fact this is so long that the most usual outcome is a 3-gamma 'pick-off' annihilation caused by the intervention of a second electron. Pick-off annihilation is increased by the presence of polar molecules such as water. Thus the technique can be a powerful measure of moisture ingression into technological resin-based materials such as aerospace components. Figure 3 shows schematically a conventional laboratory-based positron lifetime system, and figure 4 typical positron lifetime spectra with the 'pick'off' components appearing as experimental tails in the lifetime distributions determined. Mean 'pick-off' lifetimes are typically $\sim$1ns. By sandwiching a ^{22}Na positron source between two identical samples and positioning the source-samples assembly between two end-on scintillator detectors, the 1.3 MeV marker γ-ray, emitted with the positron by the source, can be used to provide the 'start' signal whilst the 'stop' signal is given by detection of one of the annihilation γ-rays. This measurement geometry is clearly unsuitable for NDE applications - a collimated beam is needed with a β-energy loss

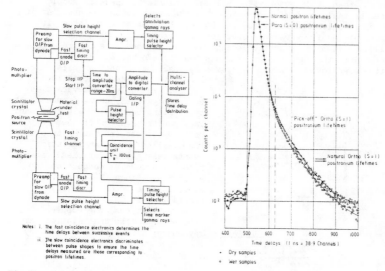

Fig 3. Schematic diagram of
        lifetime measurements

Fig 4. Typical positron
        lifetime spectra

scintillator through which the positron must pass to provide the 'start'
pulse. Such a system incorporating a powerful ^{68}Ge positron source has been
developed at Harwell to provide a penetrating positron beam for future
lifetime NDE measurements[10].

Two other technological applications of positrons have developed: one
based on the use of low-energy positron beams[4]; the other an imaging system
based on tomography. Positrons from a radioisotope have a continuous spread
of energies upto a maximum value, in common with all β-decay spectra. This
is useful for penetrating into sample material but not for studying the
surface. However certain materials have negative work functions for positron
emission. This means that positrons from a source on entering the material
and thermalising within a small distance from the sample surface are
re-emitted at the work function energy. Thus, by suitable use of electric
and magnetic fields a low-energy monochromatic positron beam can be produced.
Such beams are becoming a powerful diagnostic tool for investigating surface
structures and are beginning to rival EELS for some applications.
Unfortunately both sample and beam must be enclosed inside a vacuum and this
precludes the use of low energy beams in NDE applications for the foreseeable
future.

PET is a new NDE technique[5] which has been developed for measuring
fluid flow in engines, fluidised beds and other industrial equipment. In
PET, a small quantity of a short-lived positron emitting isotope is injected
into the fluid to act as a label. Each positron thermalises within a short
distance in the fluid and then annihilates as in condensed matter. The two
nearly collinear annihilation gamma rays are identified by their simultaneous
recording in two position sensitive detectors, one on each side of the
subject. The gamma ray flight path from each pair of recordings is computed,
and the intersections of many paths provide a tomographic map of the origins
of the gamma rays and therefore of the distribution of the labelled fluid.
The 511 keV anihilation gamma ray energy is sufficient to penetrate the

structure of engines; flaws have been observed through 100mm of steel, which
is equivalent to 300mm of aluminium. Recent developments also allow the
method to be used tomographically using synthetic holographic encoding[11]
(SHE). Figure 5 shows typical results for the oil distribution (the lighter
areas) in a bearing test rig and illustrates the value of overlaying an

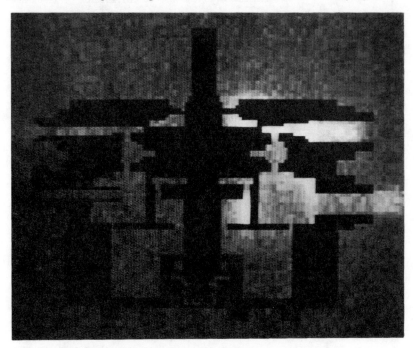

Fig 5.   Positron tomograph of oil in a bearing test rig
         (courtesy of University of Birmingham, England).

accurately scaled drawing of the subject on the radio-label image. The
tomograph and overlay represent the central slice of the rig which was 150mm
in diameter.

POSITRON LINESHAPE ANALYSIS IN NDE

Positron lineshape analysis applications to the NDE assessment of
plastic or radiation damage in metals and alloys is based on the change in
lineshape parameter, S, as the vacancy defect concentration is increased
according to a simple trapping model. The lineshape parameter S is defined
in figure 6. Note that as the line narrows, so S increases. Figure 7 shows
a sample calibration curve of change in S-parameter, ΔS, vs plastic strain as
determined from measurements on a 'tensile pull' sample of mild steel. The
change in positron S'parameter can be related to a trap concentration 'C'
where C is the fraction of lattice sites associated with vacancy
defects[12]. If τ_o is the mean positron lifetime in a reference annealed
sample and 'μ' is the trapping rate constant for the defect population, then

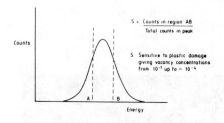

Fig 6. Definition of lineshape
S-parameter

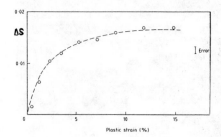

Fig 7. Calibration curve for
mild steel

the model predicts that:

$$\frac{S - S_o}{S_{sat} - S} = \mu C \tau_o \qquad (1)$$

where S_o is the reference S value for an annealed sample of the same
material, S_{sat} is the saturation S-value when all positrons are trapped at
vacancy defects (applicable when $C \geq 10^{-4}$), and S is the measured S value for
material with trap concentration C. Although absolute concentration values,
C, are important for some applications, such as quantitative determinations
of radiation damage, it is frequently sufficient to use calibration curves
such as figure 7 to assess plastic damage directly from ΔS values.

Assessment of plastic deformation

A simple case for assessment of plastic deformation is a 4 point bend
specimen shown in figure 8. Here a mild steel bar 25mm thick and 51mm wide
has been deformed beyond the elastic limit using a four point bending device.
The stress has produced a permanent curvature of approximately 500mm radius.
The variation of plastic damage along the radial direction across the wide
face of the sample has been measured using a contact source geometry and the
results are shown in figure 8. The S values increase from the neutral axis
in the centre to the inner and outer regions of the bar. For measurements
towards one end of the bar (which is relatively unstressed) the minimum value
of S in the centre is 0.4840 and this has been taken as the reference value
characteristic of the material in the starting condition (this is still
probably above the S value for an annealed sample). All scans across the bar
gave the same minimum value of S indicating that the bending has not produced
any non-uniform damage along the neutral axis of the bar. In the stressed
section, the compressive plastic strain towards the inner radial edge and
tensile plastic strain towards the outer radial edge have produced similar

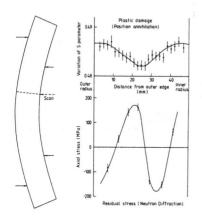

Fig 8. Plastic damage and
residual stress variation
across a four point bend
steel bar.

changes in S (ΔS = 0.0055)
relative to the centre value.
These measurements were made using
the 'contact source' geometry
suitable for NDE measurements as
shown schematically in figure 2.
The final bar geometry gives a
neutral axis radius of 400 mm.
The bar width of 51 mm implies
6.5% tensile and compressive
strains at the extreme outer and
inner radii respectively. The
yield strain for mild steel is
≈ 1500 με and this is reached in
tension or compression just 0.6 mm
from the neutral axis. Thus the
minimum S-values in the centre of
the bar are not for unstrained
material. Indeed the rise in S,
ΔS of 0.006 between centre and
extreme radii suggests that the
strain at the centre of the bar as
measured by the source is ≈ 2%
with respect to the fully
annealed calibration sample.

Figure 8 also shows the
residual axial stress variation
across the bar[13], as measured
using high resolution neutron diffraction. The oscillating stress
distribution arises to provide equilibrium between the stresses in the
central regions of the bar which apply relaxation forces and stresses in
the plastically deformed inner and outer regions which cannot relax. It is
clear that the greatest residual stresses do not occur in the same place as
greatest plastic damage. Thus complementary measurements of residual stress
and plastic damage distributions can provide important information on the
stress history of a component where this is uncertain, since the plastic
damage measurements indicate where greatest stress has been applied in the
past.

A further application of plastic deformation assessment lies in the
assessment of machine processes such as shot-peening. Preliminary results on
aluminium by the authors suggest that, by use of both ^{22}Na and ^{68}Ge sources,
sensitivity to peening levels can be maintained throughout the range of
peening grades.

Detection of incipient fatigue damage

The electric power, aerospace, aeroengine, offshore and automotive
industries all have need of a reliable means to monitor incipient fatigue
damage - particularly where high-cycle fatigue is prevalent. The electricity
and air industries have specific NDE requirements, for which the positron
lineshape technique is suitable.

Figure 9 shows a mild steel test piece 19 mm thick which has been
fatigued until a crack of 8mm length has developed. Using a 1mm contact
source, an extensive array of points has been scanned to one side of the
fatigue crack. Figure 9 gives a contour plot of the plastic damage close to
the crack where a 2mm spacing has been used. The lineshape measurements are

138

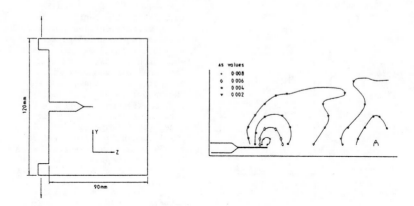

Fig 9.   Plastic damage in a cracked fatigue test specimen

expressed in terms of ΔS, giving the difference between the S value at the
measured point and a reference value measured in an undamaged region well
away from the crack. The plastic damage is greatest near to the crack face
with the maximum close to the crack tip. The plastic damage falls to a
minimum 22mm from the crack tip but then increases towards the edge of the
specimen. Figure 9 also indicates the size of the plastic zone in the
fatigued region.

    Residual stress measurements[13] (using neutron diffraction) and plastic
damage measurements were made along the crack line across the sample. Figure
10 compares the magnitudes of the stresses parallel and perpendicular to the
crack with the plastic damage variation. It is clear that both stress and
damage distributions oscillate but that the two sets of oscillations are not
in phase except at the crack tip itself. Thus such oscillating plastic
damage and residual stress may be symptomatic of high cycle fatigue damage.

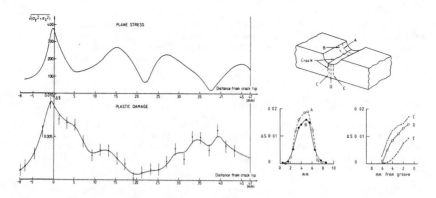

Fig 10.   Stress and plastic damage   Fig 11.   Measurement of fatigue
          variations along the crack   damage on a notched
          line in figure 9   IMI318 test piece

Note that the range in S values measured gives a maximum of $\Delta S \approx 0.008$ and it is assumed that saturation levels of plastic damage exist close to the crack tip.

A further illustration of incipient fatigue damage detection is given in figure 11 on a notched test specimen of IMI 318 aerospace alloy[14]. The S-parameter variation around the notch locates the position of greatest plastic damage and this is seen to build up as the number of fatigue cycles is increased until cracking takes place. The positron lineshape technique can detect incipient fatigue damage prior to the appearance of any discernible plastic deformation.

Applications to creep damage assessment

In many industrial plants, pressure tubes and vessels, as well as other components under high temperature and pressure are subject to degradation due to creep. NDE methods to detect and assess the extent of creep damage have become of increasing need to maintain plant and extend service life. At present the most extensively used technique is that of surface replication and the onset of creep damage is characterised by the appearance of micropores and of precipitate-coarsening. Interpretation of surface replication results can be difficult and several genuine NDE techniques are currently being explored[15]. These include x-radiography, ultrasonics, magnetic or electric methods, and positron annihilation lineshape analysis. Changes in the microstructure inherent to normal service life seem to have significant effects on the results of any individual test. It may prove necessary to combine the results of different techniques to define the creep state of components with any certainty. Figure 12 displays the results of both ultrasonic velocity and positron lineshape measurements on an aerospace CrMoV steel tensile pull sample subjected to controlled creep conditions

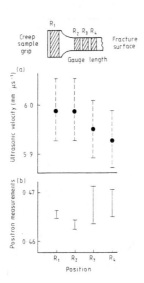

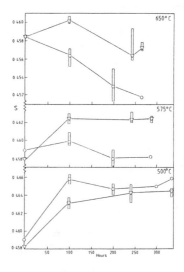

Fig 12. Positron and ultrasonic measurements of
creep damage in CrMoV
steel test specimen

Fig 13. Study of creep
damage in 2¼ CrlMo
steel (0 = creep,
∇ = control)

until failure has occurred (70 MPa for 274 hours at 675°C). The ultrasonic velocity decreases (as predicted by theory) towards the creep fracture surface where the greatest creep damage has occurred. The ultrasonic technique can be affected by the presence of strong texture or residual stresses. The positron lineshape method is not so affected by texture or stress but the sensitivity to plastic damage under creep conditions can be reduced by thermal annealing out of vacancy defects or by their migration to join coarser micropores. Nevertheless some rise in S parameter is seen as the fracture surface is approached. Clearly it is even more important for creep damage assessment than for fatigue damage assessment[16] in industrial components to fully characterise the behaviour of the component material/service environment combination.

Figure 13 shows the effects of creep damage on positron S-parameter at different temperatures for a 2¼ CrlMo steel. Some of the variations in S are complex and not fully understood. The samples show an initial increase in S value vs time at 500°C and 575°C, but a slight rise followed by a fall at 650°C. The samples subjected to creep show S-value variations which fall below those for the aged samples at 575°C or above, and at 650°C there is a continuing decrease in S value. These effects are believed to be due in part to the formation of carbide precipitates associated with creep in addition to micropores. Viable NDE measurements would seem to be possible either above 650°C, or substantially below 500°C (in the 'no creep' regime), provided calibration curves are first determined.

Figures 14 a and b are calibration curves on tensile pull samples of isochronal α-iron. Figure 14b shows that the initial increase in S value vs plastic strain (elongation) after annealing can be reduced to zero by one hour's annealing at temperatures above ∿600°C. (Two samples are shown which had each been cold rolled to 60% reduction in thickness). The annealing out

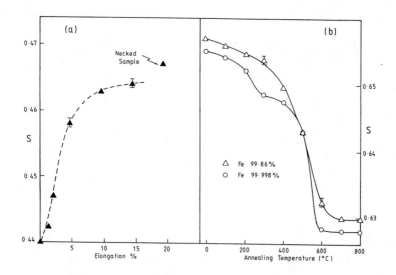

Fig 14a.   Specpure α-iron rods   Fig 14b.   Isochronal annealing
           deformed in tension   effects on S value
           after annealing at
           700°C for 200 hours

of creep damage to which positron annihilation is sensitive takes place at roughly the temperatures at which the creep characteristics become complex in 2¼Cr1Mo steel. We conclude that positron lineshape analysis should be one of the techniques considered by industry for creep assessment, but fully realistic calibration under service conditions is required, as in all cases where the precipitate microstructure changes in parallel with the appearance of vacancy defect plastic damage.

TRANSPORTABLE POSITRON LINESHAPE SYSTEM FOR NDE APPLICATIONS

All the experimental results presented in this paper have been obtained in the research laboratory. However for some time a transportable positron annihilation lineshape system has been under development at Harwell[13]. The scale of the counting electronics has been reduced to a single moderately sized crate, and the multichannel analyser used in the laboratory for lineshape analysis has been replaced by channel group selectors. The channel group selectors are set up in advance so as to give an S value directly from the ratio of accumulated counts in different energy windows.

Fig 15. Transportable lineshape system for NDE measurements

Figure 15 shows the transportable lineshape system now available for NDE measurements. The personal computer system allows S value outputs to be displayed as counting progresses. A run can be terminated when sufficient counts have been obtained to give a stable S value. The compact intrinsic germanium detector can be used to take measurements at any atitude. It is mounted normal to the surface of the component being studied with the two attached 'feet' in contact with the surface. A small sealed contact source of ^{22}Na or ^{68}Ge is then positioned in contact with the sample surface midway between the feet. A reproducible measurement geometry is thus maintained and reliable S values can be determined in less than 10 minutes per measurement. As a result, 'in the field' NDE mapping of plastic damage to ∿1mm spatial resolution in metal and alloy components is possible within realistic timescales. Future improvements are likely to consist of an even more compact electronic and computing system and remote handling cradles for individual applications.

CONCLUSIONS

Positron annihilation is both a laboratory tool for solid state physicists and materials scientists, and a technological probe for material components. Whereas angular correlation studies and low energy positron spectroscopy are likely to remain laboratory based, positron lineshape and lifetime measurements, as well as PET have the potential for NDE applications. Of these positron lineshape NDE technology in particular is now available for use in monitoring plastic deformation, (and irradiation damage), high cycle incipient fatigue damage and, in some cases when combined with other NDE methods: creep damage.

ACKNOWLEDGEMENTS

The work reported here has been supported by the Metrology and Standards Requirement Committee of the UK Department of Trade and Industry and by the UKAEA Underlying Research Programme for which UK Crown Copyright is reserved. Dr T D Beynon of the University of Birmingham is acknowledged for supplying the information on PET.

REFERENCES

1.   S. Berko, in Positron solid-state physics, edited by W. Brandt (North Holland Publishing Company, Amsterdam, 1983) pp. 64-145.

2.   K. Mackenzie, in Positron solid-state physics, edited by W. Brandt (North Holland Publishing Company, Amsterdam, 1983) pp. 196-264.

3.   W. Brandt, Scientific American 23 (1), 34 (1975).

4.   C.D. Beling and M Charlton, Contemporary Physics 28 (3), 241-266 (1987).

5.   M.R. Hawkesworth, M.A. O'Dwyer, J. Walker, P. Fowles, J. Heritage, P.A.E. Stewart, R.C. Witcomb, J.E. Bateman, J.F. Connolly and R. Stephenson, Nucl. Instr. Meth. A253, 145-157 (1986).

6.   C.F. Coleman, NDT International 10 (5) 227-234 and 235-239, (1977).

7.   A.E. Hughes, Materials in Engineering 2 (34) (1980).

8.   L. Granatelli and K.G. Lynn, Brookhaven National Laboratory report, BNL-28795 (1981). Also see: O. Brummer and G. Dlubeck, Mikrochimica Acta (Wien) Supp 11, 187-204 (1985).

9.   A. Dupasquier, in Positron solid-state physics, edited by W. Brandt (North Holland Publishing Company, Amsterdam, 1983) pp. 510-564.

10. C.F. Coleman, UK Patent application no. GB 2179829A, (July 1987).

11. T.D. Beynon, M.R. Hawkesworth, T.R. Matthews and M.A. O'Dwyer, Nucl. Instr. and Meth. - in press (1988).

12. R.N. West in Positrons in solids edited by P. Hautojarvi (Springer-Verlag, Berlin, 1979) pp. 89-139.

13. A.J. Allen, C.F. Coleman, S.J. Conchie, M.T. Hutchings and F.A. Smith in Proc 4th European NDT Conference, London, 1987, (in press) (Pergamon, Oxford 1988); also Harwell report MPD/NBS 331 (1987).

14. A.E. Hughes, C.F. Coleman and F.A. Smith in Review of progress in quantitative nondestructive evaluation, edited by O. Thompson and D.E. Chimenti (Plenum Publishing Corporation, New York, 1982) I pp. 661-664.

15. R.L. Smith, Metals and Materials 3 (4) 187-191. (1987).

16. A.J. Allen, D.J Buttle, C.F. Coleman, F.A. Smith and R.L. Smith EPRI report NP-5590 (Electric Power Research Institute, Palo Alto, California, 1987).

UTILIZATION OF ULTRASONIC MEASUREMENTS TO QUANTIFY AGING-INDUCED MATERIAL MICROSTRUCTURE AND PROPERTY CHANGES

STEVEN R. DOCTOR, STEPHEN M. BRUEMMER, MORRIS S. GOOD, LARRY A. CHARLOT, T. THOMAS TAYLOR, DONALD M. BOYD, JOHN D. DEFFENBAUGH, AND LARRY D. REID
Pacific Northwest Laboratory, P. O. Box 999, Richland, WA 99352

ABSTRACT

Many structural components are exposed to thermal environments that promote aging-induced changes in material microstructure and properties. The extent of these changes may lead to a reduction in the design safety factors and, as a worst case, premature failure of the component. In order to illustrate how nondestructive measurements can be used to indicate these changes, ultrasonic measurements were conducted on stainless steel. Specimens from a high-carbon, Type 304 stainless steel were subjected to various thermal treatments to introduce differing precipitation morphologies. Specimens were examined using transmission electron microscopy to quantify the size, density, and distribution of the precipitates formed. Ultrasonic measurements were performed to assess sensitivity to the change in precipitation levels. Shear-wave birefringence provided the most consistent measurements for tracking precipitate morphology changes. Birefringence velocity differences increased as precipitate size and density increased on grain boundaries and as additional precipitates formed in the matrix. The ability to nondestructively monitor the precipitation process is important since susceptibility to environmentally assisted cracking and low-temperature embrittlement correlated with this microstructural change.

INTRODUCTION

This work evaluates the potential of nondestructive measurements to assess material properties and property changes. The ability to make these types of measurements would benefit the safe and economical utilization of structures. Many industries such as electric power production employ large safety factors to insure that the structures will retain their integrity over their design life. Furthermore, as many structures approach their design lives, it is important to know whether they must be retired due to material degradation or if they can be used safely for an extended time period. With accurately measured material properties, fracture mechanics analysis can be performed to determine if the defect sizes that need to be detected change as the structure ages (degrades). The significance and impact of defects can be more accurately assessed and unnecessary repairs can be eliminated. The most direct and useful means to obtain the needed data is to use nondestructive methods to make in-situ measurements of the material properties at all importart locations on the structure.

Nondestructive ultrasonic methods were utilized to measure material microstructural and property changes in stainless steel. Well characterized material was thermally aged to induce various precipitate distributions and susceptibilities to environmental cracking. Direct correlations are made between ultrasonic birefringence measurements and changes in stainless steel microstructures.

PRODUCTION AND CHARACTERIZATION OF CONTROLLED MICROSTRUCTURES

The stainless steels represent a simple system where precipitate distributions can be controlled either inter- or intragranularly. Intergranular precipitation of chromium-rich carbides prompts a local depletion of chromium at grain boundaries. This phenomena is commonly referred to as "sensitization" and results in a material condition susceptible to intergranular corrosion and stress corrosion cracking (SCC). Two high-carbon Type 304 stainless steel heats (Table I) were selected for initial precipitation experiments. Heat compositions were comparable to piping specifications in operating boiling water reactor power plants. Extensive corrosion and SCC tests had been performed on these heats in various heat-treated conditions [1-3]. Thus, comparisons between NDE measurements could be made not only to precipitate microstructures, but also to cracking susceptibility.

TABLE I. Bulk Composition of 304 Stainless Steel Heats

Heat	C	Cr	Ni	Mo	Mn	Si	N
C6	0.062	18.5	8.8	0.2	1.7	0.4	0.06
C7	0.072	18.5	9.3	0.4	1.7	0.5	0.04

Three sets of specimens were heat treated to produce a wide range of microstructures. Precipitate distributions were documented by transmission electron microscopy (TEM), and corrosion/SCC susceptibility determined using the electrochemical potentiokinetic reactivation (EPR) test [4]. This test measures the degree of sensitization (i.e., chromium depletion) in a stainless steel and can be directly correlated with IGSCC susceptibility in high-temperature water environments [3]. Solution-annealed material was compared to mill-annealed material for one set. This enabled the effect of different grain sizes to be evaluated as well as the possible effect of rapid quenching which was done from the solution annealing temperature.

Solution annealing was performed at 1100°C for 1 hour followed by a water quench. Material grain size increased from about 50 to 110μm due to this anneal. Individual specimens (typically 2.5 mm x 2.5 mm x 9.1 mm) were then heat treated at 700°C for times ranging from 1 to 150 hours. This time was the duration in a pre-heated furnace and was followed by air cooling. Microstructural changes are illustrated in Figure 1. The predominant microstructural change with heat treatment is in the size, density, and distribution of carbide precipitates. Initial precipitation is confined to grain boundaries (Figure 1b). As heat treatment time increases ($\geq$20 h), intergranular carbides increase in size and intragranular carbides begin to form. After long times ($\geq$100 h), very large carbides are found both at grain boundaries and in the matrix. A minor change in material microstructure that occurred along with carbide precipitation was a variation in dislocation and stacking fault density. As carbides grew to relatively large sizes (after about 10 h at 700°C), dislocation density appeared to increase and additional stacking faults were observed. These changes were difficult to quantify since significant grain-to-grain differences were present. In order to ensure that no significant bulk changes were occurring with heat treatment, x-ray diffraction analysis was also performed. Diffraction data gives texture information from the relative intensity of lines, which can be used to construct inverse pole figures. Patterns obtained from heat treated specimens showed only a slight enhancement of (111) lines and slight suppression of (220) lines. Line intensities among samples examined were essentially unchanged, i.e., no texture change was observed. Therefore, the slight texturing that is typical of wrought stainless steel was not altered by the heat treatment.

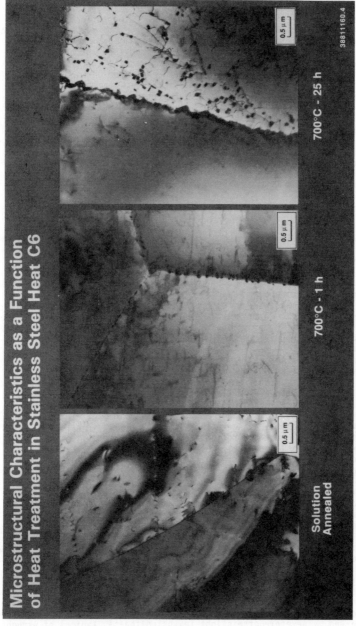

145

Microstructural Characteristics as a Function of Heat Treatment in Stainless Steel Heat C6

Solution Annealed

700°C - 1 h

700°C - 25 h

0.5 μm

0.5 μm

0.5 μm

38811160.4

FIGURE 1. TEM photos of three different heat-treated samples: a) solution-annealed sample, b) after 1 hour of heat treatment at 700°C, and c) after 25 hours at 700°C.

Degree of sensitization (as measured by the EPR test) has been previously reported [2,3] to increase rapidly in these heats with heat treatment time at 700°C. The solution-annealed material with no precipitated
carbides (Figure 1a) is not sensitized. Sensitization levels roughly
track the increase in intergranular carbide precipitation showing low
levels of sensitization after 1 hour at 700°C and reach a maximum degree
of sensitization after about 75 hours. Stainless steel specimens begin
to "desensitize" after long heat treatments as the bulk carbon content in
the matrix, and the driving force for further precipitation, decreases
[5]. As a result, a specimen aged for 75 hours will be more susceptible
to intergranular corrosion and SCC than specimens aged at 1 or 150 hours.

NONDESTRUCTIVE MEASUREMENTS

Extensive research has been performed over the years with ultrasonics
to measure materials. Acoustic velocity, attenuation, and related frequency
effects are the primary ultrasonic measurements which have been used to
characterize materials. The development of the fundamental theory of
acoustics goes back many years, but one of the better references is the
two-volumne book by Auld [6]. This fundamental theory relates the influence
of some material changes to wave velocity. The practicality of making
meaningful velocity measurements must also be considered. For example,
the effects of the transducer coupling layer on the wave velocity measurement has been shown by Vincent [7] to be in many cases much larger than
the subtle differences caused by the material property changes. Thus,
caution must be used in making velocity measurements.

Wave velocity is sensitive to several material conditions. Two material properties which have received extensive research are texture and
stress. Research has shown that birefringence measurements are related
to the stress, texture, and yield strength or fracture toughness [8-12].
This work has built both a theoretical basis for dealing with texture
anisotropy and stess, and an experimental validation of the theory in
some materials.

Other work in aluminum by Rosen et. al. [13,14] has shown that wave
velocity changes as the alloy was aged at different temperatures with a
maximum wave velocity change of 0.71%. These studies also examined heat-
induced precipitation levels of several aluminum alloys. The measured
wave velocity changes were quite small and required accurate thickness
measurements. In our review of approaches, birefringence was selected as
more attractive because it provides a self calibration and did not require
the need to know the thickness.

Longitudinal and shear velocity measurements were initially made on
each specimen. This was done as a simple bench mark for each specimen.
The shear mode created some problems because the velocity varied based on
the orientation of the probe. Shear velocity (time-of-flight) was then
taken at two orthogonal orientations where the maximal and minimal values
occur. A 12-mm diameter, 2.25-MHz transducer was coupled to the specimen
using a viscous "honey" type couplant. The values recorded were the
extremely small time shifts of the signal between the two transducer
orientations. The third back surface echo was used to obtain the time
shift of the birefringent measurement. The third echo was used because
of optimum signal-to-noise ratio. The plotted birefringence data showed
some very interesting relationships. Figure 2 shows the results obtained
on the furnace-annealed samples for heats C6 and C7. The data trend was
repeated by both specimen sets that had received the same heat treatment
although they came from a different heat of material. Making contact
shear measurements is quite difficult, and yet we found this general trend
was repeated by several personnel making the measurements in a blind
fashion. Also contained in Figure 2 is a plot of the change in the degree
of sensitization for each heat.

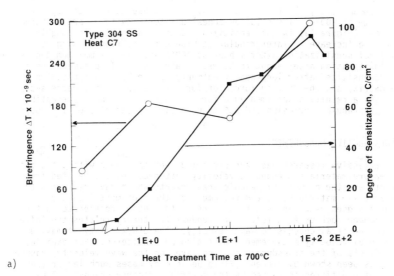

a)

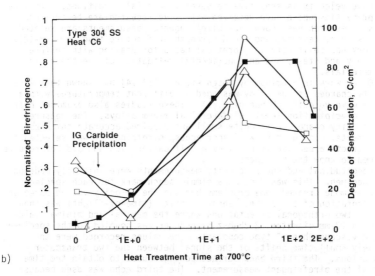

b)

FIGURE 2. Plots of the acoustic birefringence (open data points with each symbol representing results from different inspections) versus hours of heat treatment at 700°C and the degree of sensitization (filled in data points) for: a) heat C7 with absolute birefringence values and b) heat C6 with normalized ($\frac{V_R - M_M}{V_R}$ x 100) 0-1% birefringence values.

DISCUSSION OF RESULTS

The two solution-annealed sets showed a significant variation in birefringence with heat treatment where carbide precipitation and sensitization occurs. Birefringence initially decreases slightly, but then increases sharply after longer times. Differences do not increase with increasing aging after about 25 hours at 700°C. This variation is remarkably similar to the change in degree of sensitization as shown in Figure 2.

The primary microstructural change occurring is the precipitation and growth of carbides as shown in the TEM photos (Figure 1). During the initial aging, carbides are predominantly intergranular and then begin to precipitate intragranularly after about 20 hours. The increased birefringence at 10 hours of heat treatment appears to be related to the growth of intergranular carbides. Carbide nucleation occurs in very fine, semi-coherent particles during the first hour of aging. The reduction in birefringence between the solution-annealed specimens and the 1-hour specimens may be due to a change in material residual stress. Water quenching from solution-annealing temperatures can promote a higher stress state due to the rapid quench and the supersaturation of carbon in solid solution. The subsequent increase in birefringence may result from growth of semi-coherent and incoherent carbides. This process can produce significant stresses as indicated by an increase in local dislocation densities and in stacking faults in the microstructure.

Wave acoustic velocity is a function of stress that changes in the specimen due to residual stress, carbide formation, and chromium depletion along the grain boundaries, as well as the increase in dislocation density and stacking faults. The influence of the carbide formation also creates a two-phase material resulting in a global change of the velocities even though the carbides are extremely small.

The influence of texture was considered and was the rationale for conducting the X-ray diffraction analysis. Since texture change can cause a predictable shift in the birefringence, the texture in the specimens needed to be known. X-diffraction showed that texture did not change and, therefore, could not cause the measured birefringence changes.

In summary, the birefringence demonstrated a positive correlation with precipitation in solution-annealed stainless steel. At this time, it is not known whether the NDE technique sensitivity directly results from the presence of the carbides or from a characteristic that is dependent on aging and carbide growth. This analysis is continuing. A complication to these results has been identified through preliminary tests on mill-annealed materials. Contrary to the solution-annealed specimen sets, birefringence did not significantly change with aging. We are certain that these results do not invalidate the positive correlation for the solution-annealed specimens, but it may have a direct impact on the mechanism producing birefringent differences.

CONCLUSIONS

This study shows very promising correlations for ultrasonics to measure very subtle changes in the microstructure of 304 stainless steel created by heat-treatment-induced precipitation. The heat treatment did not change the texture of the specimen and, thus, could not account for the changes measured. The difference between the furnace-annealed and the mill-annealed specimens could not be explained. However, this looks like a measurement method that needs to be further developed and understood for measuring precipitate changes in materials.

REFERENCES

1. S. M. Bruemmer, L. A. Charlot, and B. W. Arey. 1988. "Sensitization Development in Austenitic Stainless Steel: Correlation Between STEM-EDS and EPR Measusrements," Corrosion, Vol. 44, No. 6, p. 328.
2. S. M. Bruemmer, L. A. Charlot, and D. G. Atteridge. 1988. "Sensitization Development in Austenitic Stainless Steel: Measurement and Prediction of Thermomechanical History Effects," Corrosion, Vol. 44, No. 7, p. 427.
3. S. M. Bruemmer. 1988. "Grain Boundary Composition Effects on Environmentally Induced Cracking of Engineering Materials," Corrosion, Vol. 44, No. 6, p. 364.
4. W. L. Clarke, R. L. Cowan, and W. L. Walker. 1978. ASTM STP 655, American Soecity for Testing and Materials, Philadelphia, PA, p. 99.
5. S. M. Bruemmer, "Quantitative Modeling of Sensitization Development in Austenici Stainless Steel," Corrosion/89, National Association of Corrosion Engineers, Houston, TX, in press.
6. B. A. Auld. 1973. Acoustic Fields and Waves in Solids. John Wiley and Sons, New York, New York.
7. A. Vincent. 1987. "Coupling Film Thickness Effect on Absolute Ultrasonic Velocity Measurments in Solids," in Nondestructive Characterization of Materials II. Ed. J. F. Bussiere, J. P. Monchalin, C. O. Ruud, and R. E. Green, Jr. Plenum Press, New York, New York, pp. 697-706.
8. J. C. Albert, O. Cassier, B. Chamont, M. Arminjon, and F. Goncalves. 1987. "Application of Polarized Shear Waves to Evaluate Anistropy of Steel Sheets." in Nondestructive Characterization of Materials II. Ed. J. F. Bussiere, J. P. Monchalin, C. O. Ruud, and R. E. Green, Jr. Plenum Press, New York, New York, pp. 515-521.
9. R. B. Mignogna, P. P. Delsanto, A. V. Clark, Jr., R. B. Rath, and C. L. Vold. 1987. "Ultrasonic Measurements on Textured Materials," in Nondestructive Characterization of Materials II. Ed. J. F. Bussiere, J. P. Monchalin, C. O. Ruud, and R. E. Green, Jr. Plenum Press, New York, New York, pp. 545-553.
10. G. Canella and M. Taddei. 1987. "Correlation between Ultrasonic Attenuation and Fracture Toughness of Steels," in Nondestructive Characterization of Materials II. Ed. J. F. Bussiere, J. P. Monchalin, C. O. Ruud, and R. E. Green, Jr. Plenum Press, New York, New York, pp. 261-269.
11. S. S. Lee, J. F. Smith, and R. B. Thompson. 1988. "Acoustoelastic Response from the Velocities of Ultrasonic Plate Modes," 9th International Conference on NDE in the Nuclear Industry. Ed. K. Iida, J. E. Doherty, and X. Edelmann, April 25-28, 1988, Tokyo, Japan, pp. 133-136.
12. R. B. Thompson, J. F. Smith, and S. S. Lee. 1985. "Inference of Stress and Texture from the Angular Dependence of Ultrasonic Plate Mode Velocities," NDE of Microstructure for Process Control. Ed. H. N. G. Wadley, American Society for Metals, pp. 73-79.
13. M. Rosen, E. Horowitz, S. Frick, R. Reno, and R. Mehrabian. 1983. "An Investigation of the Precipitation-Hardening Process in Aluminum Alloy 2219 by Means of Sound Wave Velocity and Ultrasonic Attenuation," Materials Science and Engineering, 53:163.
14. M. Rosen, L. Ives, C. Ridder, F. Biancaniello, and R. Mehrabian. 1985. "Correlation between Ultrasonic and Hardness Measurements in Aged Aluminum Allow 2024," Materials Science and Engineering, 74:1.

NDE OBSERVATIONS OF HYDROGEN EFFECTS IN 4340 STEEL

IN-OK SHIM AND J. G. BYRNE
Department of Metallurgy and Metallurgical Engineering, University of Utah,
Salt Lake City, Utah 84112-1183, U.S.A.

ABSTRACT

Cathodic charging of hydrogen into 4340 steel produces drastic deterioration of mechanical properties such as the notch tensile strength, depending on
current density, charging time, and the poison used as an inhibitor of hydrogen recombination and bubble formation at the specimen surface. Fusion of
charged samples permitted hydrogen content to be measured and revealed a
linear increase in the latter with charging current density. The change in
the Doppler spectrum sharpness parameter $\Delta(P/W)$ deviates negatively with
cathodic charging time but inversely with the charging current density (fugacity). Thus a low-fugacity hydrogen charging appears merely to screen existing
positron trap sites, whereas a high-fugacity charge also creates new defect
traps. This would result in a lesser net screeninig effect which would be
seen as a smaller negative deviation in $\Delta(P/W)$ as observed. For cumulative
charging, however, one finds oscillatory behavior in (P/W) as in earlier cumulative charging experiments with Ni. This behavior indicates that hydrogen
screening of positron traps was followed by new positron trap site creation
and that this sequence of events occurred repeatedly and in that order.

INTRODUCTION

Much of the information on the behavior of positrons in solids can be
obtained from the gamma rays emitted during the annihilation of the positron
with an electron. Initially, positron annihilation spectroscopy was applied
to investigate electronic structure [1,2] such as electron density and momentum distribution. Subsequently, it has been used more for defect studies
[3-7]. In the present work, the damage of interest is that produced by hydrogen in steel. For this there is background work in which the response of
positrons to the presence of hydrogen provides a unique and useful response
[8-15] for nondestructive testing. In essence, an increase in defect density
in a solid normally causes an increase in the sharpness of the Doppler shifted
gamma-ray energy distribution resulting from the annihilation of positrons
injected into the solid. Hydrogen cathodically or otherwise injected into a
solid can do two things. It can generate new defects and/or screen defects
from detection by positrons, i.e., either increase or decrease respectively
the peak-to-wings (P/W) [16] parameter which is used to describe the shape of
the positron Doppler distribution. The balance between the frequencies of
these two tendencies governs the total tendency of (P/W). In addition, when
hydrogen escapes from a defect trap, (P/W) will increase again, since the
defect then once more becomes a potential trap for positrons.

EXPERIMENTAL PROCEDURE

The material used was AISI 4340 steel supplied by the U.S. Air Force in
the heat-treated and double-tempered condition with tensile strength in the
range of 260-280 ksi.
Cathodic charging was preceded by mechanical and electropolishing and was
done in a hydrogen recombination poisoned 1 N solution of H_2SO_4 either continuously or discontinuously and cumulatively. For the discontinuous cumulative
case, the poison was thiourea, and the current densities ranged from 5 to 100
mA/cm^2. In continuous charging, samples were charged for different times

under either high-fugacity (10 mA/cm^2 with As$_2$O$_3$) or low-fugacity (1 mA/cm^2 with thiourea) conditions. The positron Doppler measurement system was the same as described elsewhere [12-14].

RESULTS AND DISCUSSION

Table 1 lists the surface crack observations for various charging conditions and the two poisons used. As$_2$O$_3$ caused more severe cracking during and/or after cathodic charging than did thiourea. From Table 1, critical current densities for surface cracking seem to be about 0.5 mA/cm^2 and 10 mA/cm^2 for cathodic charging with 1 N H$_2$SO$_4$ containing As$_2$O$_3$ and thiourea respectively. Figure 1 shows how drastically the notch strength falls off with charging time for two current densities and As$_2$O$_3$ as a poison. More detailed

Table I. Hydrogen-Induced Cracking Observation after Hydrogen Charging

Charging Condition		Poisons in Electrolyte, 1 N H$_2$SO$_4$	
Current Density	Time	1 g/ℓ thiourea	5 mg/ℓ arsenic trioxide
100 mA/cm^2	10 min	many cracks	many cracks
80	10	2 cracks	many cracks
60	10	no cracks	many cracks
30	10	no cracks	many cracks
10	10	no cracks	many cracks
10	3 hr	several cracks	many cracks
1	10 min	no cracks	several cracks
1	3 hr	no cracks	--
1	12 hr	no cracks	--
0.5	10 min	--	2 cracks
0.1	10 min	--	no cracks
0.1	3 hr	--	no cracks

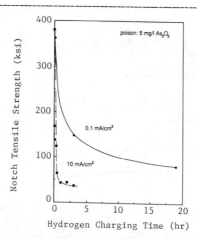

Figure 1. Notch tensile strength decrease with charging time in As$_2$O$_3$ poisoned solution.

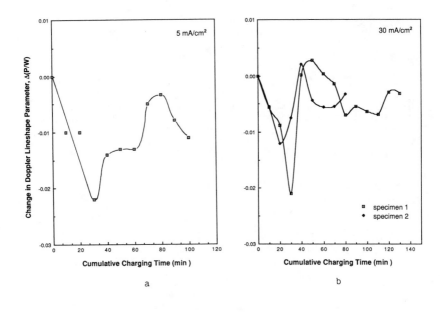

a

b

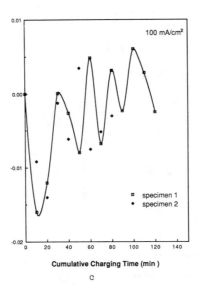

c

Figure 2.   (P/W) versus cumulative charging time for a) 5 mA/cm^2, b) 30 mA/cm^2, and c) 100mA/cm^2 in electrolyte poisoned with thiourea.

154

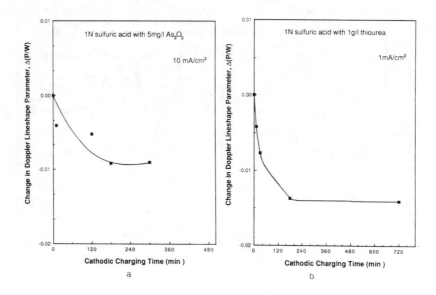

Figure 3. (P/W) versus continuous charging
time for a) high-fugacity and b)
low-fugacity charging conditions.

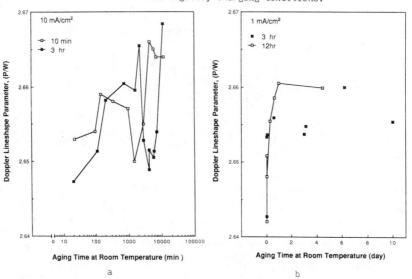

Figure 4. (P/W) versus time of aging at room
temperature for a) high-fugacity and
b) low-fugacity charging conditions.

mechanical and hydrogen analysis data [17] will, for lack of space, need to be published elsewhere; however, it is of interest to note that the maximum hydrogen content found immediately after charging was 9 wt. ppm in our ultra high strength AISI 4340 steel.

Figures 2a, b, and c all show for discontinuous cumulative charging at current densities of 5, 30, and 100mA/cm^3, a common negative starting deviation in (P/W) from its initial value for the ultrahigh-strength condition. In every case, this is followed by an increase, most often not into the positive region of Δ(P/W). This is then followed by further oscillations, especially for the highest-fugacity situation in Fig. 2c. Similar oscillations were observed earlier in nickel by Kao et al. [9].

When the charging was continuous up to the time chosen for observation, the shape of the Δ(P/W) versus time response was initially similar to that shown in Figs. 2a, b, and c in that P/W first decreased. This is shown in Figs. 3a and b for both high- and low-fugacity charging conditions respectively. The magnitude of the decrease in (P/W) is larger for the low-fugacity charging than for the high-fugacity case. Let us first address a decrease in (P/W) in general and then its magnitude. A decrease in (P/W) is ordinarily interpreted either as a lowering of the defect density or its apparent lowering due to the screening effect of hydrogen trapped at defects, thereby rendering those defects temporarily invisible to positrons [8-16]. This occurs because in a hydrogen-free sample there is an attractive negative potential of a defect for a positron. This results in the trapping of the positron at the defect and eventually its annihilation with an electron in the defect environment. Such an electron most often is a lower-energy conduction type, since the defect is by its nature deficient in ion core(s) and their higher-energy electrons. These lower-energy electron-positron annihilations put less of a Doppler shift on the gamma ray energies which accompany annihilation and hence sharpen the Doppler shift distribution about the 511 keV central value for an annihilation with a stationary electron. If hydrogen is trapped at the defect first, before the approach of a positron, then this reduces or "screens" the attractive potential of the defect for the positron. The electronic state of hydrogen in metals is still debated, but there are three possibilities: natural atomic hydrogen, a proton screened by conduction electrons, and anionic hydrogen. Hydrogen as an atom or proton screened by conduction electrons dissolves in transition metals interstitially [18]. Trapped hydrogen (tritium) at defects has been observed directly by high-resolution radiography [19,20]. Thus we can rationalize now both increases (new defect creation) and decreases in an initial value of (P/W) due to hydrogen. Let us now address the magnitude of a decrease.

The larger decrease in (P/W) in Fig. 3b for a lower-fugacity charging than represented by Fig. 3a for a higher-fugacity charging can be explained as follows. The lower-fugacity charge produces fewer new defects than the more severe charging. The competitiion between positron trapping at these new defects (which increases P/W) and hydrogen trapping at defects (which decreases P/W) is therefore dominated by the hydrogen screening effect giving a larger magnitude of the negative deviation in (P/W) than for the more severe charging case. The fact that both Fig. 3a and Fig. 3b shows initial negative deviations is because the material is initially in a very highly defected state, so that the hydrogen screening of defects is in general easier than the generation of new defects by hydrogen in an already hard material. Similar situations have been observed earlier in an fcc system [12].

Now, if we look back at Figs. 2a, b, and c, we can contrast the oscillatory behavior after the initial decrease with the behavior in Figs. 3a and b in which the initial decrease of (P/W) remains stable. As in Figs. 3a and b, the largest initial decrease in (P/W) among Figs. 2a, b, and c is for the lowest-fugacity charge and becomes gradually smaller for the intermediate and most severe charges described in Figs. 2b and 2c for the reasons just given earlier. The fact that oscillations occur during discontinuous but cumulative charging experiments probably is due not only to the synergism between defect creation and screening but also the influence of hydrogen detrapping which, like positron trapping at defects, raises (P/W).

Finally, it should be noted that aging at room temperature after charging can produce an increase in (P/W) with time for both high- and low-fugacity charging as hydrogen detraps from defects. Figure 4a shows that, after about seven days relaxation at room temperature, a specimen charged for three hours reflects more damage than one charged only ten minutes. Both curves in Fig. 4a show an intermediate decrease which may mean that some hydrogen is retrapping at defects during its diffusion from the interior to the surface. In contrast, the specimens receiving a low-fugacity charge show in Fig. 4b only a continual increase. It was found [17] that the 3-hr charge gave a full recovery of mechanical properties upon aging, indicating this low-fugacity charge created few defects. However, the higher-fugacity conditions described in Fig. 4a is known (see Table I) to produce even irreversible defects such as cracks, hence the very high endpoint values on the curves in Fig. 4a.

CONCLUSIONS

1. In discontinuous cumulative charging, (P/W) decreased initially for AISI 4340 steel in a hardened condition due to hydrogen as a screened proton trapping at defects, reducing the probability for positron trapping at defects.

2. Following the above-mentioned initial decrease of (P/W), it then increased followed by repeated decrease/increase oscillations which increase in frequency with the fugacity of the charging. The increases are ascribed to both new defect generation and hydrogen detrapping from defect traps, but the significance of the frequency of oscillation is unclear.

3. In continuous charging, (P/W) also decreased initially but remained stable at the lowest level reached as charging time increased. The lack of interruptions seems to preclude the positive oscillations in (P/W) which occur in discontinuous cumulative charging.

4. The magnitude of the decrease in (P/W) which is attained during continuous charging is lower for lower-fugacity charging than for high. This is because some of the hydrogen in the high-fugacity case probably creates new defects, thereby cancelling more of the (P/W) reduction due to defect screening by the rest of the hydrogen introduced.

5. Aging at room tmeperature for about one day after continuous low-fugacity charging causes (P/W) to continuously increase. For longer charging time, the value of (P/W) reached is slightly larger.

6. Aging at room temperature after high-fugacity charging gives an intermediate decrease in (P/W) as some hydrogen retraps at defects during its diffusion to the surface from the specimen interior.

7. Aging at room temperature for seven days after high-fugacity charging produces higher (P/W) values than the uncharged levels. Final (P/W) values were larger for 3 hrs charging than for 10 min charging, presumably because of the observed greater number of irreversible defects such as surface cracks which were creasted in the longer time.

REFERENCES

1. R. N. West, Adv. Phys. 22, 405 (1973).

2. A. T. Stewart, in Positron Annihilation, edited by A. T. Stewart and L.
    O. Roellig (Academic Press, New York, 1967), p. 17.

3. A. Seeger, in Frontiers in Materials Science, edited by L. E. Murr and C.
    Stein (Marcel Dekker Inc., 1976), p. 177.

4. R. N. West, in Positrons in Solids, Topics in Current Physics, vol. 12,
    edited by P. Hautojarvi (Springer-Verlag, Berlin, 1979), p. 89.

5. J .G. Byrne, in Dislocations in Solids, edited by F. R. N. Nabarro
    (North-Holland Pub. Co., 1981), p. 263.

6. A. Seeger, Appl. Phy. 4, 183 (1974).

7. R. M. Nieminen and M. J. Manninen, in Positrons in Solids, Topics in
    Current Physics, vol. 12, edited by P. Hautojarvi (Springer-Verlag,
    Berlin, 1979). p. 145.

8. F. Alex, T. O. Hadnagy, K. G. Lynn and J. G. Byrne, AIME Conference on
    Effect of Hydrogen on Behavior of Materials (1975), 642.

9. P. W. Kao, R. W. Ure, and J. G. Byrne, Phil. Mag. 39A, 514 (1979).

10. J. G. Byrne and F. Alex, U.S. Patent No. 4,064,438 (Dec. 1977).

11. S. Panchanadeeswaran and J. G. Byrne, Scripta Met., 17, 1329 (1983).

12. J. J. Kim and J. G. Byrne, Scripta Met., 17, 773 (1983).

13. Yi Pan and J. G. Byrne, Materials Science and Engineering, 74, 215
    (1985).

14. Yi Pan and J. G. Byrne, Materials Science and Engineering, 84, 195
    (1986).

15. I. R. Harvey and J. G. Byrne, Review of Progress in Quantitative NDE,
    presented at Review of Progress in Quantitative NDE, U.C. San Diego, Aug.
    1-5, 1988 (in press).

16. M. L. Johnson, S. Saterlie and J. G. Byrne, Met. Trans. 9A, 841 (1978).

17. In-Ok Shim, Ph.D. Thesis, University of Utah, 1988.

18. R. A. Oriani, in Fundamentals of Stress Corrosion Cracking, edited by
    R. W. Staehl, J. Hochmann, R. D. McCright, and J. E. Slater (NACE-5,
    NACE, 1977), 225.

19. P. Lacombe, M. Aucouturier, and J. Chene, in Hydrogen Embrittlement and
    Stress Corrosion Cracking, edited by R. Gibala and R. F. Heheman (ASM,
    Cleveland, 1984), 79.

20. T. Asaoka, in Metal-Hydrogen Systems (Pergamon Press, Oxford, 1982), 197.

A STUDY OF MICRODEFORMATION AND CREEP-FATIGUE DAMAGE
USING ACOUSTIC EMISSION TECHNIQUE

YUNXU LIU[*] AND XINGREN LI[**]
Jilin Institute of Technology, Changchun, 130012, China

*Professor and President
**Faculty member of Materials Engineering Department

ABSTRACT

This paper deals with the relationship between the plastic deformation damage and microstructure by means of Acoustic Emmission. The plastic deformation behavior of AISI 4340 steel of various microstructures was investigated in both the tensile and creep-fatigue testings with a view to providing new insights into properties of high performance steel. Based on Theory of Damage Mechanics, a creep-fatigue law was derived and formulated. The reason of early failure and the service life prophecy of high strength steels was studied.
The damage micromechanism of four stages was studied by the optical microscope, scanning electron microscope, and microhardness tester. It seems that the nucleation and the growth of the voids at the martensite-ferrite interface is the dominant mechanism of damage. The monitoring of Acoustic Emmission indicated that the plastic deformation did not appear in the circulating hardening stage. But in the circulating softening stage, the accumulation of the plastic deformation and the creep-fatigue damage become more and more severe. The total energy of Acoustic Emmission was successfully applied to measure the degree of the damage caused by the plastic deformation.

EXPERIMENT

Acoustic Emmission Technique is an avariable method for measuring the microdeformation resistance (Fig.1-1). The effect of microstructures on the microdeformation was investigated by means of the tensile test. It was found that the soft phases, such as free ferrite and upper bainite, were damage origins, the resisrance of microdeformation decreased significantly as the amount of soft phases increased (Fig.1-2), and the lower the strength, the lower the microdeformation resistance. The experimental results also revealed that 1-3% free ferrite decreased the microdeformation resistance largely by 40-70%.
In tensile creep testing, specimens of various microstructures performed different creep properties (Fig.2-1). The soft phase ferrite raised the creep deformation and its deformation rate obviously, and the creep deformation increased as the amount of ferrite increased. The results of the creep test also showed that the plastic deformation of the same specimen was varied in every tension. The circulating hardening appeared in all the specimens of various microstructures, but this behavior was more severe in the tension of the specimens containing free ferrite (Fig.2-2). Some new explanations about the role of soft phases was submitted.
It was shown by the creep-fatigue test that the fatigue damage resulted from the accumulation and the development of the plastic microdeformation. In the cycling testing of constant stress amplitude, it was found firstly that the plastic deformation and its accumulation caused the damage of materials and increased with the number of cycles (Fig.3-1). The existing of soft phases in materials made the plastic deformation appear earlier and the damage more severe (Fig.3-2). In the same testing, it was also found that the four stages of creep-fatigue damage took place successively

in the view of microregion. The damage micromechanism of four stages was studied by the optical microscope, scanning electron microscope and microhardness tester. It seems that the dominant mechanism of damage is the nucleation and the growth of the voids at the martesite-ferrite interface. The monitoring of Acoustic Emmission indicated that the plastic deformation did not appear in the circulating hardening stage. But in the circulating softening stage, the accumulation of the plastic deformation and the creep-fatigue damage become more and more severe (Fig.3-3). The total energy of Acoustic Emmission was successfully applied to measure the degree of the damage caused by the plastic deformation.

Investigation on the behavior of AISI 4340 steel was carried out in the plastic deformation cycling. The results showed that the ferrite deformed and the martensite-ferrite interface debonded. Under the condition of the load cycling, it could be said that the free ferrite plays the role of micronotches. Stress concentration occurs at the tips of these notches and voids initiate very easily at the martensite-ferrite interface. The degree of damage is controlled by the complex phenomenon of the void growth. The growth and linking of the voids ultimately cause the failure of materials. The service life of materials mostly loses in the development of these multiple void system.

Continum Damage Theory was proposed by Kachanov for the creep damage and modified by Lemaitre for the fatigue damage. According to Theory of Damage Mechanics, authors of the present paper introduce a scalar parameter D, called the creep-fatigue damage. For convinience, 0 (zero) is assigned to the initial undamaged state and 1 to the failure state so that D varies in the range from 0 to 1.

$$D = 1 - \Delta\sigma/\Delta\sigma_0 \qquad (1)$$

$$dD/dN = (1 - D)^{-p}/(1 + p)N_f \qquad (2)$$

where p is a property of the material. It is dependent on the applied loading condition and can be easily determined by the constant amplitude test. $\Delta\sigma/\Delta\sigma_0$ is the relative change in stress amplitude. N_f is the number of fracture cycles. N is the number of stable cycles. The service life N_s

$$N_s = 1 - (1 - N/N_f)^{1 + p} \qquad (3)$$

This formular fits the experimental results very well. The availible service life of high performance parts can be calculated by this formular.

SUMMARY

It is possible to predict the result of the variable amplitude test based on the damage relationship obtained in the constant amplitude testing. A number of experiments showed that the analytical estimation for the service life conformed favorably to the experimental results. The physical observation of the creep-fatigue showed clearly that the mechanism of the creep-fatigue was controlled by the growth and development of the multiple void system. The concepts of the classical fracture mechanics are extremely difficult or even impossible to be applied in the present cases because of the complex phenomena of the void interaction. The proposed theory of the continuum damage can offer better estimation of the service life for

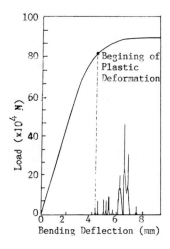

Fig. 1-1. The relationship between deformation and Acoustic Emmission.

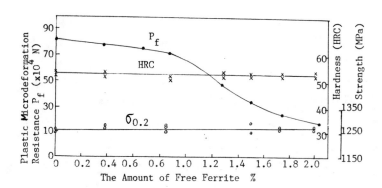

Fig.1-2. The amount of free ferrite vs. the resistance of plastic deformation.

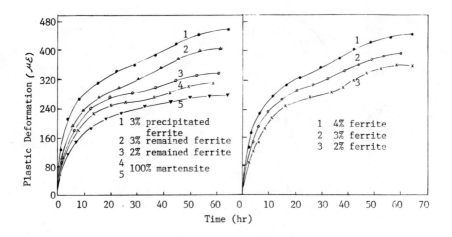

Fig. 2-1. The effects of various microstructure
and free ferrite on the creep property.

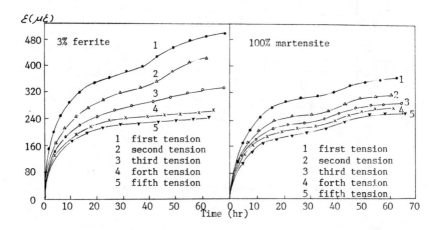

Fig. 2-2. The difference of creep property between
ferrite specimen and martensite specimen

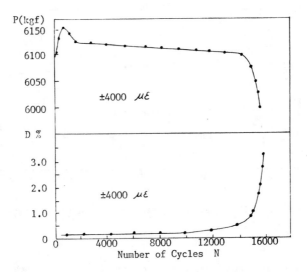

Fig. 3-1. The relationship between the damage and cycles.

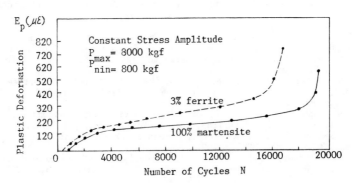

Fig. 3-2. The plastic deformation accumulation
of different microstructure specimen.

164

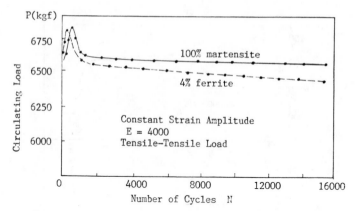

Fig. 3-3. The effect of ferrite on creep-fatigue damage

AISI 4340 steel. Only simple testing procedures are needed which were developed and can be widely used in material charaterization. It seems that the application of this model will improve the service life prediction of materials with very little extra work.

REFERENCES

1. J. Lamaitre, and A. Plumtree, "Application of Damage Concepts to Predict Fatigue Failure", ASME J. of Eng. Materials and Technology, Vol. 101, July 1979, p 284-292.
2. K.N. Smith, P. Watson, and T.H. Topper, "A Stress-Strain Function for the Fatigue of Metals", J. of Materials, Vol. 5, No.4, Dec. 1970, p 767-778.
3. M.S. Starkey and P.E. Irving, "The Influence of Microstructure on Fatigue Crack Initiation in Spheroidal Graphite Cast Iron", Proc. Int. Symp. on Low Cycle Fatigue Strength and Elasto-plastic Behavior of Metals, Stattgart, 1979.
4. G.N. Gilbert, "Factors Relating to the Stress/Strain Properties of Cast Iron", J. of the British Cast Iron Res. Association, Vol. 6, No. 11, 1957, p 546-588.
5. J.W. Fash and D.F. Socie, "Fatigue Behavior and Mean Effects in Grey Cast Iron", Int. J. of Fatigue, Vol. 4, No. 3, 1982, p 137-142.
6. K. Sadananda and P. Shahinian, "Creep Crack Growth in Alloy 713", Met. Trans. A, Vol. 8A, 1977, p 439.
7. R.C. Boettner, C. Laird, and A.J. Meevily, "Crack Nuceation and Growth in High Strain Low Cycle Fatigue", TMS-AIME Trans., Vol. 233, 1965, p 379.
8. B.K. Min and R.Raj, "A New Method for Fatigue Testing at Elevated and Room Temperatures", J. of Test and Evaluation, September, 1977.
9. M. Chrzahowski, "Use of the Damage Concept in Describing Creep-Fatigue Interaction and Prescribed Stress", International Journal of Mechanical Science, Vol. 18, 1976, p 69-73.
10. D.K. Wilsdorf, "Dislocation Behavior in Fatigue", Materials Science and Engineering, Vol. 39, 1979, p 231-245.

ASSESSMENT OF CREEP DAMAGE BY NDT

Herbert H. WILLEMS
Fraunhofer-Institute for Nondestructive Testing, University, Building 37,
D-6600 Saarbrücken 11, Federal Republic of Germany

ABSTRACT

Residual lifetime analysis of power plant components requires information on the degree of degradation of the material. In case of high-temperature creep, material damage can be related to cavity formation or to the accumulated creep strain. At present, only metallographic replication technique is widely used for in-service inspection to detect creep cavitation. Other NDT-techniques like ultrasonic velocity and electrical resistivity, which have potential for detection of low pore concentrations, are being developed for practical application. The use of capacitive strain gages yields encouraging results for the long-time monitoring of creep deformation. Potential and limits of these NDT-techniques together with industrial needs for creep damage assessment are reviewed.

INTRODUCTION

The lifetime of high-temperature components in power plants which operate under creep conditions is limited by material degradation due to the development of creep damage. In order to ensure safe and economic operation, reliable procedures for the assessment of residual lifetime have to be established. In addition to parameters like material properties, loading conditions and component features, such procedures have to take into account the actual state of the considered material, i.e. appropriate NDT-techniques for damage assessment are required. If such techniques are available, retirement-for-cause strategies can be applied and the full life capacity of a component could be used instead of replacing components after the conservatively calculated lifetime. Significant life extension would be possible in most cases because conservative lifetime calculations involve lower bound material properties as well as considerable safety margins. This is of great economical importance for fossil power plants reaching the end of design life.

The nondestructive assessment of early creep damage requires techniques which are sensitive to small microstructural changes. Presently, only the metallographic replication technique is successfully used for component inspection. The potential of some other techniques which showed promising results under laboratory conditions and which are now being developed for practical application is discussed in the paper.

MATERIAL DEGRADATION BY CREEP

The strain-time behavior of a metallic material subjected to a constant load at temperatures higher than about half the melting temperature is usually described by a creep curve with primary, secondary and tertiary stage (Fig. 1). The primary and secondary creep range are characterized by dislocation and phase reactions. Depending on the type of steel, the formation of micropores at the grain boundaries can be observed by means of light microscopy during secondary creep in low-alloy ferritic CrMo-steels used for pipes and tubes and during the transition from secondary to tertiary creep in CrMoV-steels used for rotors, blades and casings, respectively /1/. Pore growth, coalescence of pores, microcrack

166

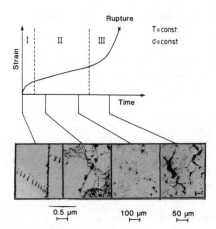

Fig. 1 Schematic representation of strain-time behavior and evolution of
microstructure during creep.

formation and crack growth lead subsequently to an increasing creep rate
causing finally fracture by plastic instability at a certain loss of
effective cross sction. Creep damage in these steel types can be classified
by the cavitation pattern (isolated pores, pore chains, microcracks and so
forth). In some other types of steel like 12 Cr-steels used for turbine
rotors and blades, no creep cavities have been observed even near the
fracture area /1/. In such cases, local strain measurements now rendered
possible for long-time applications by using capacitive strain gages /2/
offer the possibility of determining the material exhaustion, which is
defined as the ratio of consumed lifetime to total lifetime. From a
metallurgical point of view, material exhaustion is a result of the
consumption of deformation potential by creep. For a given material
relationships between the damage state and the exhaustion grade have to be
established in order to perform reliable lifetime estimations. Some
phenomenological models like the Kachanov-Rabotnov model /3/ relate the
density change due to cavity formation to the material exhaustion. Such
relationships have to be evaluated for all the used materials.
   Long time heating also affects the carbide structure of the steels
mentioned and carbide precipitation as well as carbide coarsening takes
place. As a consequence, material softening is observed and hardness
measurement has therefore been suggested to follow the degree of
degradation. But since the softening process is mainly thermally
conditioned, no direct relationship between damage and hardness can be
expected. This is also supported by experimental results /4/. Since
material softening will cause an increasing strain rate at a given load,
hardness measurement may yield valuable information on the actual strength
properties of the considered material.
   According to practical experience, creep damage in piping components
generally starts from the outer surface /5/. Whether creep damage will
spread over larger areas of a component or will be locally restricted
depends essentially on the actual stress state which is determined by
geometrical conditions, structural inhomogeneities and the applied loads.
In the case of weldments or fittings, comparatively high stress gradients
are present. Then the damage development will be locally restricted and
forthcoming surface cracks can be traced by conventional NDT-tchniques
before reaching critical length. The safety-risk potential is negligble
because leakage-before-fracture behavior can be assumed in case of

METHOD	INFORMATION OBTAINED	RESOLUTION	REMARKS
REPLICA 1) INSPECTION	MICROSTRUCTURE (MICROPORES, CARBIDES)	$\simeq 0.1$ µm	- SPOT CHECKING, - COLD STATE REQUIRED
CAPACITIVE 2) STRAIN GAGE	PLASTIC DEFOR- MATION, CREEP RATE	$\pm$ 30 µm/m Drift: < 0.2 µm/md	- CONDITION MONITORING - SELECTED POSITIONS - THERMAL CALIBRATION REQUIRED - EXPERIENCE FOR > 20.000h ON COMPONENTS
ULTRASONIC 3) VELOCITY	MICROPORE CON- CENTRATION (VOL. CONTENT)	$\simeq 2 \times 10^{-3}$	- FAST MEASURING TIME - UNDAMAGED REFERENCE MATERIAL REQUIRED - ONLY LABORATORY EXPERIENCE UP TO NOW
ELECTRICAL 3) RESISTIVITY	"	$\simeq 1 \times 10^{-3}$	
3) MICROMAGNETICS	COERCIVE FORCE (RELATED TO MICROSTRUCTURE)	$\simeq 0.1$ A/cm	

TABLE 1 Nondestructive testing methods for detection of creep damage
(1) in use 2) being field tested 3) under development)

failure /5/. However, in pipe bends with low stress gradients, micropores can develop over large areas and spread through the whole wall thickness. Leakage-before-fracture can no longer be assumed and catastrophic failure cannot be ruled out /5/. To prevent this type of failure, the early detection of cavitation at the stage of micropores is necessary. The development of new NDT-techniques for creep damage assessment at our institute is especially devoted to this type of component.

NDT-TECHNIQUES FOR CREEP DAMAGE ASSESSMENT

The in-field detection of creep damage at the micropore stage is presently only possible by means of replica techniques. Some other techniques showed promising results when applied under laboratory conditions. They are now being developed for practical application. Potential, advantages and drawbacks are discussed in the following sections (see also Table 1):

Replica Technique (Surface Metallography) /6,7/

The replica technique (Fig. 2) yields direct information on the microstructure, i.e. grain size, carbide distribution, degree of cavitation etc. Selected positions are ground, polished and etched, and a replication of the microstructure is obtained by means of a triafol film softened with acetone. After drying, it can be examined with an optical microscope or with a scanning electron microscope and a resolution of $\simeq 0.1$ µm is obtained.

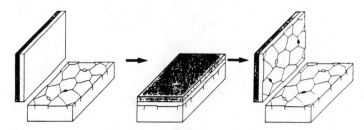

Fig. 2 Replica inspection - schematic.

Since it is essentially a spotwise inspection, a careful selection of the weakest positions of the considered component is necessary. For application a cold state of the component is required. Depending on the degree of the detected damage inspection intervals are set for the inspection of plant components.

Capacitive Strain Gage /2/

The problem of local strain measurements at higher temperatures over long times seems to be solved by using capacitive strain gages (Fig. 3). Reported longtime drifts are < 2 x 10^{-5}%/day which gives at most 0.3% over 40 years /2/. Experience over more than 20.000 h at components is now achieved. A big advantage of this technique is that it can be applied for condition monitoring of the component during operation. Individual thermal calibration of each gage is required, and as in case of the replica technique careful selection of measuring positions is necessary. Then, the measured strain and strain rate, respectively, can be used for lifetime estimation. In Fig. 3, an example of capacitive strain measurement shows the influence of the change of operation temperature on the strain rate. When continuously applied to new components the accumulated creep strain can be measured. In the case of old components, the accumulated strain can only be estimated from the actual strain rate, if the material is in the second creep range and if constant operation conditions for past operation are assumed.

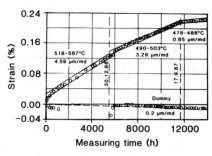

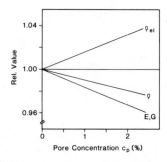

Fig. 3 Creep strain measurement on a 14 MoV 6 3 bend at different operating temperatures by means of capacitive strain gage (Ref. /2/).

Fig. 4 Relative elastic moduli E,G (in iron), electrical resistivity ρ_{el} and density ρ as a function of pore concentration c_p (theory).

Ultrasonic Velocity

Ultrasonic velocity is directly connected to the elastic moduli (Young's modulus, shear modulus) and the density of a solid material. Because these properties are affected by cavitation (Fig. 4), ultrasonic velocity was used for creep damage characterization /8,9/. So far the experience was obtained under laboratory conditions including investigations of service-exposed material from a damaged pipe bend /8/. Results (see Fig. 5) show that pore concentrations down to $\simeq 2 \times 10^{-3}$ (volume content) can be detected at a measuring accuracy of $\pm 10^{-3}$ for relative ultrasonic time-of-flight measurement. The pore concentration is obtained from the velocity change relative to the undamaged material with the same microstructure. Such reference material is available at low-loaded parts of the component itself. The advantage of this technique is the fast measuring time which enables screening of large component areas in short time. An automated system for time-of-flight measurement with a reproducibility of ± 1 ns was developed at IzfP /10/ and will be applied now for component investigations. Because creep damage in bends is starting at the outer surface, surface waves are used. The analyzing depth can be varied by changing the ultrasonic frequency. In order to avoid coupling problems, special EMAT-probes for the measurement of the ultrasonic velocity of surface waves are also under development. The main point for velocity measurement is now to show whether reliable detection of creep damage at components such as pipe bends is feasible.

Electrical Resistivity

The effect of porosity on the electrical resistivity is even larger than on ultrasonic velocity as supported by both theoretical calculations /11/ and experimental results obtained under laboratory conditions (see Fig. 5). But reliable systems for practical application are not yet available. IzfP is now developing a system based on a four-point-contact probe. Using a 16 bit ADC and signal averaging the reproducibility of voltage measurement will be better than ± 100 nV. This will allow to measure the specific electrical resistivity with an accuracy of $\simeq 10^{-3}$, which should enable the determination of pore concentrations down to 10^{-3}. The advantages of this technique are similar to those of ultrasonic velocity measurement and likewise the problem is now to verify the practial applicability.

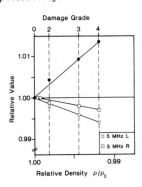

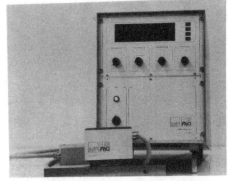

Fig. 5 Influence of creep damage on physical properties in 14 MoV 6 3 steel (● electrical resistivity, o (□) ultrasonic velocity).

Fig. 6 3MA-instrument and transducer for measurement of Barkhausen noise and incremental permeability.

Micromagnetics

From basic micromagnetics theories /12/, it is well known that non-magnetic particles (e.g. carbides or voids) in magnetizable materials disturb micromagnetic dynamic processes like Bloch-wall jumps and rotational processes. The strongest effect is expected, if the diameter of the inclusion or the pore is in the range of the Bloch-wall thickness ($\approx$50-100 nm) of these materials. In case of creep porosity an increase of the coercivity with increasing damage grade was observed for 1/2 CrMo-steel.

IzfP has developed an 3MA-instrument (Micromagnetic-, Multiparameter-, Microstructure- and Stress-Analyzer) using Barkhausen-noise- and incremental permeability analysis /13/. Fig. 6 shows the equipment and the handy transducer which consists of an U-shaped electromagnet and several sensors: a Hall-element is used to control the tangential magnetic field strength and inductive coils are used to receive the magnetic noise or to perform a small inner hysteresis loop for measurement of incremental permeability. Both techniques can be used in order to measure the coercivity at components and a reproducibility of better 0.1 A/cm is presently achieved.

CONCLUSIONS

Reliable estimation of residual lifetime of high-temperature components used in fossil power plants is only possible if the damage state of the considered material can be determined quantitatively by NDT-techniques. So far only the replica technique is successfully used in practical application provided the damage can be quantified in terms of cavitation formation with the limitation that it gives information only on the direct surface. Direct measurement of creep strain and strain rate at components during operation is achieved by means of capacitive strain gages. The practical applicability of other advanced NDT-techniques like ultrasonic velocity, electrical resistivity and micromagnetic measurements has still to be demonstrated.

REFERENCES

/1/ W. Lempp, N. Kasik, U. Feller, in: Proc. of the 2nd Int. Symp. on Non-destructive Characterization of Materials, Plenum Publishing Corp., 1987, 441-449
/2/ P. Hofstötter, Int. Conf. "Life Assessment and Extension", Conf. Proc. Session II, The Hague 1988, 125-129
/3/ F.A. Leckie, Phil. Trans. R. Soc. London. A 288 (1978) 27-47
/4/ J.L. Brinkman, L.B. Dufour, R.J. H. Baten, Int. Conf. "Life Assessment and Extension", Conf. Proc. Session II, The Hague 1988, 44-50
/5/ W. Bendick, H. Weber, ibid., Conf. Proc. Session III, 76-84
/6/ B. Neubauer, U. Wedel, ASME Int. Conf. on Advances in Life Prediction Methods, Albany 1983, 307-313
/7/ P. Auerkari, Eurotest Technical Bulletin, Dec. 1983
/8/ H. Willems, W. Bendick, H. Weber, in: Proc. of the 2nd Int. Symp. on Nondestructive Characterization of Materials (eds. J.-P. Bussière et al.), Plenum Publishing Corp., 1987, 451-460
/9/ H. Willems, 13. MPA-Seminar, Stuttgart, 8.-9. Okt. 1987, Band I
/10/ E. Schneider, R. Herzer, VDI-Berichte Nr. 679 (1988) 301-311
/11/ G. Ondracek, Z. Werkstofftech. 8 (1977) 280-287
/12/ E. Kneller, Ferromagnetismus, Springer-Verlag 1962
/13/ W.A. Theiner, B. Reimringer, H. Kopp, 3. Int. Symp. on Nondestructive Characterization of Materials, Saarbrücken, 3.-6. Okt. 1988

X-RAY DIFFRACTION DETERMINATION OF METALLURGICAL DAMAGE IN TURBO BLOWER ROTOR SHAFTS

JOHN F. PORTER[*], DAN O. MOREHOUSE[*], MIKE BRAUSS[**],
ROBERT R. HOSBONS[***], JOHN H. ROOT[***], AND THOMAS HOLDEN[***]

[*]Defence Research Establishment Atlantic, Dartmouth, Nova Scotia, Canada, B2Y 3Z7
[**]Proto Manufacturing Ltd., Oldcastle, Ontario, Canada, N0R 1L0
[***]Atomic Energy of Canada Ltd., Chalk River, Ontario, Canada, K0J 1J0

ABSTRACT

Studies have been ongoing at Defence Research Establishment Atlantic on the evaluation of non-destructive techniques for residual stress determination in structures. These techniques have included neutron diffraction, x-ray diffraction and blind-hole drilling. In conjunction with these studies, the applicability of these procedures to aid in metallurgical and failure analysis investigations has been explored. The x-ray diffraction technique was applied to investigate the failure mechanism in several bent turbo blower rotor shafts. All examinations had to be non-destructive in nature as the shafts were considered repairable. It was determined that residual stress profiles existed in the distorted shafts which strongly indicated the presence of martensitic microstuctures. These microstructures are considered unacceptable for these shafts due to the potential for cracking or in-service residual stress relaxation which could lead to future shaft distortion.

INTRODUCTION

Considerable effort has been expended at DREA on the development and application of non-destructive residual stress measurement techniques. The program was initiated as part of a research effort investigating residual stress effects on the fatigue life of welded structures. This activity has been expanded to include investigations of other potential applications of these techniques. One application has been as an aid in failure analysis; particularly in cases where a component cannot be sectioned or otherwise destructively examined. In principle, if a component has a known or anticipated residual stress distribution, the failed similar component can be examined for residual stress anomalies which may have contributed to the failure. This approach can be successful if undesirable changes in microstructure are highlighted by modifications in the residual stress profiles. One example where this approach has proven successful has been in the failure analysis of turbo blower rotor shafts used in some marine steam propulsion systems. This paper will briefly outline recent activities in qualifying residual stress determination techniques and will then focus on the application of the x-ray diffraction procedure to the turbo blower investigation.

RESIDUAL STRESS DETERMINATION TECHNIQUES

Although the neutron diffraction, x-ray diffraction and blind-hole drilling techniques are well established as residual stress determination procedures[1,2,3], a round robin was undertaken to identify limitations of the techniques and to investigate the applicability of these procedures to structures with complex non-uniform with depth residual stress distributions. Plate samples of HY80 and HY100 steel were heat treated to remove fabrication residual stress. The samples were then strain gauged and loaded in four point bending such that the zones furthest from the neutral axis had yielded plastically. This resulted in permanent deformation of the samples upon release of the bending moment. Thus, a predictable, non-uniform residual stress gradient was established in the samples. From the strain gage results and the load histories, the theoretical residual stress profiles for each sample were determined. The samples were then distributed to establishments in possession of various residual stress measurement expertise. Blind-hole drilling measurements were conducted at Defence Research Establishment Atlantic (DREA) and Metals Technology Laboratories (MTL). X-ray diffraction measurements were undertaken at Proto Manufacturing and MTL, while neutron diffraction measurements were made by Atomic Energy of Canada (AECL).

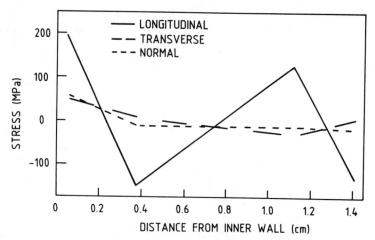

FIGURE 1: Neutron diffraction prediction of through-thickness residual stress in HY80 sample

Figure 1 shows the longitudinal, transverse and normal components of residual stress determined at positions through the wall of the bent HY80 sample, assuming bulk elastic constants. The residual strains were determined by measuring the shifts in angular position of the (110), (112), (002), (220) and (222) diffraction peaks of the ferritic material. Results for the (110), (112) and (220) strains are identical. The results conform to the model stress distribution expected for a perfectly elastic - perfectly plastic material strained beyond the yield point from about half the distance from the neutral axis to the plate surface.

TABLE 1
Longitudinal Surface Residual Stress Results (MPa)

| | HY80 | | HY100 | |
	Convex	Concave	Convex	Concave
Predicted	-268	+268	-241	+241
Neutron Diff.				
AECL	-167	+249	-	-
Blind-Hole				
DREA	-166	+227	-	-
MTL[4]	-220	+230	-226	-
X-Ray Diffrac.				
Proto	-300	+236	-276	+279
MTL[4]	-250	+260	-246	+200

The round robin results are displayed in Table 1. Although substantial variance is apparent, the differences are related to the peculiarities of the different techniques. In the blind hole drilling procedure, as a small hole is incrementally introduced into a sample, the measured surface strain redistribution allows a through thickness calculation of residual stress to shallow depths. Surface strains are translated to residual stress with depth values by the compliance matrix method. The values shown in Table 1 result from extrapolation of the stress with depth profiles to the surface and are therefore approximate. A steep stress gradient near the surface would therefore not be detectable.

X-ray diffraction allows direct measurement of surface residual strain by the measurement of the elastic atomic lattice spacing in the material crystal structure. Values shown in Table 1 were obtained with the Canmet/Proto Stress Analyser. Measurement locations were incrementally electropolished up to .007 inches and surface strain measurements recorded at each step. Values shown in Table 1 are extrapolations of the residual stress gradient to the sample surface. Proposed future round robin activities will be helpful in establishing the expected standard deviation with these procedures.

TURBO BLOWER ROTOR INVESTIGATION

DREA was contacted in January 1988 to conduct a metallurgical investigation on several turboblower rotor shafts which had bent in service. The investigation was hampered because the shafts could not be sectioned or in any way further damaged as attempts were to be made to salvage and repair the shafts for future service. Following a nital etching of the shaft surfaces, dark spots were evident as shown in Figure 2, indicating that the shafts had experienced prior bending failures and "hot spot" repairs.

FIGURE 2: Turbo blower rotor shaft showing "hot spot" repair

The hot spot repair procedure is one in which a bent shaft is straightened by applying concentrated heat to a localized spot of the shaft corresponding to the apex of the bend. The spot is heated until local plastic deformation of the material occurs, following which cooling sets up high tensile residual stresses which draws the shaft straight. The turbo shafts are fabricated from several different alloy steels, all containing combinations of chromium, nickel, molybdenum, and vanadium alloying elements. The shafts were originally heat treated to obtain a ferritic and pearlitic microstructure to acquire a good balance between strength and ductility.

Prior to identifying metallurgical changes in the hot spots, the surface residual stress distributions across the hot spots were measured with the portable CANMET/Proto Stress Analyser x-ray diffraction apparatus. Figure 3 shows a representative residual stress profile across a "hot spot" repaired surface containing two adjacent repairs. High tensile residual stresses were expected at the center of the spots, however high compressive stresses were recorded. This distribution was only possible where phase transformation effects dominated the residual stress profile creation. The formation of martensite, accompanied with a specific volume increase during cooling, would lead to the compressive residual stress measured at the hot spot centers. In the turbo blower shafts investigated, such transformations could occur if the material transformation temperature of 800 degrees C is exceeded by the concentrated heat source during a hot spot repair[5]. Detrimental effects due to the presence of such a martensitic microstructure in the shaft include a reduction in the material ductility and the potential for the relaxation of residual stress at service temperatures. The relaxation of the hot spot residual stress during service could result in a further distortion of the shaft to an unservicable condition.

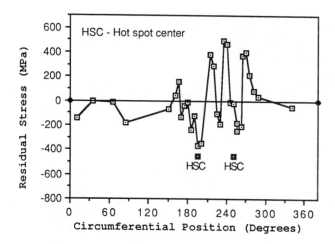

FIGURE 3: X-ray diffraction hoop residual stress results across two turbo blower "hot spots"

Due to the significance of the x-ray diffraction results and the questions raised about over-heating, permission was granted to destructively examine one of the non-repairable rotor shafts containing hot spots that exibited the presumed martensitic related residual stress distribution. The microstructure at the center of the hot spot examined is shown in Figure 4.

FIGURE 4: Turbo blower hot spot microstructure indicating a martensitic / banitic microstructure (500 X)
This metallographic examination confirmed that martensite had

been formed during the "hot spot" repair process. Consequently alternative shaft straightening procedures are being investigated. Additional studies will investigate the degree to which martensitic microstructures allow residual stress relaxation at the shaft operating temperatures. X-ray diffraction will be employed to measure surface residual stress profiles in samples containing "hot spot" repairs introduced with varying degrees of heat input. Each sample will be held at the shaft operating temperatures for several hundred hours, following which a re-evaluation of the hot spot surface residual stresses will be conducted.

DISCUSSION

The round robin of non-destructive residual stress measurement techniques has yielded some very interesting results. Neutron diffraction experiments on plastically deformed HY80 samples have shown good through thickness agreement with predicted profiles. The neutron diffraction surface residual stresses, shown in Table 2 were extrapolated from measurements taken at depths greater than one millimeter. This would explain the variance between the determined and the predicted values. A similar explaination is possible for the blind-hole drilling experiments which yielded surface residual stress results lower than expected.

Non-destructive residual stress determination techniques are under continuous refinement such that improved accuracy and flexibility are being realized. The application of portable X-ray diffraction apparatus to characterize undesirable residual stress profiles, indicative of improper microstructures has greatly aided in the turbo blower failure analysis.

CONCLUSIONS

DREA has investigated through a round robin, several non-destructive residual stress measurement techniques for use in fatique and fracture studies. X-ray diffraction residual stress measurement was successfully employed to non-destructively identify the existance of undesirable martensitic microstructures in "hot spot" repaired turbo blower rotor shafts.

REFERENCES

1. M.T. Flaman, "Qualification Studies on Sectioning Techniques for Residual Stress Measurement", CEA/CN-106-G-275, Dec. 1984.

2. B.D. Cullity, The Elements of X-Ray Diffraction, 2nd Edition, Addison-Wesly, Reading Mass., 1978.

3. C.S. Choi, H.J. Prask and S.F. Trevino, Journal of Applied Crystallography, Vol. 12, page 327, 1979.

4. G. Roy and E. Cousineau (Private communication).

5. "Flame Straightening of Turbine Rotors", W.H.Allen Engineering Department Report, Feb. 1965.

EFFECTS OF GRAIN BOUNDARY CHARACTERISTICS OF STEEL
ON MAGNETOACOUSTIC EMISSION SPECTRA

M. Namkung*, W. T. Yost*, D. Utrata**, J. L. Grainger*** and
P. W. Kushnick****
*NASA Langley Research Center, Hampton, VA 23665
**Association of American Railroads, 3140 S. Federal St., Chicago, IL 60616
***AS&M Inc., 107 Research Dr., Hampton, VA 23666
****PRC Kentron Inc., 303 Butler Farm Rd., Hampton, VA 23666

ABSTRACT

The pulse height distribution of a magnetoacoustic emission (MAE) spectrum is expected to be generally Gaussian due to its random nature. The functional form of distribution depends on the microstructure of a ferromagnet since the domain wall-lattice defect interaction produces MAE. The present study investigated the effects of grain boundary characteristics on the properties of MAE spectra obtained by external AC magnetic field-driven domain wall motions. The results show the enhancement of domain wall-defect interaction as more grain boundary disorder is introduced to HY80 steel samples. This was confirmed by the growth of a non-Gaussian-like distribution in the tail section of histograms with increased population of impurities trapped at the grain boundaries causing embrittlement. It is found that the enhancement of domain wall-defect interaction, which is responsible for generation of high amplitude MAE pulses, also tends to reduce the rate of such MAE events by limiting domain wall motions. Application of a stronger AC magnetic field should increase the count rate of high amplitude MAE signals in more embrittled samples.

INTRODUCTION

Strong local lattice strain fields are created at 90° domain walls of an iron-like ferromagnet due to spontaneous magnetostriction [1]. The strain fields at 180° domain walls are much weaker and shorter-ranged than that of 90° domain walls [1,2]. Lattice mismatch at defects also creates strong local strain fields. The domain walls and lattice defects interact with each other through the strain fields they create and the interaction is much stronger for 90° domain walls [2]. Lattice defects, therefore, act as effective potential barriers against 90° domain wall motion.

Motion of 90° walls is followed by the rearrangement of lattice strain fields that release elastic energy. A discontinuous motion of 90° domain walls, therefore, produces a sudden burst of MAE signals. Since the abrupt motion of 90° walls is due to pinning and unpinning of these walls at the defect sites, it is natural to assume that the characteristics of large amplitude MAE signals are related to the properties of the grain boundary in cases where lattice mismatch may be severe.

Our initial study has proven that differentiation between embrittled and unembrittled HY80 steel samples is possible by pulse height analysis applied to the MAE spectra obtained with a 60 Hz AC external magnetic field [3]. The results have shown broader Gaussian-like distributions of histograms in the embrittled samples which is consistent with the assumption of domain wall-defect interaction at the grain boundaries. By varying AC field frequencies and amplitudes, our recent study in the same HY80 steel samples has provided a complete verification of the above assumption [4,5]. The results also have shown that much information on temper embrittlement of these steel samples can be obtained directly from the MAE spectra.

In our earlier study, a distinct non-Gaussian-like structure has been observed in the tail section of histograms obtained for the HY80 samples. The appearance of this tail structure in repeated measurements, however, was not very reproducible, presumably due to a lack of statistical accuracy. The purpose of the present study is to investigate this tail structure with an enhanced statistical accuracy in the histograms.

EXPERIMENTS

The detailed properties and treatment history of the HY80 steel samples have been documented in Ref. 3, and only a brief description will be given here. The casting steel was quenched and tempered to obtain the desired toughness while producing a yield stress of approximately 80 ksi. During the stress relief heat treatment after the above stages, impurity atoms execute thermally activated diffusional motion in the lattice. Trapping of certain types of impurities at the grain boundaries in this process is known to cause embrittlement. Sets of test samples were obtained from the blocks heat treated at 538°C for 1, 5, 24, 50 and 100 hours to produce different amounts of temper embrittlement. A separate Charpy V-notch test has shown a rapid decrease in impact toughness from 127.5 ft-lbs for unembrittled HY80 samples to 54.5, 15.0, 9.5, 6.5 and 5.0 ft-lbs by increasing the hours of heat treatment time as given above.

Fig. 1 shows the block diagram of the MAE instrumentation used for the experiment. The magnetizing unit consists of a function generator that controls the power supply/amplifier. A digitizer was used to record the MAE spectra. Histograms were constructed directly by using a multichannel analyzer. Since the multichannel analyzer accepts positive voltage signals only, an inverting circuit was used to accumulate both positive and negative MAE pulse counts. For the present experiment, the multichannel analyzer was used to process about 100 million events for each histogram.

RESULTS AND DISCUSSION

The upper part of Fig. 2 shows the MAE spectrum obtained at a 60 Hz AC field frequency with the HY80 sample heat treated for 24 hours. The lower part of this figure shows an approximately sinusoidal waveform of the induction pickup coil output. The random noise band between the MAE peaks is due to system background noise.

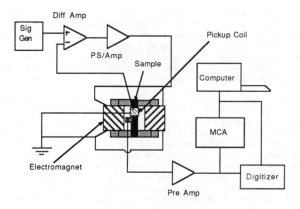

Fig.1. Block diagram of the MAE instrumentation.

The MAE envelope in Fig. 2 is seen to be asymmetric. This is because some 90° domain walls are not able to jump over the potential barriers at the grain boundaries due to strong resistance. An immediate proof of such incomplete 90° domain wall motion over the barriers is given in Fig. 3. The symmetry of the MAE envelope is seen to be recovered by increasing the AC field intensity as reflected by the peak amplitude of the pickup coil output. Whether these MAE peaks shown in Fig. 3 have reached their possible maximum amplitude or not is, however, not clear at the present stage.

Fig. 4 shows the histogram of the system background noise. The horizontal resolution of the histogram is about 2.44 millivolts and the channel 2048 corresponds to zero. The figure shows the range of background noise in the histogram to be approximately ± 250 channels about the center. Except for the region between the two peaks, the histogram shows a smoothly decaying Gaussian-like distribution indicating the random nature of the noise spectrum.

The histogram obtained for the unembrittled HY80 sample is shown in Fig. 5. Since no filtering scheme has been applied, the histogram consists of a mixture of background noise and low amplitude MAE signals in the range identified in Fig. 4. Beyond this range, the histogram maintains Gaussian-like distribution except at both ends where non-Gaussian tails are seen.

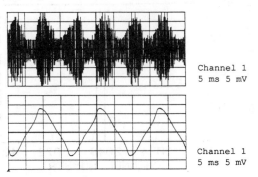

Channel 1
5 ms 5 mV

Channel 1
5 ms 5 mV

Fig. 2. MAE spectrum obtained at 60 Hz for the HY80 steel sample heat treated for 24 hrs.

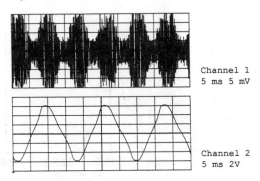

Channel 1
5 ms 5 mV

Channel 2
5 ms 2V

Fig. 3. Results of repeated measurements of Fig. 2 with an increased AC magnetic field amplitude.

The presence of such a tail structure, of course, indicates the existence of a different contributing mechanism. It is assumed that a mixture of the domain wall-defect interactions inside grains and at grain boundaries contribute to the Gaussian-like distribution. The interactions at the grain boundaries with more pronounced lattice mismatch, on the other hand, are assumed to be the sole contributor to the non-Gaussian-like distribution at the tail of the histogram.

Fig. 6 shows the histogram obtained for the HY80 steel sample which has been heat treated for 5 hours. The increased count rate of low amplitude MAE signals in this sample is seen to remove the smooth transition at the end of the background noise region. At the same time, a growth in count rate is seen at the tail sections of the histogram indicating clearly an increase in high amplitude MAE activity as the domain wall-defect interaction becomes stronger.

All the histograms obtained for the embrittled HY80 steel samples show such a distinct tail structure which is not found in the histogram of the unembrittled sample. The trend of growing count rate in the tail section of the histogram, however, is found to level off and decrease somewhat as the samples become more embrittled. Fig. 7 shows the high-amplitude portion of

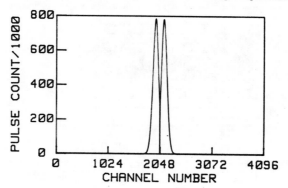

Fig. 4. Histogram representing the pulse height distribution of system background noise.

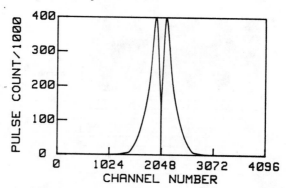

Fig. 5. Histogram obtained for the unembrittled HY80 sample at 60 Hz AC field frequency.

histograms for the first three samples. The curve for the unembrittled
sample residing in the lowest position shows an almost straight line for the
last 300 channels. Between the two curves of the embrittled samples, the
count rate in the same region is shifted upward as the heat treatment time
is increased from 1 hour to 5 hours. The same results are shown in Fig. 8.
In this figure the curves shifted downward as the heat treatment time is
increased to 24, 50 and 100 hours.

The decreased count rate in the tail section found in Fig. 8 is not a
surprise. As the grain boundaries become stronger potential barriers
against 90° domain wall motions, those 90° walls succesfully jumping over
the boundaries will generate high amplitude MAE signals. The presence of
enhanced potential barriers, however, allows fewer 90° domain walls to
execute such jumps, and consequently, results in decreased MAE activity in
the more embrittled samples. A clear evidence of such an effect has been
observed from the AC magnetic field amplitude dependence of MAE peaks in
unembrittled and embrittled samples [5]. Comparing those curves in the tail
section, it appears that the MAE peaks in Fig. 3 have not reached the
possible maximum amplitude.

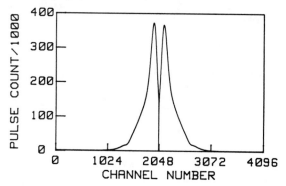

Fig. 6. Histogram obtained for the HY80 sample heat treated
for 5 hours.

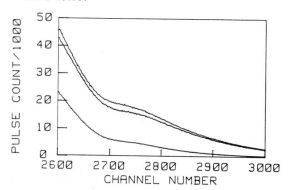

Fig. 7. Non-Gaussian tail structure of histograms for the
samples heat treated for 0, 1 and 5 hours. The curves
shift upward with increased heat treatment time.

182

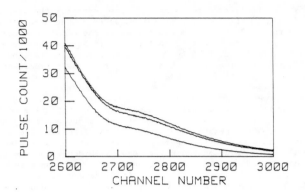

Fig. 8. Tail structure of histograms for the samples heat treated
for 24, 50 and 100 hours. The curves shift downward with
increased heat treatment time.

SUMMARY

This paper presents the initial results of MAE pulse height analysis of
unembrittled and embrittled HY80 steel samples obtained by using a
multichannel analyzer. The enhanced 90° domain wall-defect interaction in
the embrittled sample is seen to generate high amplitude MAE signals as
evidenced by the presence of a distinct tail structure in the histogram that
is not found in the unembrittled sample. Such a trend is seen to level off
and reverse slightly in the highly embrittled samples due to the enhanced
potential barriers at the grain boundaries that tend to limit 90° domain
wall motions. It is, therefore, necessary to repeat these measurements with
a stronger AC magnetic field.

ACKNOWLEDGEMENT

The authors are grateful to Mr. Robert DeNale of the David Taylor
Research Center for his support in this research and provision of test
samples with metallurgical information.

REFERENCES

1. B. D. Cullity, Introduction to Magnetic Materials (Addison-Wesley, Menlo
   Park, 1972).
2. H. Trauble, in Magnetism and Metallurgy, Vol. II, edited by A. E.
   Berkowitz and E. Kneller (Academic Press, New York, 1969).
3. S. G. Allison, W. T. Yost, J. H Cantrell and D. F. Hasson, in Review of
   Progress in Quantitative Nondestructive Evaluation 7B, 1463 (1987).
4. M. Namkung, W. T. Yost, J. L. Grainger and P. W. Kushnick, submitted to
   Review of Progress in Quantitative Nondestructive Evaluation, La Jolla,
   CA (August, 1988).
5. M. Namkung, W. T. Yost, D. Utrata, J. L. Grainger and P. W. Kushnick,
   submitted to IEEE Ultrasonics Symposium, Chicago, IL (October, 1988).

MEASUREMENTS OF STRAIN INDUCED RESISTIVITY
BY THE EDDY-CURRENT DECAY METHOD

K.THEODORE HARTWIG, JR.

Texas A&M University, Department of Mechanical Engineering, College
Station, TX 77843-3123.

ABSTRACT

The eddy-current decay method developed by Bean for electrical
resistivity measurements is well-suited for bulk metal characterization
studies. The technique can be applied to investigations of low temperature
plastic strain in pure aluminum.

INTRODUCTION

The electrical resistivity of metal is often determined by a DC
potential-difference method [1]. This involves passing a known current
through a sample with known geometry and measuring a potential difference
between carefully placed voltage contacts. The measurement is easy provided
the geometry is well-characterized and the voltage drop can be measured
accurately. The method is difficult for large samples with low resistivity.
This is often the case for bulk measurements on superconductor stabilizer
materials such as pure aluminum and OFHC copper. For such situations, the
eddy-current decay method is well suited.

EDDY-CURRENT DECAY METHOD

Bean, in 1959, described an eddy-current decay (ECD) method for
measuring electrical resistivity of metallic specimens [2]. The method was
subsequently evaluated by the National Bureau of Standards for measurements
of residual resistivity and residual resistivity ratio in bulk, high-purity
metal samples [3]. The measurement is made by determining the rate of decay
of magnetic flux from a bar placed in an external magnetic field that has
been brought rapidly to zero. The method requires no electrical contacts to
the specimen, is quick and is accurate to several percent. For a
cylindrical bar of constant cross section and uniform isotropic resistivity,
determination of resistivity is based on the relationship:

$$\rho(\Omega cm) = \frac{2.17 \times 10^{-9} \mu r^2}{\tau} \qquad (1)$$

where μ is the relative permeability, r is the specimen radius in
centimeters and τ is the time constant in seconds.

Complications that may arise in using the ECD technique are listed in
Table 1.

TABLE 1. Possible Complications in Using the ECD Method.[a]

Problems	Basis/Solutions
1) Sample dimensions too small.	In very pure materials, the electron mean free path may be comparable to specimen dimensions. Diameters larger than several millimeters are often sufficient in six nines pure metal.
2) Measurement imposed magnetoresistivity.	Detect magnitude of effect by measurements at different field levels (primary coil current levels).
3) Ferromagnetic or superconductive specimens.	Assumption of constant permeability is in error. Make measurements in a constant external field that saturates the ferromagnet or causes the superconductor to be in the normal state.

_____

[a] Resistivities that cause relaxation times from several milliseconds to seconds are measured easily. These correspond to resistivities of about 100 to 0.1 nΩcm on 6.4-mm diameter cylindrical rod.

Coil designs, measurement apparatus and detailed experimental procedures are described elsewhere [3,4].

STRAIN RESISTIVITY IN PURE ALUMINUM

A serious concern to designers of large superconductive magnetic energy storage devices is conductor strain [5]. First time cooldown and the periodic charge-discharge cycles of the magnet subject the windings to stress and strain. Components of the conductor that have a low yield strength, i.e. the pure aluminum stabilizer, may undergo plastic strain and resistivity increases depending on overall stress levels. The magnitude of the resistivity increase is related to the amount of plastic strain. Larger strains result in larger resistivity increases.

The ECD method is convenient for study of strain-resistivity effects in centimeter-diameter samples of pure aluminum. Two examples of the work are reported here to underscore how damaging low temperature strain can be to very low resistivity aluminum.

The first example relates to cooldown strain. If the structure holding the windings of a magnet composed of aluminum conductors is held rigid, then the conductor is subjected to a tensile strain of about 0.004 in/in as the magnet is cooled from room temperature to 4.2 K. The resistivity increase associated with this strain depends on the cooldown rate and the yield strength of the aluminum. The worst case is rapid cooling of low-yield strength material. Rapid cooling minimizes recovery, a time dependent annealing process. Low-yield strength aluminum experiences a larger amount of plastic strain.

Figure 1 is a plot of change in resistivity ($\Delta\rho$) versus plastic strain ($\Delta\varepsilon$) at 4.2 K for 99.999% pure aluminum with an initial residual resistivity ratio (RRR = ρ(273 K)/ρ(4.2 K)) of about 3400. For the worst case cooldown scenario, the aluminum experiences about 0.37% plastic strain (total strain of 0.4% minus elastic strain of about 0.03%). If we assume all aluminum with RRR > 1000 has the same yield strength and $\Delta\rho$ versus $\Delta\varepsilon$ relationship, then the magnitude of the cooldown strain effect is that shown in Table 2.

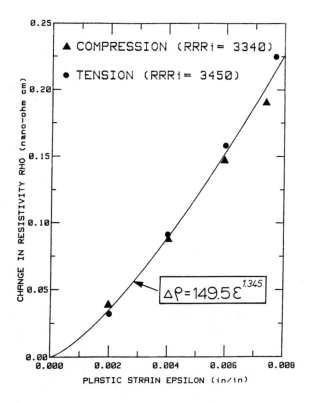

FIGURE 1. Change in resistivity versus plastic strain at 4.2K in annealed 99.999% aluminum.

TABLE 2.   Cooldown Strain Resistivity in Pure Aluminum[a]

Initial RRR	RRR after Cooldown
1000	970
5000	4290
10,000	7520

[a] Assumptions are rapid cool down, 0.37% plastic strain and same strain-resistivity relationship for all cases ($\Delta\rho$ is 0.08 nΩcm).

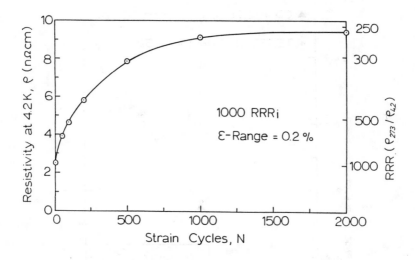

FIGURE 2.   Residual resistivity versus strain cycles at 4.2 K for 1000 RRR aluminum.   Strain range is 0.2%.

The second example is that of cyclic strain. When a magnet is taken through many charge-discharge cycles, the windings experience cyclic stress and strain. The loading environment is constant-stress-range cyclic strain. Every stress cycle subjects the conductor to additional plastic strain. The plastic strain per cycle decreases as the number of cycles increases, and the pure aluminum slowly work-hardens. After several thousand cycles, the pure aluminum becomes fully hardened and the resistivity increase per cycle drops to zero. The total change in resistivity, which may be large, depends primarily on two factors: the stress (or strain) range and the initial yield strength of the aluminum.

Figure 2 shows resistivity versus number of strain cycles at 4.2 K for 99.995% aluminum with an initial RRR of 1000. In this example, the loading environment is constant-strain-range cyclic stress, with a strain range of 0.002 in/in. Resistivity increases rapidly during the first several hundred cycles and reaches equilibrium after about 1500 cycles. Notice that the total resistivity increase is very large. The final RRR is about 260!

The phenomena of cyclic-strain resistivity increases in pure aluminum have been studied extensively [5] and are still under investigation by the author. The ECD method for resistivity characterizations has been and is being used successfully and almost exclusively in these studies.

ACKNOWLEDGEMENT

The sponsors of this research are gratefully acknowledged: the Electric Power Research Institute and Texas A&M University.

REFERENCES

1. F.R. Fickett, in Materials at Low Temperatures, edited by R.P. Reed and A.F. Clark (ASM, Metals Park, Ohio, 1983), p. 165.

2. C.P. Bean, J. Appl. Phy. 30(12), 1976 (1959).

3. A.F. Clark, V.A. Deacon, J.G. Hust and R.L. Powell, Standard Reference Materials: The Eddy-Current Decay Method for Resistivity Characterization of High Purity Metals, NBS Special Publication 260-39 (U.S. Dept. of Commerce, Nat. Bur. Standards, 1972).

4. K.T. Hartwig, in Eddy-Current Characterization of Materials and Structures, STP 722, edited by G. Birnbaum and G. Free (ASTM Publishers, Philadelphia, 1981), p. 157.

5. K.T. Hartwig and G.S. Yuan, in Cryogenic Materials '88, Volume 2, Structural Materials, edited by R.P. Reed, Z.S. Xing and E.W. Collings, (ICMC, Boulder, CO, 1988), p. 677.

IN-SITU NMR STUDY OF DISLOCATION JUMP DISTANCE DURING CREEP OF PURE AND DOPED NACL SINGLE CRYSTALS

K. LINGA MURTY* AND O. KANERT**

*North Carolina State University, Raleigh NC 27695-7909
**University of Dortmund, Postfach 500 500, 46 Dortmund 50, FRG

ABSTRACT

Nuclear magnetic resonance pulse techniques are used *in-situ* during creep of single crystals of NaCl to evaluate the contribution of mobile dislocations to spin relaxation. ^{23}Na spin-lattice relaxation rates were measured in the rotating frame ($T_{1\rho}$) during compression creep of single crystals of NaCl along [110] direction at 473K at an applied stress of 20 MPa. The relaxation rates are evaluated from the spin-echo height following $\pi/2$, locking and 67° pulse sequence. The height of the free induction decay decreased as soon as the load is applied followed by a gradual increase until the steady-state is reached, at which point a saturation value is observed corresponding to the constant steady-state creep-rate. The mean jump distance of the mobile dislocations, evaluated from the ratio of the signal heights without deformation and during creep, decreased with time/strain reaching a constant value during steady-state creep regime. The results are compared with the dislocation-dislocation spacing, subgrain size as well as the jump distance predicted from creep models. The effects of dilvalent Ca and solid solution with LiCl are examined.

INTRODUCTION

Creep deformation of materials is due to the motion of dislocations which move in the field of external and internal stresses with the latter being caused by the forest dislocations. The fundamental parameters controlling the steady-state creep behavior are the density and velocity of the mobile dislocations that appear in the fundamental Orowan equation for the strain-rate:

$$\dot{\varepsilon} = \phi \alpha = \phi \rho_m b \ v \ , \tag{1}$$

where ε is the creep-rate, α the shear strain-rate, ρ_m the density of mobile dislocations, b the Burger's vector, v the mean velocity and ϕ the geometric factor. In addition, the time/strain dependencies of these factors dictate the extent and the nature of the transient creep as well. The temperature and more significantly the stress dependencies of the dislocation velocity are important in that they characterize the specific dislocation micromechanism controlling the creep behavior of a given material such as the dislocation climb [class M] versus viscous glide [microcreep, class A].[1] Moreover, the characteristics of the transient creep [normal, inverse, etc.] are dependent on the class of the material and the nature of the transients is dictated by the time/strain dependence of the dislocation velocity [or jump distance] and/or the density of the mobile dislocations.

Recent advances in both the theoretical and experimental aspects lead to the feasibility of using the pulse NMR techniques to evaluate the dislocation jump distance during deformation[2] and we extended these to the creep behavior of NaCl single crystals. This technique is unique in that it can be used in-situ during plastic flow to non-interactively determine the characteristics of the mobile dislocations in both the steady and primary stages of creep thereby enabling an investigation of the time and strain variation during creep transients. In addition, some preliminary experiments were performed to investigate the effects of divalent impurity as well as solid solution on the jump distance.

The spin lattice relaxation rate in a rotating field H_1 [locking field], $1/T_{1\rho}$, of the resonant nuclei in the sample is enhanced due to the motion of dislocations. The resulting total relaxation rate is given by,

$$\frac{1}{T_{1\rho}} = \{\frac{1}{T_{1\rho}}\}_0 + \{\frac{1}{T_{1\rho}}\}_\perp \tag{2}$$

where the subscripts $\perp$ and 0 represent the contributions of the jumping dislocations and background respectively. The background relaxation is due to various factors including two-phonon scattering processes, atomic motion, static dislocations, etc., representing the relaxation rate in the absence of plastic deformation. The dislocation-induced relaxation rate is given by[3]

$$\{\frac{1}{T_{1\rho}}\}_\perp = \frac{\delta_Q}{H_1^2 + H_{loc}^2} \langle V^2 \rangle g_Q(L) \frac{\rho_m}{\tau_c} \tag{3}$$

Here, δ_Q is the quadrupole coupling constant, $\langle V^2 \rangle$ the mean-squared EFG of a dislocation of unit length, H_1 the locking field, H_{loc} the local field comprising of dipolar (H_D) and quadrupolar (H_Q) components, $g_Q(L)$ a geometrical factor dependent on the jump distance (L) with a value close to unity for practical conditions,[3] and τ_c the effective dislocation jump time. Combining the above equation with an appropriate mechanical equation of the dislocation motion, one may relate the relaxation rate to the jump distance. For the example where the dislocation motion is governed by the Orowan equation [Eq.1] with mean dislocation velocity given by L/τ_c, Eq.3 can be written as

$$\{\frac{1}{T_{1\rho}}\}_\perp = \frac{\delta_Q \langle V^2 \rangle}{H_1^2 + H_{loc}^2} g_Q(L) \frac{1}{\phi b L} \dot{\epsilon} \tag{4a}$$

Assuming different jump processes of type k which can be characterized by different correlation times, we find

$$\{\frac{1}{T_{1\rho}}\}_\perp = \frac{\delta_Q \langle V^2 \rangle}{H_1^2 + H_{loc}^2} \sum_k \frac{g_Q(L_k)}{L_k} \frac{\dot{\epsilon}}{\phi b} \tag{4b}$$

It is clear then that the dislocation mechanism with smallest jump distance makes significant contribution to the relaxation.

EXPERIMENTAL ASPECTS

NaCl single crystals of high purity [<7ppm] were prepared as parallelopiped samples measuring 5x5x12 mm^3 with long axis along <110> direction. To investigate the effects of impurities, similarly oriented single crystals were used with either 0.01 a/o Ca or LiCl. All test samples were carefully polished to ensure parallelism of the faces to minimize any bending during compressive loading.

A compression creep machine was designed and built for Creep-NMR experiments so that the sample and rf signal coils along with the heater assembly fit into a 4.23T superconducting magnet.[4] A Bruker SXP4-100 NMR pulse spectrometer was used for $T_{1\rho}$ measurements where the free induction signal of the ^{23}Na nuclei [resonant frequency of 47.584 MHz] was monitored following the rf pulse sequence consisting of a $\pi/2$ pulse followed by a locking field H_1 of 5G [and 100 msec long] phase shifted 90°. The nuclear magnetization is allowed to decrease in the presence of the locking field for a time τ_1 and then H_1 turned off following which the nuclear free induction decay signal is measured whose amplitude decays with time constant $T_{1\rho}$. It turned out that the signal shape slightly altered during deformation and thus the height of the free induction following the locking pulse is monitored by applying a 67° pulse[2] and noting the height of the spin echo [Fig.1]. The application of load/strain-rate causes a significant reduction in the relaxation time $T_{1\rho}$ due to the motion of dislocations. All measurements were made during creep by storing the data in a microcomputer which was also used for signal averaging as well as data analyses.

RESULTS AND DISCUSSION

We report here results on pure and doped NaCl single crystals at a temperature of about 473K. Constant loads corresponding to initial stresses ranging from 10MPa to 30MPa were used. The height of the spin-echo was monitored before applying load and during creep from which the dislocation-induced component of the spin-lattice relaxation time is evaluated. All of the tests could be performed for a maximum period of 300 seconds due to the stability of the electronics used. Nevertheless, both primary and steady-state creep regimes were spanned in all cases.

Fig.2a depicts the creep curve for pure NaCl at about 20MPa along with the time dependence of the spin-echo amplitude. The time dependence of the creep strain clearly indicated normal primary creep followed by a relatively long secondary stage. As soon as the load is applied, the amplitude of the echo decreased but started increasing as the creep strain-rate decreased during the transient stage reaching a plateau in the secondary creep regime. The time or strain dependence of the dislocation-induced relaxation times were evaluated as a function of time or strain

from the change in the signal amplitude [Fig.2b] which then
facilitated the evaluation of the jump distance using Eq.4b :

$$L = \frac{\delta_Q \langle V^2 \rangle}{H_1^2 + H_D^2 + H_Q^2} \frac{g_Q}{\phi b} \varepsilon(T_{1p})_\perp \tag{5}$$

where H_{loc} is replaced by the dipolar and quadrupolar terms. In
evaluating L, the following values were used for the various
terms in this equation,[2-4]

$\delta_Q \langle V^2 \rangle = 2.6 \times 10^{-9}$ $G^2 cm^2$, $g_Q = 1$, $H_1 = 5$ G, $H_D = 0.9$ G, $H_Q = 1$ G,
$\phi = 0.5$, $b = 3.98 \times 10^{-8}$ cm.

Thus evaluated L is depicted as a function of time in Fig.3 along
with strain obtained from the results in Fig.2a. It is clear that
during the primary creep region where the strain-rate decreased
with strain, the dislocation jump distance also decreased and
both of them reached constant values in the secondary creep
regime. It should be noted that the dislocation jump distances
are very small in the secondary creep regime, of the order of
23b.

It is now instructive to compare the dislocation jump
distance with some of the substructural features relevant to
creep. Before discussing these correlations, we note that at the
test conditions considered here, creep in NaCl is controlled by
dislocation climb mechanism akin to that noted during high
temperature creep of metals and nonmetals. The jump distance
found here is noted to be 1000 times less than the subgrain size
and 100 times less than the dislocation-dislocation separation
predicted at the applied stress of 20MPa.[4]

One of the classic descriptions of the steady-state creep
was put forward by Weertman,[5] the so-called 'Pill-Box' model
which successfully predicted both the temperature and stress
dependences of steady-state creep-rate at temperatures above
about $0.4T_m$. Accordingly, the lead dislocations in the pile-ups
generated from Frank-Read sources on the parallel slip planes
interact with each other thereby hindering their further
movement. The lead dislocations of opposite sign then climb
towards each other and get annihilated. Once each of the lead
dislocations climbs for annihilation, a pair of new dislocations
is produced resulting in the steady state between the rate of
dislocation generation and recovery. The climb distance of the
lead dislocations required for mutual annihilation is given by,[4,5]

$$h = \frac{Gb}{4\pi\sigma} = 99\,b = 0.04\,\mu m \tag{6}$$

which compares with the experimental value of 23b. This implies
that climb of edge dislocations as described by Weertman is the
dominating process controlling creep of NaCl and as far as we are
aware, this is the first time that a quantitative measurement of
the jump distance of climbing dislocations during creep has ever
been reported.

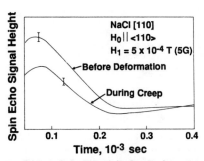

Fig.1. Spin Echo Signals Before Deformation and During Creep.

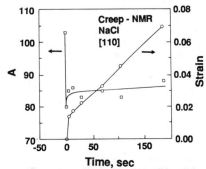

Fig.2a. Time Dependence of Creep Strain and Spin Echo Height

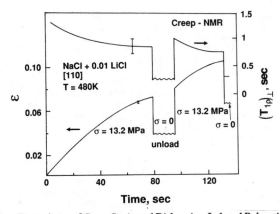

Fig.2b. Time Dependence of Creep Strain and Dislocation-Induced Relaxation Rate

Fig.3. Creep Strain and Dislocation Jump Distance Vs Time

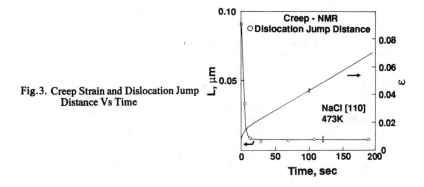

Effect Of Impurities

We made a limited study of the effects of divalent Ca and solid solution with LiCl on the dislocation jump distance in single crystals of NaCl. Testing was made at a nominal temperature of 473K but the stresses were varied to obtain similar strain-rates. Following is a compilation of the experimental results at a constant creep-rate of 3×10^{-4} sec^{-1} :

Material	τ, MPa	$\langle L \rangle$, μm
NaCl	20.1	0.007±0.001
NaCl+0.01Ca	25.2	0.008±0.002
NaCl+0.01LiCl	13.2	0.023±0.003

Thus, we note that Ca addition made the material slightly stronger while LiCl made it much softer. The dislocation jump distance is relatively unaffected by Ca whereas LiCl additions resulted in distinctly large dislocation jump distance [by about 4 times] in comparison to that in pure NaCl. The reasons for these observations are at best speculative at the present time.

It is known that Ca addition increases the hardening of the material due to elastic distortions created by Ca atoms displacing Na on the NaCl lattice. LiCl is expected to be a solid solution in NaCl in which case the creep mechanism might change from dislocation climb [similar to pure metals and ClassII alloys] to viscous glide typical of ClassI alloys where no distinct subgrains are formed. In ClassI alloys, glide controlled creep dominates due to the interaction of the dislocations by the solute atmospheres; in the present case, the atmospheres are formed by Na$^+$ and Li$^+$ ions and these atmospheres result in decreased rates of dislocation glide. It is possible then that the active dislocation sources are spaced farther than in the case of climb controlled creep such as in NaCl and pure metals. Other possibilities exist in that the addition Of LiCl might influence the shear modulus which, if increases due to solute additions, results in increased dislocation jump distance. Moreover, the stacking fault energy of the material in general decreases due to alloying which might result in decreased number of active slip systems or increased jump distances for dislocation annihilation by climb.

ACKNOWLEDGEMENTS

We wish to express our sincere appreciation to D.Begert who made the initial measurements in NaCl and to Messrs. R.Munter, U.Schlagowski and M.Backens for assistance and discussions. One of us [KLM] is grateful to DAAD for a visiting fellowship which made this work possible.

REFERENCES

1. T.T. Fang, R.R. Kola and K.L. Murty, Met.Trans. 17A, 1447 (1986).
2. O. Kanert, Physics Reports, 91, 183 (1982).
3. W.H.M. Alsem, J.Th.M. De Hosson, H. Tamler and O. Kanert, Phil. Mag. 46, 451 (1982).
4. K.L. Murty, D. Begert, R. Munter and O. Kanert, presented at the 1988 ASM World Materials Congress, Chicago, Ill, Sep. 1988.
5. J. Weertman, Trans. ASM, 61, 681 (1968).

A STUDY OF HIGH TEMPERATURE DAMAGE PROCESSES USING MICRORADIOGRAPHY

J. E. BENCI AND D. P. POPE
Department of Materials Science and Engineering, University of Pennsylvania, 3231 Walnut St., LRSM, Philadelphia, PA 19104.

ABSTRACT

Synchrotron radiation and microradiographic techniques were used to study the development of creep damage in notched tensile samples. The creep damage in these samples was recorded using microradiography. The density and distribution of creep damage was measured from the microradiographs using an image analysis system. The results from the image analysis can be compared to damage predictions from finite element models of the damage process to determine the quality of these models.

Notched tensile samples of copper, iron and a low alloy steel were subjected to slow strain rate tensile tests at 500°C or 700°C. The tests were interrupted after various fractions of the creep lives had been expended. 1 mm thick longitudinal sections were then removed from the center of each sample for microradiography using electro-discharge machining.

Creep damage in the copper alloy was concentrated in a fairly narrow band around the plane of minimum cross-section in the samples. This is in stark contrast to the results from iron and the low alloy steel. The creep damage in these materials developed at fairly sharp angles to the notch or crack plane. These results show that the damage process in iron and this steel is controlled by the equivalent stress while the formation of damage in copper is controlled by the maximum principal or hydrostatic stress.

INTRODUCTION

Radiography was first used to investigate the internal structure of materials shortly after the discovery of x-rays[1,2]. However, the technique has never gained widespread use in the area of materials research because of the limitations in resolution and thickness of sample which could be investigated with reasonable exposure times. The availability of synchrotron radiation has generated new interest in the technique. Synchrotron sources produce very intense, highly collimated white radiation. Since the beam is so well collimated, large source to specimen distances are possible with very little loss in intensity. Large source to specimen distances result in excellent resolution, with the main factor limiting the resolution of the technique being the resolution of the recording medium. The intensity of the beam allows for the investigation of relatively thick specimens (1 to 2 mm thick) with exposure times of several minutes. A monochromater can be inserted into the beamline to obtain any wavelength from about 0.3 to 2.0Å. Since the sensitivity of the technique depends on the absorption coefficient of the material, which is a function of wavelength, the sensitivity achieved in an experiment can be chosen by picking the incident radiation energy. The drawback of a monochromater is the enormous intensity loss and the resulting increase in exposure time. White radiation was therefore used for all of the experiments described in this paper.

The details involved in estimating the resolution and sensitivity of the technique are discussed elsewhere[3-5]. The resolution for these experiments, defined here as the minimum separation of features in a direction perpendicular to the beam which can be resolved, is about 1.1μm for 1 mm thick samples of copper, iron and steel. The sensitivity, defined as the minimum feature dimension in the direction parallel to the beam which can be detected, is about 15 to 20μm for 1 mm thick samples assuming 10% differences in contrast can be seen on the x-ray film. A further discussion of the technique of microradiography and some of its possible applications is given in ref. 6.

EXPERIMENTAL PROCEDURE

The starting material for this study was OFHC copper alloy 101 (99.99% pure) obtained in the form of 19 mm diameter rod. The material was annealed at 800°C at an O_2 pressure of 1 atm for 4 hrs to saturate it with oxygen, and then water quenched. This material condition will be referred to as Oxygen Saturated, or O.S., copper. O.S. copper exhibited totally intergranular fracture when subjected to slow strain rate tensile testing at temperatures between 200 and 700°C. The intergranular fracture was due to the nucleation, growth and eventual coalescence of grain boundary cavities under the applied stresses, as determined by SEM micrographs of the fracture surfaces.

Axisymmetric notched tensile samples with two different notch geometries, A and B, were machined from the O.S. copper. The A geometry is a relatively blunt notch, while the B geometry is a fairly sharp, deep notch. Slow strain rate tensile tests were performed on these samples at 500°C in air with a constant applied displacement rate of 2.11×10^{-5} mm/s. Identical samples of both notch geometries were interrupted after various test times. Longitudinal center slices 1.0 to 1.5 mm thick were then removed from the gauge section of each sample using EDM. These slices were mounted and polished on both faces to obtain the smooth and parallel surfaces necessary for radiography. Radiographs of the creep damaged samples were then obtained using white synchrotron radiation and high resolution plates for the recording medium. Selected radiographs were examined using a light microscope equipped with an image analysis system. The area fraction of creep damage as a function of position was plotted for these radiographs. A more thorough description of the procedure for these experiments is given in ref. 3.

RESULTS AND DISCUSSION

Figure 1(a) is a low magnification radiograph of a notch A sample which was interrupted at $0.75\sigma_{max}$ while the stress was still increasing. The small white features about 100µm long in the radiograph are failed grain boundary facets, or microcracks. Figures 1(b-d) are slightly higher magnification radiographs of notch A samples which were interrupted at the max. eng. stress, $0.9\sigma_{max}$ and $0.5\sigma_{max}$, respectively, after the stress had reached its maximum value. The microcracks in these radiographs appear black. Figure 2(a) is a low mag. radiograph of a notch B sample which was interrupted at $0.75\sigma_{max}$ while the stress was still increasing. Again, the small white features are microcracks. Figures 2(b-d) are slightly higher mag. radiographs of notch B samples which were interrupted at the max. eng. stress, $0.9\sigma_{max}$ and $0.4\sigma_{max}$, respectively, after the stress had reached its maximum value. Figure 3 is a higher magnification view from the radiograph in fig. 1(c) which demonstrates the detail which can be imaged with microradiography. The important features to note from the radiographs is figs. 1 and 2 are that the creep damage for both notch geometries developed in a fairly narrow band around the plane of minimum cross-section, and that the regions with the highest concentration of damage in the lightly damaged samples appear to occur in the interior of the samples, and not at the sample surface.

The results from the O.S. copper samples are in stark contrast to results from a previous investigation of creep damage formation in high purity iron and a 1.0 Cr, 1.25 Mo, 0.25 V steel[5]. Single Edge Notched Tension (SENT) samples of the iron alloys and Double Edge Notched Tension (DENT) samples of the low alloy steel were subjected to high temperature, slow strain rate tensile testing. The intergranular creep damage in these systems appears to nucleate at the notch surfaces, not ahead of the notch, and the damage forms at sharp angles to the plane of the notch in both materials.

Finite element modeling of the slow strain rate tensile tests carried out on the axisymmetrically notched O.S. copper samples was performed using the ABAQUS finite element code. The modeling was performed for an applied dispacement rate of 8.47×10^{-6} mm/sec as compared to the experimentally applied dispacement rate of 2.11×10^{-5} mm/sec. This modeling took into account elastic and creep deformation but did not take into account any possible stress redistributions due to the presence of creep damage. In the small damage limit, stress redistribution due to damage should be minimal and the damage contours are expected to closely follow the appropriate stress contours responsible for the formation of damage. Figure 4 shows the results from the finite element calculations for the A and B notch geometries for a particular test time, 4.0 hrs for notch A and 1.5 hrs for notch B. The maximum value of the equivalent stress occurs at the notch surface for both geometries and the contours develop out at

197

an angle to the plane of the notch. However, the maximum values of the max. principal and hydrostatic stresses occur subsurface from the notch for both notch geometries and the contours develop out ahead of the notch in the notch plane.

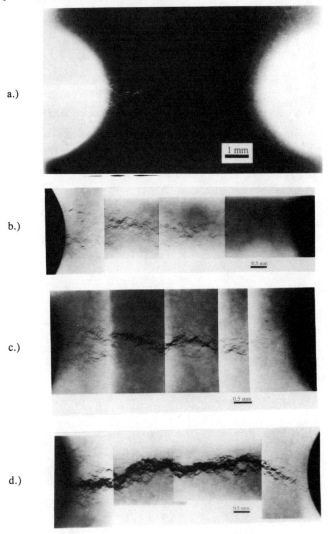

Fig. 1 Radiographs from the notch A geometry O.S. copper samples which were interrupted at (a) 0.75 σ_{max} while the stress was still increasing, (b) σ_{max}, (c) 0.9σ_{max} after the stress had reached its maximum value and (d) 0.5σ_{max} while the stress was decreasing. The white features in (a) and the black features in (b) - (d) approximately 100µm long are failed grain boundary facets or microcracks.

198

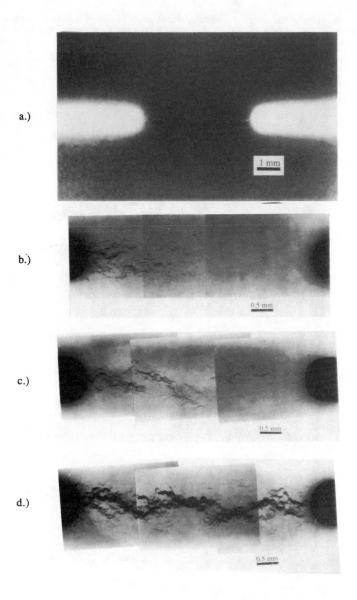

Fig. 2 Radiographs from the notch B geometry O.S. copper samples which were interrupted at (a) 0.75 σ_{max} while the stress was still increasing, (b) σ_{max}, (c) 0.9σ_{max} after the stress had reached its maximum value and (d) 0.4σ_{max} while the stress was decreasing.

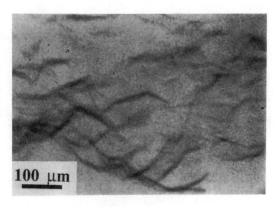

Fig. 3 Microradiograph from an area near the center of the radiograph in fig. 1(c) illustrating the detail which can be resolved with microradiography using white synchrotron radiation.

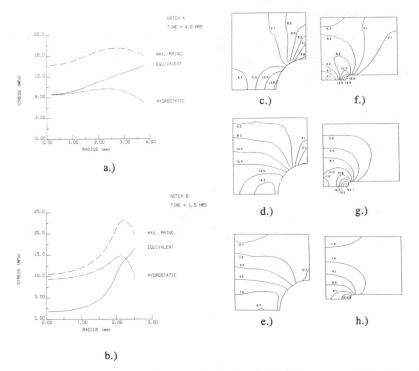

Fig. 4 Finite element results for the stress distributions in the O.S. copper notch A and notch B sample geometries undergoing constant displacement rate tensile testing for 4 hrs. and 1.5 hrs., respectively, at 8.47×10^{-6} mm/sec and 500°C. The equivalent, max. principal and hydrostatic stresses as a function of position in the plane of the notch (z=0) are shown for the (a) notch A and (b) notch B geometries. The stress contours for the equivalent, max. principal and hydrostatic stresses are shown in (c), (d) and (e), respectively, for the notch A geometry and (f), (g) and (h), respectively, for the notch B geometry, in units of MPa.

Comparing the finite element results to the results from microradiography, the damage for O.S. copper correlates well with a maximum principal or hydrostatic stress controlled damage process. If one assumes that the stress distributions around the notches in SENT and DENT sample geometries will resemble those from the axisymmetric notched geometries, the formation of damage in iron and this low alloy steel is predominantly controlled by the equivalent stress.

CONCLUSIONS

Significant advances in the technique of radiography have been made through the use of synchrotron radiation allowing the use of microradiography to investigate creep damage processes. The combination of microradiography and finite element analysis can provide much information about the mechanisms of creep damage formation. The creep damage process in O.S. copper is predominantly controlled by the maximum principal and/or hydrostatic stress components. Therefore, the development of damage is a diffusion controlled process. The equivalent stress has a strong influence on damage development in iron and CrMoV steel. Cavity growth in these materials must be due to plasticity.

ACKNOWLEDGEMENTS

This work was funded through the synchrotron topography project beamline, X-19C, of the National Synchrotron Light Source, Brookhaven National Laboratory, which is supported by the United States Department of Energy under grant DE-FG02-84ER45098. Part of this work was performed at the Stanford Synchrotron Radiation Laboratory which is supported by the Department of Energy, Office of Basic Energy Sciences. The use of LRSM facilities at the University of Pennsylvania supported by the NSF MRL program under grant DMR85-19059 is also acknowledged. The authors would like to thank Dr. R. J. Fields of the National Bureau of Standards for his assistance with the image analysis.

REFERENCES

1.   G.J. Burch, Nature, Lond., **54**,111 (1896).
2.   C.T. Heycock and F.H. Neville, J. Chem. Soc., **73**, 714 (1898).
3.   J.E. Benci and D.P. Pope, Mater. Sci. Eng., in press.
4.   J.E. Benci, E.P. George and D.P. Pope, Mater. Sci. Eng., **A103**, 97 (1988).
5.   J.E. Benci and D.P. Pope, Metall. Trans., **19A**, 837 (1988).
6.   D.K. Bowen, in "Applications of Synchrotron Radiation to the Study of Large Molecules of Chemical and Biological Interest," R.B. Cundall and I.N. Muro, eds., Daresbury Laboratory Proceedings DL/SCI/R13, p. 37.

PART IV

Polymers and Composites

THE USE OF TIME-TEMPERATURE SUPERPOSITION TECHNIQUES TO FORECAST CREEP BEHAVIOR IN ENGINEERING PLASTICS

B. J. Overton, C. R. Taylor and J. W. Shea, AT&T Bell Laboratories, 2000 Northeast Expressway, Norcross, Georgia 30071

ABSTRACT

A complete understanding of the time-temperature dependent, viscoelastic mechanical properties of polymeric materials is essential for their successful application to engineering designs. We describe a computer-based system for acquiring, reducing and storing the mechanical properties of polymers, from which the useful forms of the mechanical characteristics may easily be reproduced and compared. We provide several examples illustrating the variety of ways in which these properties are utilized to forecast the performance of polymeric components of engineering designs under extreme field conditions.

INTRODUCTION

The role of polymeric materials in engineering applications continues to increase throughout industries where metals, wood and natural fibers were historically the only useful choices. The advantages in cost, weight, ease of fabrication, and often strength are accepted today, and the development of specialized polymers with specific properties is accelerating.

A thorough understanding of the nature of polymer behavior is essential for designing with plastics, the stress-strain response of polymers being fundamentally different from that of many engineering materials. For example, polymeric materials exhibit viscoelastic creep under loading, where the gradual deformation (or recovery) of a polymer-based part or component can render the part nonfunctional if the change is beyond a critical limit. Fortunately, tools are available to obtain accurate determinations of the short- and long-term mechanical characteristics of polymers, both filled and neat.

We will present a brief synopsis of the principles of time-temperature equivalence in the viscoelastic mechanical behavior of polymers and describe techniques grounded in this relationship for obtaining data with which the long-term/high temperature performance of a polymer system can be predicted with accuracy. We then discuss several instances of application of such information to the selection of materials for specific uses, where the reliability of dimensions and/or strength through years of service under extreme conditions is essential.

TIME TEMPERATURE SUPERPOSITION

The simplest useful model showing the mechanical response of a viscoelastic material under a constant strain is the Maxwell model[1], a single element mechanical analog of which is illustrated in Figure 1. For

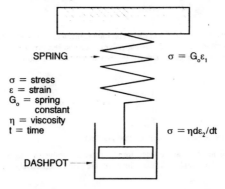

SPRING $\quad \sigma = G_o \varepsilon_1$

σ = stress
ε = strain
G_o = spring constant
η = viscosity
t = time $\qquad \sigma = \eta d\varepsilon_2/dt$

DASHPOT

FIGURE 1.   MAXWELL MODEL OF A VISCOELASTIC MATERIAL

Mat. Res. Soc. Symp. Proc. Vol. 142. ©1989 Materials Research Society

the single Maxwell element the time dependent stress is described by

$$\sigma(t) = \sigma_o \exp(-t/\tau_o) \tag{1}$$

where t is time and σ_o is the initial stress at $t = o$. τ_o is a characteristic relaxation time constant defined as $\dfrac{\eta_o}{G_o}$ where η_o is the viscosity associated with the dash pot and G_o is the spring constant, Figure 1. Dividing through by the strain, equation (1) becomes

$$G(t) = G_o \exp(-t/\tau_o) \tag{2}$$

For uncrosslinked polymers, represented by a large number of Maxwell elements in parallel, the time dependent modulus $G(t)$ is given by

$$G(t) = \sum_{i=1}^{N} G_i \exp(-t/\tau_i) \tag{3}$$

or in integral form,

$$G(t) = \int G(\tau) \exp(-t/\tau) \, d\tau \tag{4}$$

where $G(\tau)$ is a function defining the distribution of relaxation times (that is, the contributions to stress relaxation associated with relaxation times between τ and $d\tau$).

The relaxation modulus, $G(t)$, for polymers is functionally dependent on temperature as well as time. Returning for illustration to the single Maxwell element described by eq. (2), the effect of temperature can be attributed to the temperature dependence of the dashpot's viscosity, Figure 1. (This viscosity is incorporated into $\tau_o = \eta_o / G_o$). From eq. (2) the value for $G(t)$ at time t_o and temperature T_o is

$$G(t) = G_o \exp(-t_o/\tau_o) \tag{5}$$

At some other time and temperature, the equation is

$$G(t)' = G_o \exp(-t/\tau) \tag{6}$$

If $t/\tau = t_o/\tau_o$, or $t = t_o (\tau/\tau_o)$, then $G(t) = G(t)'$ and the moduli are equivalent. The ratio τ/τ_o is termed the "shift factor", a_T, between temperatures T_o and T. For a polymer, all the relaxation times, τ, shift by the same ratio when there is a temperature change, so that eq. (4) may be modified to show

$$G(t, T_o) = G\left[\frac{t}{a_T}, T\right] = \int G(\tau) \exp\left[\frac{-t}{\tau a_T}\right] d\tau \tag{7}$$

The principle of time-temperature superposition is based upon this finding.

For the dynamic measurements such as obtained for the materials above, it can be shown that the elastic and viscous components of the complex shear modulus G^* are given, respectively, by the generalized Maxwell model as

$$G'(\omega) = \int G(\tau) \left[\frac{\omega^2 \tau^2}{1 + \omega^2 \tau^2} \right] d\tau \tag{8}$$

and

$$G''(\omega) = \int G(\tau) \left[\frac{\omega \tau}{1 + \omega^2 \tau^2} \right] d\tau \tag{9}$$

where ω is the frequency of sinusoidal oscillation in radians per second. Here again, the effect of a temperature change is to multiply τ by the factor a_T. Because of the forms of eqs. (8) and (9), this is equivalent to multiplying ω by a_T. By analogy to eq. (10),

$$G'(\omega_o, T_o) = G'(\omega a_T, T) \tag{10}$$

The shift factors, a_T, are determined empirically by manipulation of isothermal data as shown in Figure 2. The curves represent isothermal measurements of log $G'(\omega)$ as a function of log ω (which could just as easily have been log $G(t)$ as a function of log time). Choosing a temperature T_o as a reference temperature, the other curves are shifted along the frequency axis to superimpose on the reference temperature data, forming a smooth curve over an extended frequency range (a "master curve") as shown in the bottom half of the figure. The shift factor, log a_T, for each set of data is found by the distance on the frequency axis the set is moved relative to the reference set.

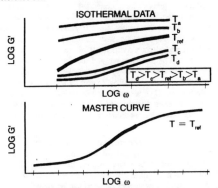

FIGURE 2. SHIFTING ISOTHERMAL MODULUS-FREQUENCY DATA TO FORM A MASTER CURVE OF MODULUS vs REDUCED FREQUENCY

Having obtained the dynamic modulus as a function of frequency by this procedure, the data may be converted to the often more useful time dependent, or relaxation, modulus by the approximation[2][3]

$$G(t) = G(\omega) - 0.40 \, G''(0.40 \, \omega) + 0.014 \, G''(10 \, \omega) \text{ where } \omega = 1/t \tag{11}$$

or to other viscoelastic properties via similar relationships.

MEASUREMENT TECHNIQUES

A number of instruments exist for determining the mechanical response of a material through a range of oscillatory frequencies and temperatures. In this work we utilize two Rheometrics™ units, the RMS 7700 and the RDS II dynamic spectrometers, both capable of a frequency range of 0.01 to 100 per second. Materials may be characterized in shear via torsional bar or parallel plate measurement or in tension if they are available in film form. A typical characterization procedure is as follows.

At 25°C, the elastic component G' of the dynamic shear modulus G^* and the phase angle δ between the elastic and viscous responses are measured directly at 16 oscillatory frequencies between 0.1 and 100 per second, inclusive. The viscous component G'' is calculated from the relationship

$$\tan \delta = G'' / G'$$

for each measurement point. The measurements are repeated through a range of lower temperatures down to about 50°C below the glass transition temperature of the material and then through a range of higher temperatures up to the limit of the instrument or until the material becomes fluid-like in behavior. One may continue measurements on the material in the fluid state if the parallel plate or cone and plate configurations are used.

As described above, the isothermal plots of log(modulus) versus log(frequency) can be overlaid and shifted horizontally by hand to obtain the shift factors for the frequency axes of all of them relative to a reference temperature, usually 25°C. We have found it expedient to utilize a computer-based system to construct the master curves for the mechanical characteristics, store them in a database as mathematical descriptions, and calculate properties such as creep and stress relaxation as required.

COMPUTER ANALYSIS

The data from the Rheometrics™ instruments may be placed in digitized form directly on floppy disk via interactive programs run through desk-top computers. We built a system based on a desk-top computer with which the isothermal modulus-frequency data, converted to ASCII format, may be represented graphically on-screen and shifted to obtain the temperature dependent shift factors $\log a_T$. A very small vertical shift is often necessary to account for rubber elasticity and microscopically confined regions of yielding in the specimen. The program applies the shift factors to the isotherms, combining them to construct a master curve of the elastic and viscous components of the modulus versus reduced frequency for the polymer, as illustrated in Figure 2. The master curve and the shift factor versus temperature curves are fit to 14th order and 7th order polynomials, respectively, the coefficients of which are stored permanently on a hard drive. Utilizing the relationships between the time, frequency and temperature dependent forms of the mechanical properties, such as given in eq. (11), we are able to produce from the polynomial descriptions of the master curve and shift factors other, more directly useful properties such as the relaxation of stress under constant strain (the time-dependent or relaxation modulus) and the temperature dependent creep behavior of the material.

At present we have about 500 materials in our database, cross-referenced under 5 major descriptors for locating materials for specific applications. This sort of availability of data has been invaluable. For instance, if a class of polymer is chosen for an important property like processability, but the particular resin first tried proved unsatisfactory in some mechanical criterion, other grades are on the database so that those candidates that are superior in the mechanical properties required can be identified immediately.

Examples of the usefulness of these predictive tools are discussed below.

APPLICATIONS

AT&T Bell Laboratories designed a small communication wire connector for joining small pair count outside plant telephone cables. The connector contained two phosphor-bronze contact elements with characteristics to splice a full pair of wires together firmly. The body and cap of the connector were molded from polycarbonate. In the design, each phosphor-bronze contact element is cold-staked into the body of the connector utilizing a concave, punch-like mandrel to shear and cold-form the polycarbonate to fill the slot over the center bar of the contacts as shown in Figure 3. Cold-staking was the method of choice for economic reasons. The nature of this operation left a diametric seam line in the staked plastic, which was the focus of attention for proving the reliability of the structure.

Cold-forming the polycarbonate left frozen-in stresses which could affect the dimensional stability of the part. For the connector, the question was one of survival through warehouse conditions without deformation that would hinder the alignment of the contacts. (Once the connector is employed, the interaction between the body and cap hold the contacts firmly, and the stake is no longer of concern.) Hot storage conditions might allow an unacceptable degree of creep-recovery in the stake, specifically in that the seam in the stake might open and let the contacts loosen.

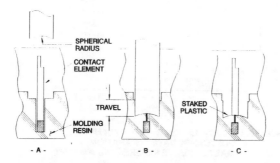

**FIGURE 3. COLD STAKING SEQUENCE FOR SMALL PAIR
COUNT CONNECTOR**

The stored stress in the cold-stake would result in viscoelastic recovery of the plastic towards its original shape. The rate at which this occurs is temperature dependent, a dependence with is known through the polycarbonate's dynamic mechanical properties measured with the Rheometrics™ instrumentation. This means that gap separation in the stake seam could be measured as a function of time at several elevated temperatures, and the acceleration due to the high temperatures could be determined with a log a_T curve, Figure 4, from the mechanical properties master curve for the polycarbonate.

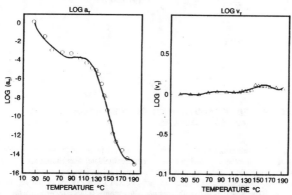

FIGURE 4. HORIZONTAL (LOG a_T) AND VERTICAL (LOG v_T) SHIFT FACTORS vs
TEMPERATURE FOR SPLICE CONNECTOR POLYCARBONATE RESIN.
REFERENCED AT 25°C.

It was conservatively estimated that the five year storage of connectors in Arizona could expose the parts to 1.5 years at 70°C,[4] and this became the benchmark against which to measure. Gap separation was determined in the seam as a function of time at 110, 130 and 140°C. At lower temperatures the separation which occurred within the 4 week period of the study was too small to be obtained with accuracy. At the same time, the force required to push the contact from its stake was obtained as a function of aging time at 40, 55, 80, 110 and 130°C. The gap separation results were shifted relative to 110°C using the shift factors in Figure 4. The reduced data, Figure 5, indicate that at 110°C the gap changes very little out to 1 year. The shift factor for 110°C relative to 70°C is about 10 (1 on the log scale in Figure 4), meaning that at 70°C the gap would remain very tiny for 10 years. In addition, force to push out the contact was found to be constant with aging temperature up through 130°C, so the cold-stake was shown to be reliable with a large safety factor even under the most extreme storage conditions.

208

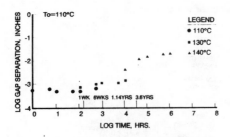

FIGURE 5.   COLD-STAKED CONTACT ELEMENT,
POLYCARBONATE SEPARATION
MASTER CURVE REFERENCED AT 110°C

A second application of time-temperature superposition to forecast creep behavior was in the choice of materials for an edge card connector designed to integrate plug-in components onto PC boards. As shown in schematic in Figure 6, the connector consists of a series of copper alloy contacts placed so that the insertion

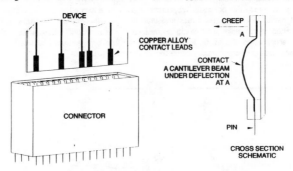

FIGURE 6.   EDGE CARD CONNECTOR DESIGN FOR MOUNTING
PC BOARD DEVICES

of an edge card, a board with conductors at proper intervals along its edge, would complete the circuits desired while leaving open all others. Filled poly(butylene terephthalate) was chosen for the body because of durability and ease of fabrication. However, the first PBT resin made into prototypes underwent creep with time at slightly elevated temperatures due to an inward-directed force exerted at the top of the connector walls by the contacts. The space between the copper alloy contacts was reduced to the extent that the system could be shorted out if there were an extended time between power-up and the insertion of an edge card terminated device. With knowledge that this PBT resin would be unsatisfactory, other PBT resins were characterized and compared with the initial one in creep compliance at 85°C, a temperature used to simulate the effects of the most extreme field conditions in Arizona. The data are shown in Figure 7. PBT A is the original material, with B and C as the preferred replacements. It had been determined experimentally that a creep compliance of 8 EE-11 (dynes/cm.sq.), or -10.1 on the log scale, was the critical value above which the material would allow the contacts to short. From the temperature dependent shift factors, it was calculated that 190 hours at 85°C are equivalent to 20 years in Arizona for PBT A, but the material's creep compliance rises above the critical value in less than 20 minutes at 85°C. PBT B and PBT C are similar in behavior. For both materials 150 hours at 85°C is equivalent to 20 years in Arizona. Based on the data in Figure 7, it is predicted that neither material would creep in the connector structure to allow shorting of the contacts through this lifetime. After considering a number of other factors including cost and availability,

PBT B was chosen for use in the connector.

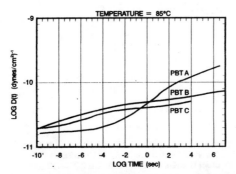

FIGURE 7. CREEP COMPLIANCE OF FILLED POLYBUTYLENE TEREPHTHATE MOLDING MATERIALS FOR EDGE CARD CONNECTOR

A last example involves the choice of molding material for an AT&T lightguide rotary splice connector, depicted in Figure 8. The considerations included very high dimensional stability throughout a service life at elevated temperature, as well as fabrication demands such as a very high degree of precision in the molded part's geometry. The first choice was a polycarbonate material. Although this resin performs excellently in ordinary field conditions, under certain specialized post-processing fabrication procedures a polycarbonate connector may deform.

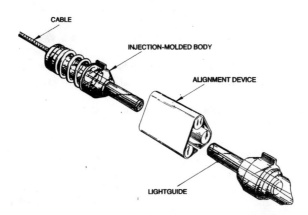

FIGURE 8. AT&T LIGHTGUIDE ROTARY SPLICE CONNECTOR.

It was found that a rotary splice might see 180°C for up to 30 minutes during the most severe of the specialized processes, so a criterion was set that an alternate connector material, while meeting the needs already enumerated, must also have limited stress relaxation in one hour at 180°C. A polysulfone and a polyethersulfone were identified which were good prospects, and their mechanical characteristics were obtained by time-temperature superposition techniques as described above. The time dependent or relaxation moduli of these materials are compared with that of the polycarbonate in Figure 9. The polyethersulfone was selected as best meeting all the requirements, particularly limited stress relaxation at 180°C, for the connector. The performance of this material in the rotary splice was excellent throughout all post-processing and aging studies.

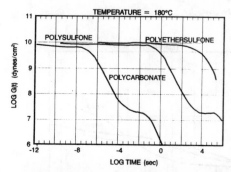

FIGURE 9. TIME DEPENDENT (RELAXATION) MODULUS FOR MOLDING RESINS CONSIDERED FOR USE IN THE AT&T LIGHTGUIDE ROTARY SPLICE CONNECTOR

SUMMARY

We have demonstrated the necessity of a complete understanding of the time and temperature dependent viscoelastic properties of polymeric materials for projecting their performance in engineering applications. We have described a computer-based data acquisition and storage system which allows rapid determinations of all important viscoelastic parameters from the polynomial expressions of the master curves and shift factors for the characterized materials, and we have shown by example the value of a full understanding of polymer mechanical properties for solving engineering problems in reliability forecasting.

References

1. I. M. Ward, *Mechanical Properties of Solid Polymers*, p. 89, (New York, Wiley, 1983) 2nd Edition.

2. J. D. Ferry, *Viscoelastic Properties of Polymers*, p.90 (New York: Wiley, 1980), 3rd Edition.

3. L. E. Nielsen, *Mechanical Properties of Polymers and Composites*, p. 157 (New York: Marcel Dekker, Inc. 1974).

4. "The National Atlas of the United States of America," United States Department of the Interior Geological Survey, Washington, D.C., 1978.

NOV/bjo.14

QUANTITATIVE NDE APPLIED TO COMPOSITES AND METALS

JOSEPH S. HEYMAN, WILLIAM P. WINFREE, F. RAYMOND PARKER, D. MICHELE
HEATH, CHRISTOPHER S. WELCH*
NASA, Langley Research Center, Nondestructive Measurement Science
Branch, Hampton, VA 23665
*The College Of William And Mary, Williamsburg, VA 23185

ABSTRACT

This paper reviews recent advances at LaRC in quantitative measurement
science applied to characterizing materials in a nondestructive environment.
Recent demands on NDE have resulted in new thrusts to achieve measurements
that represent material properties rather than indications or anomalies in a
background measurement. Good physical models must be developed of the
geometry, material properties, and the interaction of the probing energy
with the material to interpret the results quantitatively.

In this paper are presented NDE models that were used to develop
measurement technologies for characterizing the curing of a polymer system
for composite materials. The procedure uses the changes in ultrasonic
properties of the material to determine the glass transition temperature,
the degree of cure, and the cure rate. A practical application of this
technology is a closed feedback system for controlling autoclave processing
of composite materials.

An additional example is in the area of thermal NDE. Thermal diffusion
models combined with controlled thermal input/measurement have been used to
determine the thermal diffusivity of materials. These measurements are
remote, require no contact with the material under test and thus have
interesting promise for NDE applications.

INTRODUCTION

Today, the field of NDE is in a rapid state of flux. Recent advances in
electronics, computers and physical models have opened the door to a more
comprehensive professionalism that embraces a multidiscipline of sciences
and technologies. In the past, NDE was thought of as an after-the-fact
technology, used only when a part was built and ready for "inspection" .
Today, in contrast, NDE plays an important role in the research of
developing new materials, of developing process controls to evaluate and
produce precursors of new materials, of fabrication to produce structures,
of final quality assurance, and of recertifying structures for service.

The new emerging science of NDE is married to the word quantitative,
along with the acronym QNDE. Significantly, by measuring real physical
properties, NDE has advanced from a detection process to one of evaluation.
It is critical to understand the underlying physics of the observed effects.
For example, in an ultrasonic examination of a composite material, a fall-
off might be detected in the measured signal strength during a scan of a
given material. The important question is what physical property caused
that measurement and will that property affect the performance of the
material. The analysis is not trivial and involves a broad understanding of
the measurement interaction physics, materials science, and the projected
use of the material.

The measured signal drop, perhaps, is caused by an increase in attenuation of the material. Attenuation is caused by absorption (conversion to heat), by scattering (reradiation of the wave), by refraction (coherent redirection of the wave), and by mode conversion (changes in propagation mode). Most of these processes are frequency dependant which makes the analysis rather complex. Absorption, for example, can vary linearly with frequency, as a square power, or as the fourth power, to mention a few of the more likely interactions. In addition, as the acoustic intensity in the wave increases, such as at the focus of a transducer, nonlinear effects begin to play a more important role and lead to harmonic generation and the formation of shock waves both of which effect attenuation. Such nonlinear interactions can significantly alter the requirements for measuring as well as the analysis of the data.

Inaddition, there are critical situations that involve unusual attenuation-like phenomena; these situations are not uncommon and require a high level of expertise to analyse correctly. For example, the fall off in signal strength may have nothing to do with real attenuation. It can be caused by an artifact of the measurement process itself. Most transducers which convert acoustic waves into an electrical signals, are larger than a wavelength. Furthermore, most transducers are phase sensitive. Therefore, if one part of the transducer "sees" a wave of different phase than another part of that transducer, the result will be phase cancellation. This usually results in a measurement error which appears as a signal drop-off. Such phase cancellation occurs in materials of nonplaner geometry, velocity anisotropy, or inhomogenity. Unless one is measuring parallel, homogeneous isotropic solids with normal incidence of perfect plane waves, this can be serious!

That sounds like a bleak situation for ultrasonics NDE. In reality, what that means for ultrasonics, and for most other measurement energies, is that NDE requires an analysis professionalism equal to that in any of the sciences or engineering practices. Clearly, velocity variations in a sample tell a different story than scattering attenuation. If only simple measurements are taken, one cannot differentiate the two phenomena. NDE must not be a study in anomalies, but rather a scientific analysis of material properties. In short, the simple measurements/analysis are not adequate for complex situations.

The exciting future for QNDE, is to be able to identify the mechanisms that cause the measurement and to link those mechanisms to a real property change in a material. That opens the door to communications with the materials scientist, a door that has been partially closed in the past by the inability to link the measurements to physical properties.

In this paper, two areas of QNDE are reviewed that have benefited from recent advances. Those areas are ultrasonic characterization of the curing of composites and thermal NDE applied to composites and to complex geometries, such as the solid rocket motor (SRM). Each of these advances is based on physical models of the interactions; models that identified meaningful measurements that resulted in improved understanding of the phenomena.

COMPOSITES PROCESSING

An important problem facing practical use of composite materials on a broad scale is cost. Costs are driven by the raw material, labor intensive fabrication, and by poor yield. The use of QNDE can reduce costs by permitting automation of process through reliable sensors and through increased yield through process control feedback. A large composite wing section can cost in the $100,000 range before it is even cured. The part must be placed in an autoclave for curing where it undergoes a heating and pressurization schedule inside a vacuum bag to insure compaction during cure.

The autoclave process first raises the temperature causing a decrease in the viscosity of the resin permitting it to flow easily. Any excess is removed by a bleeder cloth surrounding the part. Further into the cycle, the viscosity begins to increase as the molecular length of the polymer chains increases. It is necessary to increase the pressure in the autoclave during this time to insure compaction of the composite, to reduce the amount of gas volitiles and the size of any porosity present. As the process continues, the viscosity further increases until the polymerization is complete.

Usually, the only monitoring in an autoclave is of time, temperature and pressure. Is that adequate to insure good parts? Yes, if all the starting material is uniform and known, if the part cures at the same rate over its geometry, if the part cures at the same rate for variations in thickness, and if the autoclave environment is uniform. The last factor is easily controlled. Are all the other factors controlled? Is the starting resin of the same chemistry as the sample test run weeks (or years) earlier to determine the schedule? Did the resin sit at room temperature for a period of time changing its initial degree of cure?

Some say proper monitoring is accomplished by placing thermocouples on the part itself. Such monitoring proceedures are risky in that they measure only the environment in which the part sits. It is more quantitative to measure properties of importance, in the part itself, such as viscosity or degree of cure. A sensor is needed to control the autoclave so that pressure is not applied too early in the curing cycle thus forcing out too much resin, or too late, unable to compact the material. Either error could result in an inferior part. There are physical relationships between the degree of cure and ultrasonic parameters of the resin. These are shown in the next section, to be possible inputs for feedback control of autoclaves improving the reliability, yeild, and cost effectiveness of composites.

Relationship Between Polymer Cure And Ultrasonics

The longitudinal velocity is related to the degree of cure using the principle of additive moduli. The principle of additive moduli, which relates the bulk moduli of an organic liquid to a sum of contributions from the different molecular groups, was first found for liquids by Rao[1], then expanded on by Van Krevelin[2] and has been shown to be applicable for solids[3]. The analytical expression for relating the moduli (K) to the contributions of different molecular groups is given by the expression

$$K(t) = \rho \ (\Sigma C_i \ (t) \ R_i \ / \ (\Sigma C_i \ (t) \ V_i \ )^6 \eqno(1),$$

where C_i are the concentrations, R_i is the Rao function and V_i is the molecular volume for each molecular group and the sum is over all molecular groups. From this expression, the velocity as a function of degree of cure can be shown to be[4]

$$V(t) = [\ (\ S_2 + \alpha(t) \ S_3 \ )^6 \ (\ 1 + 4/3 \ A \ ) \ -4/3 \ A \ S_2^6]^{1/2} \eqno(2),$$

where S_2 and S_3 are sums of Rao functions., A is an experimentally

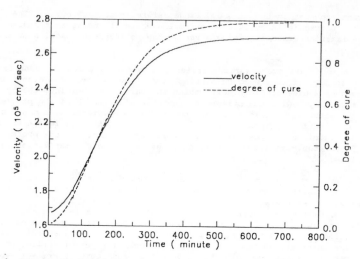

Figure 1. Longitudinal velocity measured during cure two part resin system with the degree of cure calculated from the velocity using equation (2).

determined constant, and $\alpha(t)$ is the degree of cure.

Experimental Results - Ultrasonics

The experimental configuration consists of a parallel plate cell with a 20 Mhz transducer bonded to one of the outside faces. The cell is placed in an oven to control the temperature of the resin during the cure reaction. A test material is prepared by mixing resin and curing agent and then introduced the mixture into the cell. The raw data is obtained from the digitized pulse-echo ultrasonic signals. The data is processed to improve the signal-to-noise and to remove the effects of the transducer/cell from the measurements of the propagation in the resin. The signal processing is based on a physical model of the test cell. The model adjusts the acoustic propagation parameters of the model to determine a best estimate of the velocity, the attenuation, and the frequency dependance of the attenuation. Based on a criteria of least squares error, the processing uses all the data rather than just the leading edges or peak amplitudes, as is usually done in simple time-of-flight measurements. This approach significantly improves the resolution and accuracy of the data. Velocity as a function of cure time found for a typical run and the degree of cure as calculated by equation (2) is shown by figure 1.

The time dependence of the degree of cure depends on the reaction kinetics of the system. Its time dependence can be changed by changing the temperature of the reaction or the concentration of the reactants. One of the reaction rate constants, K, has a functional dependence on the initial epoxide concentration and the temperature given by the expression

$$K = A_2 \, C_i^2 \, \exp \, (-E_g/RT) \tag{3},$$

where A_2 is a constant, C_i is the initial epoxide concentration, E_g is the

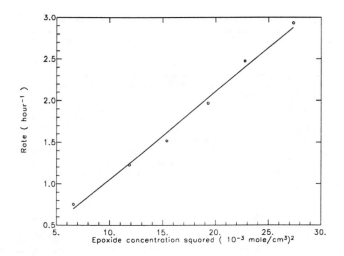

Figure 2. Dependence of the reaction rate on concentration of epoxide for epoxide-amine resin system.

activation energy, R is the universal gas constant and T is the absolute temperature. The functional dependence of K on reactant concentration and temperature can be found from the acoustically measured degree of cure and a comparison made with equation (3).

Different mixtures of the reactants are prepared to test the dependence of K on the concentration. The rates as measured acoustically for these mixtures are shown in figure 2 plotted against the initial epoxide concentration squared. A line has been drawn to facilitate visualizing the

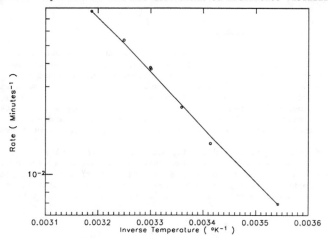

Figure 3. Reaction rate measured as a function of temperature for an epoxide-amine resin system.

linear relationship between the rate and the initial epoxide concentration squared as predicted by equation (3). As can be seen from this figure there is good agreement between the measured values of K and reaction kinetics.

The dependence of the reaction rate on temperature is also given by equation (3). The longitudinal velocity was measured at several different temperatures and, K was calculated using the method described above. For ease of comparison with theory, the log of the rate is plotted against the inverse of the absolute temperature in figure 3. With the data, a linear least squares fit of the data is plotted. As can be seen from figure 4, there is good agreement between the temperature dependence of K and the temperature dependence predicted by reaction kinetics. The slope of the linear least squares fit can also be used to calculate the activation energy of the system. We find an activation energy of 13.9 kcal/mole, which is well within the range expected for cure of aliphatic and aromatic amines/epoxy systems.

PHYSICAL MODELS AND THERMAL NDE IN COMPOSITES

Another technology that has seen rapid and dramatic changes is the field of thermal NDE. Ever since the availability of infrared cameras, it has been easy to use thermal energy to qualify physical situations. One of the early thermal demostrations was an image of an empty theater shortly after a show. Each seat stored thermal information so that one could identify which seats had been occupied during the show. Such simple applications led many researchers to explore this straight-forward technology. The easy experiments were done, and thermal NDE was placed on a shelf for appropriate applications.

However, instead of using thermal information by the seat of the pants as was done in the theater experiment, the current activity is similar to that in X-Ray at the time tomography was discovered. Physical models of heat diffusion are being developed using the heat equation, finite element analysis, and other computational tools. The models are used to invert the data and predict internal structure properties.

Experimental Results - Thermal NDE

In general, the use of thermal energy for quantitative NDE requires extremely tight contol of all the experimental variables. We assume that the sample is at a uniform initial temperature, is then exposed to an initial heating period and then is observed as a function of time during the cooling/diffusion time. The parameters that are involved in the measurement include the energy (assumed optical) focused onto the sample , the sample's absorption coeficient, the sample geometry, diffusivity, and emissivity, and the thermal properties of the medium in which the test is performed.

Simplifing assumptions are usually appropriate for specific examinations. We assume that the input heating profile is a line produced by scanning a laser beam on the sample surface. For thin plate samples [5], such that the temperature profile through the thickness is uniform after heating, one need only solve a one dimensional heat equation of the form

$$T(x,t) = (4\pi kt)^{-1/2} e^{-bt} \int_{-\infty}^{\infty} T(x',0) \exp\left(-\frac{(x - x')^2}{(4kt)}\right) dx'$$

(5)

where b is the fractional heat loss rate at the plate surface and k is the diffusivity of the plate. In the general analysis, the measured temperature profile just after the pulse, is used as the initial condition, $T(x',0)$, for the model. Usually, the temperature profile starts as a gaussian, further simplifying the problem so that

$$T(x,t) = T_0 \, e^{-x^2/a^2} \tag{6}$$

where T_0 is the peak amplitude of the initial heating profile and a is its half width. The analysis shows that if the initial heating profile is of the form of a gaussian, then the time evolution remains a gaussian simply spreading in time such that the half width follows the relationship $c^2 = a^2 + 4kt$. Thus for this case, the square of the half width is used to directly calculate the diffusivity. A comparison of the two diffusivity calculations is given in Table 1.

A specific application of this approach is to determine the fiber orientation of graphite/epoxy composites. Again a heating line is produced on the sample with a scanned laser. The diffusivity is determined as before, but this time as a function of angle with respect to the fibers in the composite sample. The diffusivity is much greater along the carbon fibers than in the matrix, resulting in a sinusoidal variation of the measured diffusivity as the heating line is rotated. Figure 4 shows the diffusivity variation as well as its offset providing quantitative vector analysis of the thermal properties of the composite material under examination.

Table 1. Comparison of in-plane diffusivity measurements made in various materials using the two analysis techniques described in the text.

	Stainless Steel	Brass	Aluminum (2024-T6)	Graphite- Epoxy 10[0]
Literature Values	0.042	0.337	0.518	N/A
General Analysis Results				
Mean Diffusivity (cgs)	0.0417	0.350	0.581	0.0326
Standard Deviation	0.0004	0.005	0.011	0.0006
Mean Loss Rate (%/s)	1.04	13.73	19.71	1.39
Standard Deviation	0.08	0.88	0.94	0.07
Chi-Square Range	1.10 −1.55	1.56 − 2.99	8.9 − 16.3	0.59 − 1.07
Gaussian Analysis Results				
Mean Diffusivity (cgs)	0.0413	0.374	0.577	0.0356
Standard Deviation	0.0005	0.004	0.014	0.0004
Mean Loss Rate (%/s)	0.72	10.31	13.91	0.92
Standard Deviation	0.04	0.46	0.55	0.04
Correlation Range	.99967 −.99975	.99865 −.99973	.99790 −.99928	.99912 −.99996

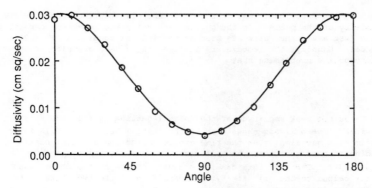

Figure 4. Comparison between measured diffusivity as a function of
sample orientation and a least-squares offset cosine curve. The sample is a
thin plate of unidirectional graphite-epoxy. The maximum and minumum values
of measured diffusivity give the principal values of the two-dimensional
diffusivity tensor, while the angle corresponding to the maximum gives the
direction of the principal axis, here aligned with the graphite fiber axis.

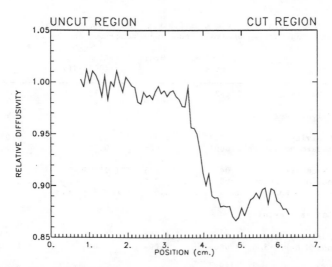

Figure 5. Indication of cut fiber damage in graphite-epoxy sample using
diffusivity measurements. In this example, the step change in underlying
diffusivity values clearly delineates the region in which fibers were
deliberately broken by cutting.scan, clearly revealing the drop in thermal
diffusion caused by the internal damage.

a   b

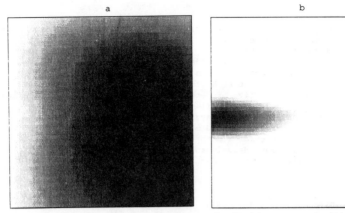

Figure 6. Improvement in contrast obtained by processing thermographic
data with the Winfree filter. a. Thermographic data averaged over 100
frames with contrast expansion. b. Same data as a following filtering and
application of threshold level.

Another application of thermal NDE, is to determine the internal damage
in a composite caused, for example, by impact damage. Since fibers provide
most of the tensile strength of these materials, it is important to
characterize broken fiber damage. A composite sample was fabricated with
internal broken fibers. The fibers were cut with a razor prior to curing so
there would be no damage to the matrix. After curing, the sample was
scanned with the diffusivity system to see if the broken fibers could be
detected. Figure 5 shows the data from that scan, clearly revealing the
drop in thermal diffusion caused by the internal damage.

An improved two-dimensional physical model has been developed which
permits delaminations in samples of Space Shuttle Solid Rocket Motors (SRM)
to be detected in thermograms even in the presence of edge cooling effects
and uneven heating [6]. The geometry of the SRM consists of an outer steel
skin, followed by a series of insulation layers and then rubber like fuel.
By considering heat flow into the insulation and fuel layer behind the steel
skin to be that of a distributed sink, a convolution filter was designed
which discriminated anomalies in the sink, associated with delaminations,
from variations in observed temperature associated with heat flow within the
steel skin. The filter design resulted from a physical model analysis of
the SRM which underscored the importance of flux imaging. This filter is
the central core in in the analysis shown in Figure 6. Figure 6a shows a
thermogram of a test sample with the Solid Rocket Motor geometry of 0.5
inches of steel painted on the face and adhesively bonded to 0.1 inch of
insulation and several inches of inert propellant. A delamination was formed
on the left side of the sample by including a triangular "pull-tab" in the
assembly and withdrawing it following the cure cycle. The thermogram was
obtained during sample cooling subsequent to applying uniform heating to the
steel surface. The temperature distribution in the thermogram is dominated
by remnants of uneven heating and uneven cooling caused by sample edge
effects. The same data are shown in Figure 6b following processing with the
filter. The dark, triangular shape prominant in this figure delineate the
edges of the delamination closely. Thus, the primary effect of the
filtering operation has been to raise the desired signal above the unwanted
noise.

CONCLUSIONS

Project managers want to see a new generation of NDE that will help them make cost effective and safe decisions   The present-day technology, as practiced, does not fully achieve that goal. Often, in a readiness review, NDE raises questions, rather than provides answers. There are exciting steps taking place in many labs today in the direction of making NDE more quantitative. A major bridge to be built in the technology, is one that links the physical nondestructive property measurements with the engineering requirements. That bridge will be built by a multidiscipline team that brings together such professions as physicists, materials scientists, chemists, computational scientists and electrical engineers. When that bridge is built, it will provide a path joining the fields of measurement science, NDE, finite element analysis and fracture / failure mechanics. It will result in a safer, more cost effective tomorrow.

REFERENCES

1. R. Rao, J. Chem. Phys. 9, 682 (1941).

2. D. W. Van Krevelen, Properties of Polymers. Correlations with Chemical Structure, (Elsevier, Amsterdam, 1972).

3. B. Hartman and G. F Lee, J. Appl. Phys. 51, 5140 (1980).

4. W. P. Winfree, F.R. Parker, Review Of Progress In Quantitative Nondestructive Evaluation, 5B, ed D. O. Thompson and D. E. Chimenti, 1055, Plenum, New York, 1986)

5. C. S. Welch, D. M. Heath and W. P. Winfree, J. Appl. Phys. 61, 895 (1987).

6. W. P. Winfree, C. S. Welch, P. H. James, and E. Cramer, 88 Review Of Progress In Quantitative Nondestructive Evaluation, to be published

DIFFICULTIES IN EVALUATING AGING IN HIGH VOLTAGE UNDERGROUND CABLES FROM IN SITU MEASUREMENTS

JEAN-PIERRE CRINE
Institut de recherche d'Hydro-Québec (IREQ), C.P. 1000, Varennes, P.Q., Canada, J0L 2P0.

ABSTRACT

The techniques used to evaluate HV cable aging during service are reviewed. It is concluded that none gives a satisfying evaluation of cable condition.

INTRODUCTION

In urban areas high-voltage lines (≥ 15 kV) tend to be more and more underground. The cables used are made of a metallic conductor insulated nowadays with polyethylene (PE) or crosslinked PE (XLPE) which are inexpensive and have excellent dielectric properties. The various cable lengths are connected together by joints which are, more than often, the weak links of the network. After installation (including joining) the cable section is electrically tested in order to detect short circuits, poor or incomplete joining and sometimes gross defects in the cable itself. Usually, this test consists in applying a high DC voltage (~ 3 times the operating voltage) for a given time duration. It is essentially a go/no go test and it is notoriously known for being unable to reliably evaluate the cable condition.

Under service conditions cables are subjected to electrical, thermal and mechanical stresses and they are often surrounded by water or moist soil. The combination of water and high electric field leads to a cable degradation process known as water treeing. This phenomenon is still poorly understood but it is known to considerably reduce the cable life. When one considers the replacement price of the cable, the cost of the cable removal and installation (including manpower and equipment) it appears that the unexpected replacement cost of prematurely failed cables is in the hundreds of million $ per year. Thus, there is definitely an economic incentive to improve HV cable life and to reduce replacement and repairs as much as possible. This could be achieved by manufacturing improved cables and also by regularly testing the condition of cables in operation. This raises the question of what are the best tests to evaluate cable aging without disconnecting or destroying cables.

To be able to determine if cables have aged and by how much, condition monitoring is necessary. The selected monitoring techniques are subjected to many constraints:
- They should be nondisruptive.
- They should be sensitive to subtle changes in insulation condition as well as to gross defects.
- The interpretation of their results should be simple and should not require high levels of training for technicians and engineers.
- They should be safe for the equipment and the operators.
- They should not be cumbersome (portable whenever it is possible) and not too expensive.

All these requirements imply that no one monitoring technique is expected to be able to completely evaluate the conditions of all cables in underground network.

The objective of this paper is to examine the various tests currently used to evaluate cable condition. It is shown that most nondestructive tests are not very informative regarding cable aging. Some recently developed techniques and methods are also discussed. However, no simple method has yet been found for performing nondestructive in situ tests of HV cable condition.

EXISTING TESTS USED TO EVALUATE CABLE CONDITION

Two types of tests are currently performed to evaluate cable aging: physico/mechanical tests (essentially destructive tests) and electrical tests (made in situ but that can be potentially destructive).

Physico/Chemical/Mechanical Tests: Density of insulation, tensile strength, elongation, oxidation content, water content, etc. These tests are made in laboratory after cable aging either in service or in laboratory (to pass acceptance tests, for example). Although their results are extremely informative on cable aging processes, they cannot be performed in situ during cable operation.

Electrical and Dielectric Tests: Such tests have been used by the electrical industry for decades [1]. One common feature of all these tests is that the change rather than the absolute value is considered significant.

DC Insulation Resistance: The cable insulation resistance (IR) between each conductor and ground is measured under a voltage of 500 or 1000 V. Many commercial instruments are available. A single reading is of little value and results should be compared with other cables or results obtained with the same cable at various time. This parameter is extremely sensitive to temperature and water content in the insulation which makes its interpretation extremely difficult. In fact, this test gives significant results only with newly installed cables. In other words, it is of limited interest for evaluating cable aging [2, 3].

Sometimes a normalization factor known as the polarization index (PI) is calculated as [2, 3]:

$$PI = \frac{IR \text{ reading at } 10 \text{ min}}{IR \text{ reading at } 1 \text{ min}}$$

It is hoped that the PI value is more or less independent of T which allows a direct comparison of results obtained at various T [2]. The interpretation is that $PI > 1$ is acceptable whereas $PI < 1$ is not (because it means a high leakage current). In any case, small local damage cannot be easily detected by this test.

DC High Voltage Test: As already stated, this test is useful after installation but it is of limited value for cable monitoring since it detects essentially short circuits and gross defects (bad joint, for example).

Impulse high voltage tests: Impulse tests are sometimes performed in order to simulate lightning. If a cable survives several impulse voltages, it is hoped that its condition is still good [4]. This is a poor indication of cable condition.

AC High Voltage Tests: Several AC high voltage tests can be performed. One consists in applying 3 times the operating voltage for a given time and to detect partial discharges. If the cable survives to high AC voltage (usually 3 times V_0 for at least 15 min) it means that it is still in good condition. If done regularly, this could be a useful method to evaluate aging. However, this is extremely time consuming and the interpretation of partial discharge signal is often awkward. This is essentially used as an acceptance test in manufacturing plant.

Another test consists of superimposing signals (with other frequency than 60 Hz) and to detect their leakage to ground. This is particularly useful for fault detection and location. This is, of course, of no value for diagnostic measurements.

Another test performed under AC voltage is to measure the capacitance and the dielectric losses (tan δ) of the cable. The measurement is prone to electrical noise and may be difficult to perform in some locations. However, recent equipment and computer-assisted techniques are now available. It should be noted that dielectric losses are sensitive to many parameters (insulation morphology, water content, oxidation, etc.) which can make the interpretation of results somewhat difficult. Recent results have shown that tan δ can be sensitive to water treeing [5].

AC breakdown tests are not useful as monitoring tests, though they may be very useful to understand cable aging. In recent years, other techniques (TDR, TDS, step voltage, etc.) have been used and they are discussed below. From this brief review, it appears that it is nearly impossible to obtain a good picture of cable aging from the currently used nondestructive tests.

NEW NONDESTRUCTIVE TESTS TO EVALUATE CABLE CONDITION

Partial Discharges Detection Tests:

Although the detection of partial discharges in dielectrics is nothing new, it has gained recently a new impetus with the advent of computer-assisted techniques that can make the interpretation of results somewhat simpler [6-8]. Cable insulations and joints contain defects which may act as the sites of partial electrical discharges in presence of high electric stress. These discharges will further accelerate the local degradation of the insulation which eventually leads to premature cable failure. It is therefore important to be able to locate the discharge site and also to determine its origin. It seems possible to precisely locate pinholes and some protrusions in cables with modern equipment [6-8]. There are several commercial partial discharge detectors and the major difficulty with this technique is the noise   (especially for measurements in the field). Another difficulty is the interpretation of   signals and much work remains to be done before this issue could be resolved. Nevertheless, this test has a promising potential for monitoring cable condition.

Detection of voids by microwaves:

Voids and small defects in cable insulation could possibly be detected by the microwave cavity perturbation technique. However, laboratory tests have shown that the sensitivity of this technique is too low for the detection of small defects in cables [9].

TDR Tests

Time-domain reflectometry (TDR) consists in measuring the impedance of a transmission line (i.e. the cable) under high frequency (MHz to GHz) [3, 10]. Reflections caused by discontinuities in the cable (voids, protrusions, end of line) affect the impedance and the location of the discontinuity can be precisely estimated from the velocity of wave propagation. This is a nondestructive technique that can be used in situ since it requires low-voltage pulses that can be superimposed on the power frequency voltage. The primary disadvantage is the lack of discrimination between impedance discontinuities caused by artifacts (joints, for example) and significant damage due to cable aging. Another difficulty is the noise level that can severely impede the sensitivity of the technique. We have recently shown that high water contents (> 500 ppm) in XLPE insulation could be detected by TDR [10]. It is not yet clear if the presence of water will significantly affect the signal associated with other defects. For TDR to be useful, it will be necessary to perform differential measurements, that is to make measurements over a long period of time and compare the results with these obtained in the previous measurement. A commercial instrument has been recently developed [11] but it remains to be seen whether the technique is really useful for cable monitoring or not.

TDS Tests

In time-domain spectrometry (TDS) one measures the response (in the time-domain) of a dielectric subjected to a step-voltage excitation [3, 12]. The response is measured over long period durations ranging from 10 μsec up to 3000 sec. After conversion of the time-domain signals to frequency-domain, the response can then be expressed in equivalent frequency range. The technique is extremely sensitive, and is especially useful in the very low frequency (long time) range of $10^{-3} - 10^{0}$ Hz. It is in this range that interfacial effects (at protrusions or voids) can be detected. Only limited data is available but results appear to be promising. However, TDS has several drawbacks: cable length will limit the maximum frequency obtainable (but only low frequencies are interesting so this is not a major problem); the cable must be well shielded (to exclude the environment) and it should be disconnected from any parallel load; finally, only average quantities distributed all over the cable length are measured.

Other New Tests

A method recently proposed consists of applying a step DC voltage and to record the time evolution of the transient current [13]. The relaxation time of this current should be somehow associated with cable aging, especially water treeing. The great difficulty in this experiment is to differentiate water outside trees from water in water trees. Apparently this can be done but the technique is still in its infancy and much work remains to be done before it would have some practical application.

An instrument called "Indenter", currently tested by EPRI, seems to give indentation values (i.e. something related to the material's tensile strength) that are related to cable aging [14]. More tests are needed but this portable instrument may have some interest as a field-tester of the mechanical properties of cables in service.

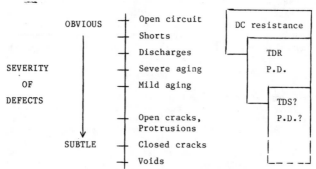

Figure 1: Summary of cable defects and nondestructive tests (from ref. [3]).

CONCLUSIONS

This brief review of existing or newly developed nondestructive tests for evaluating cable aging indicates that none is really satisfactory. Although short circuits or gross defects can be sometimes detected, subtle defects and mild aging are not yet easily detectable, as shown in Figure 1. However, TDS, TDR and partial discharge detection (with computer-assisted signal processing) are interesting candidates for future developments. More research and development is obviously needed to improve the reliability of nondestructive testing of underground high voltage cables.

REFERENCES

[1] Details can be found in "Proc. of Workshop on Power Plant Cable Condition Monitoring," EPRI Report EL/NP/CS-5914-SR, Palo Alto, July 1988.
[2] P.H. Reynolds, in Ref. 1, p. 18-1.
[3] F.D. Martzloff, in Ref. 1, p. 25-1.
[4] R.A. Hartlein, V.S. Harper and H.W. Ng, IEEE PES Summer meeting, paper 88-SM 516-7 (1988).
[5] M.C. Michel and J.C. Bobo, J. Appl. Polym. Sci. 35, 581 (1979).
[6] M.S. Mashikian, in Ref. 1, p. 22-1.
[7] G.C. Stone and S.A. Boggs, IEEE Trans. Elec. Insul. 17, 143 (1982).
[8] W.L. Weeks and J.P. Steiner, IEEE Trans. PAS 104, No. 7 (1982).
[9] "Non Destructive Evaluation of XLPE Cable by Microwave Cavity Perturbation," Can. Elec. Ass. Report 185D354, Montréal (1987).
[10] T.K. Bose, M. Merabet, J.-P. Crine and S. Pélissou, IEEE Trans. Elec. Insul. 23, 319 (1988).
[11] R.D. Meininger, in Ref. 1, p. 19-1.
[12] F.I. Mopsik, Rev. Sci. Instr. 55, 79 (1984); IEEE Trans. Elec. Insul. 20, 319 (1985).
[13] E. Evers and M.G. Kranz, Proc. 2nd ICPADM, p. 955, Beijing (1988).
[14] J.B. Gardner and T.A. Shook, in Ref. 1, p. 20.1.

DYNAMIC CURE AND DIFFUSION MONITORING IN THIN ENCAPSULANT FILMS

David R. Day and David D. Shepard
Micromet Instruments, Inc.
26 Landsdowne St., Suite 150
Cambridge, MA 02139

ABSTRACT

Recent developments in the area of microelectronics now enable the fabrication of microdielectric sensors that can follow the drying, curing, and diffusion phenomena in thin films. This paper first investigates the response of the microdielectric sensor to cure in a thin epoxy film. The multi-frequency loss factor data are reduced to a single "ion-viscosity" response curve representative of the change in ion mobility during the cure reaction. The dynamic dielectric data are compared to Tg as a function of time as determined by differential scanning calorimetry (DSC) on several partially cured samples. The dielectric response is shown to be extremely sensitive to the entire cure process. A similar epoxy film is exposed to alternately wet and dry environments while the dielectric response is monitored. The diffusion coefficient of water in the cured resin is estimated through the dynamic change in permittivity using a simple Fickian model and the results are shown to be in good agreement with the literature values. Finally, changes in dielectric properties of a photoresist during UV exposure are presented.

INTRODUCTION

Standard analytical techniques are often difficult or impossible to apply in measuring the properties of thin films. Recent developments in the area of microelectronics now enable the fabrication of microdielectric sensors that can follow drying, curing, and diffusion phenomena in thin films through monitoring the dielectric quantities of permittivity and loss factor. By analyzing the multi-frequency loss factor response, the ionic conductivity can be extracted and related to viscosity and Tg during a curing reaction[1]. After cure, the permittivity of a material may be measured to determine not only if the material is cured to the proper state, but also to dynamically monitor the influence of diffusing substances such as water vapor. This work investigates the dielectric response through all these stages with the cure of an epoxy resin and the subsequent exposure to moisture. A final investigation into the dynamic dielectric response of a UV initiated solid state reaction in a photoresist is also discussed.

EXPERIMENTAL

Thin films were applied to microdielectric sensors (Figure 1) by placing either a small drop of epoxy on the active area (2 x 3 millimeters) or by lightly pressing a small sheet of photo resist film onto the active area. Since the microdielectric sensors measure only the first 12 microns of material, all films used were at least that thickness. The dielectric response was measured from 0.1 to 10,000 Hz with a Micromet Instruments, Inc. Eumetric System II Microdielectrometer.

The epoxy studied consisted of stoichiometric quantities of EPON 828 (diglycidylether of bisphenol-A, DGEBA) and diamino diphenyl sulfone (DDS). The components were heated and mixed until complete dissolution occurred. The mixture was then cooled and stored in a sealed container in a freezer until used. Epoxy samples were placed on microdielectric sensors and in DSC pans just before use. During isothermal cure at 200°C, the DSC pans were removed at various times while the dielectric response was continuously monitored. A similar procedure was used during non- isothermal cure. A DuPont DSC 9900 was used to determine the glass transitions of the various cured samples.

Moisture influence was monitored by arranging sensors with cured epoxy films in a near 100% humidity chamber at ambient temperature and subsequently in a dry (silica gel) chamber. The dielectric response was monitored until equilibrium was attained for both moisture uptake and drying.

The photoresist material was placed on a sensor under a 254 nm mercury vapor light. The UV light was switched on and off as dielectric properties were continuously monitored.

RESULTS AND DISCUSSION

Reaction Monitoring

During the cure of the thermoset, two dielectric properties, the permittivity (e'), and the loss factor (e") were measured. The ionic conductivity or its inverse, ionic resistivity, can be extracted from multi-

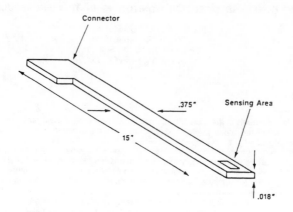

FIGURE 1. Schematic diagram of microdielectric sensor used for monitoring response from .1 to 10,000 Hz.

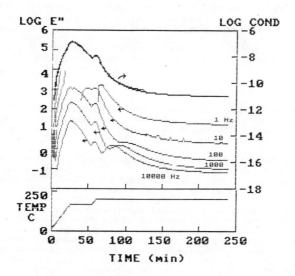

FIGURE 2. Dielectric loss factor and extracted conductivity (Siemens/cm) during non-isothermal cure of an epoxy.

frequency dielectric loss factor data (Figure 2)[2]. This single frequency independent curve is useful for analyzing property changes during cure. Figure 3 shows the increase in resistivity of the DGEBA/DDS system during isothermal cure at 200°C. Also shown is the change in glass transition temperature during isothermal cure as determined by differential scanning calorimetry. If a cross plot is constructed of Log Resistivity vs Tg a near linear relationship is observed for this material until the Tg approaches the cure temperature to within a few degrees (Figure 4). Using this relationship it is possible to predict Tg during non-isothermal cure[3]. In order to predict Tg values based on changes in conductivity, the Log (cond)-Tg relationship is assumed to be linear and will subsequently be referred to as the "linear Tg model." Using this model, the Tg during cure may be determined simply by:

$$Tg = \frac{Log\ (Cond) - 0\%\ Log\ (Cond)}{100\%\ Log\ (Cond) - 0\%\ Log\ (Cond)} * (Tg\ 100\% - Tg\ 0\%) + Tg\ 0\%$$

[1]

In order to use Equation 1, the Tg must first be measured at 0% and 100% cure. In addition, the Log (Cond) at 0% and 100% cure must be known. If non-isothermal cures are run, then the Log (Cond) at 0% and 100% cure must be known as a function of temperature. These data are determined from the heat-up and cool-down data from a cure at 200°C shown in Figure 5. The temperature dependence of the 0% and 100% Log (Cond) is represented by the linear functions indicated below and shown by the dotted lines in Figure 5:

$$0\%\ Log\ (Cond) = 0.016 * T - 9.5$$
$$100\%\ Log\ (Cond) = 0.053 * T - 21.4$$

[2]

The linear approximation of the Log (Cond) temperature dependence does not fit well for this material at temperatures below 100°C. However, this is not critical since significant reaction does not occur until the temperature exceeds 100°C.

By combining equations 1 and 2, the Tg may be calculated from the measured Log (Cond) and temperature during non-isothermal cure. Figure 6 shows the result of using the linear-Tg model on the same data from Figure 2. This method yields calculated Tg values that are in good agreement with Tg values measured by DSC (solid circles in Figure 6). It should be noted that this model requires a resin that does not change ionic concentration level during cure, that has only one dominating reaction mechanism, and that does not have a conductivity that is influenced by loss of volatiles at the temperature of interest. While many resins fit these criteria, many other resins, such as polyamic-acids and phenolics do not. However, even if the linear Tg model cannot be applied, the dielectric data still provide valuable information about the dynamics of the curing process.

Diffusion Monitoring

After completion of reaction, the epoxy coated sensors were allowed to equilibrate at room temperature. The sample was then placed into a 100% humidity environment. The dielectric response was monitored over the course of many hours as the moisture was taken up by the film. The change in dielectric constant measured at 10,000 Hz is shown in Figure 7. These data show the dielectric constant slowly increasing and leveling in time, due to the increasing concentration and saturation with water. At low levels of moisture uptake, the change in dielectric constant is often linearly related to the moisture concentration[4]. Using Fick's law of diffusion, the following equation may be derived[5]:

$$\frac{C - C0}{C1 - C0} = 1 - \frac{4}{\pi} * \sum_{n=0}^{\infty} \frac{(-1)^h}{2n + 1}\ EXP\frac{-D\ (2n+1)^2\ \pi^2\ t}{4L^2} * COS\frac{(2n + 1)\ \pi\ X}{2L}$$

[3]

where:
    C = concentration at distance x into the film
    C1 = concentration at surface (saturated)
    C0 = initial uniform concentration (zero)
    D = diffusion coefficient
    t = elapsed time
    L = film thickness

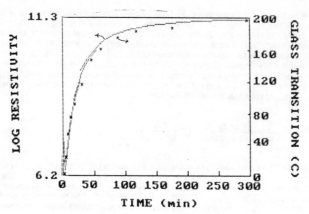

FIGURE 3. Tg (measured by DSC) and Log(resistivity) (cm/Siemen) during isothermal cure.

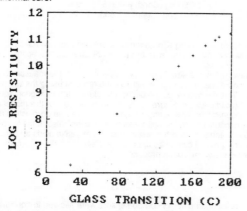

FIGURE 4. Cross plot of Tg and Log(resistivity) (cm/Siemen) during isothermal cure (data from Figure 3).

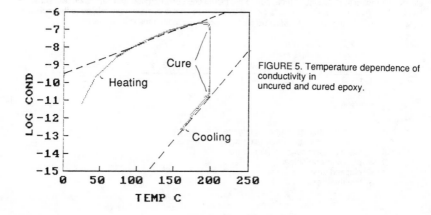

FIGURE 5. Temperature dependence of conductivity in uncured and cured epoxy.

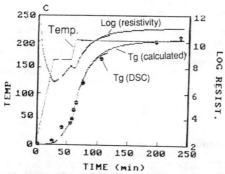

FIGURE 6. Actual Tg and calculated Tg during non-isothermal cure using linear-Tg model (data from Figure 2).

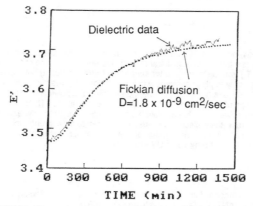

FIGURE 7. Change in dielectric constant (10,000 Hz) of epoxy coating during moisture uptake. Dotted line is best fit Fickian diffusion curve where D=1.8x10 cm/sec.

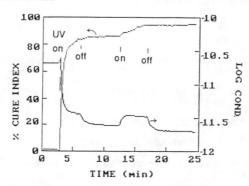

FIGURE 8. Change in conductivity and temperature corrected "dielectric cure index" during intermittent UV exposure to a photoresist.

X = distance from back of surface of film
(n need only be carried to about 20 for convergence of summation)

Using the above equation, the moisture level can be calculated at the back surface of a film (at the sensor/film interface) as a function of time if the Diffusion coefficient and the film thickness is known. Since the diffusion coefficient is not known, several values were tried until the calculated "Fickian" curve best matched the observed change in dielectric constant. The dotted line in Figure 7 is a best fit Fickian diffusion curve for the concentration at the back surface of a thin infinite sheet. The diffusion constant determined by this fit is equal to 1.8x10 cm /sec. This compares to a literature value of about 1.6x10 cm/sec measured by weight uptake[6].

Photoreaction

Figure 8 show the specific conductivity (1/specific resistivity) monitored during the exposure of a photoresist to UV radiation. The response is influenced both by the reaction as well as temperature fluctuations caused by the UV lamp turning on and off. A recently developed technique[7] was used to reduce the data to the dielectric cure index which is temperature independent. The resulting data show a monotonic increase in the degree of cure. After the first exposure, reaction is observed to continue for about 60 seconds. During the second exposure a small degree of further reaction is observed. Using these data it is relatively easy to determine the reactivity and necessary exposure time required for this material.

CONCLUSIONS

A relationship between the glass transition temperature and dielectric response was demonstrated during an epoxy curing reaction. The near linear relationship enabled the use of the "linear Tg" model to dynamically determine Tg from dielectric and temperature information during non-isothermal cure. Although this model may only apply to systems with a single dominant reaction mechanism, the dielectric response is sensitive to all types of reactions including silicones and polyimides. The dielectric constant was shown to be a function of moisture diffusion in an epoxy encapsulant. The rate of change in dielectric constant over several hours was used to calculate a diffusion constant for the material that agreed well with other published data. Finally, the dielectric response was shown to be a strong function of a solid state photoreaction in photoresist. After removing the temperature influence, the photoreaction was observed to continue several seconds after removal of UV excitation.

REFERENCES

1A. S.A. Bidstrup, N.F. Sheppard Jr., S.D. Senturia, Proceedings of the Society of Plastics Engineers Annual Tech. Conf. (ANTEC,45th), May 4th, pg. 987 (1987)

1B. J.Gotro & M.Yandrasits, Proceedings of the Society of Plastics Engineers Annual Tech. Conf. (ANTEC,45th), May 4th, pg. 1039 (1987)

1C. J.W. Lane, J.C. Seferis, & M.A. Bachmann, Polym. Eng. Sci., 26(5), pg.346 (1986)

2. D.R. Day, Poly. Eng. Sci., 26, 5, 362(1986)

3. D.R. Day and D.S. Sheppard, Proceeding of the 43rd Annual SPI/Composites Conference (Cincinnati), Feb. 1 (1988)

4. D.D.Denton, J.Kamou, & S.D.Senturia, Proceedings of the 1985 Int. Symp. on Moisture & Humidity, Wash.D.C., April 15, pg.505 (1985)

5. J. Crank,Mathematics of Diffusion, Oxford Univ. Press (1956)

6. J.B. Enns, J.K. Gillham, J.Appl.Polym.Sci., 28, 2831(1983)

7. D.R.Day, Proceedings of the Society of Plastics Engineers Annual Technical Conference (ANTEC), May 4th, pg.1043 (1987)

GEL POINT DETERMINATION IN CURING POLYDIMETHYLSILOXANE
POLYMER NETWORKS BY LONGITUDINAL ULTRASONIC WAVES

A. SHEFER*, G. GORODETSKY**, AND M. GOTTLIEB*
*Chemical Engineering Department, Ben Gurion University, Beer
Sheva 84105, Israel
**Physics Department, Ben Gurion University, Beer Sheva 84105,
Israel

ABSTRACT

The formation of well characterized poly-dimethylsiloxane networks has been studied by means of an acoustic
interferometer. The hydrosilation reaction used to form these
networks was followed by infra red spectroscopy. The relative
changes in velocity of the ultrasonic longitudinal waves
propagating through the system are found to be very sensitive
to gelation and to the density of crosslinks. Due to our
simultaneous kinetic and ultrasonic studies it was possible to
relate the changes in acoustic properties of the curing system
directly to the cure state.

INTRODUCTION

Polymer chains may be linked together to form intricate
three dimensional network structures [1,2]. At some point
during the process leading to the formation of the network the
material undergoes a well defined transition from a liquid
(melt) state into a solid or solid-like (gel) state. This
critical transition point occurs at the so called gel point
and the material at this point is commonly referred to as the
critical gel. The formation of these network structures is of
great practical and theoretical interest. Yet, many of the
aspects related to the structure-property relationship and its
dependence on the extent of crosslinking reaction are still
unclear [3,4,5,6]. The determination of the exact state of
cure, i.e. the state of the system relative to the gel point
and the extent of chemical reaction past this point is of
considerable importance in the manufacture and processing of
elastomers and thermosetting materials. Several techniques
have been suggested for the on-line determination of the state
of cure. These include capillary flow [7], falling ball
viscometry [8,9], small amplitude low frequency oscillatory
deformation [10,11,12], and ultrasonic measurements
[13,14,15,16]. Only the Winter-Chambon [12] method seems to be
of general nature applicable to a large variety of chemical
systems. But, it raises serious questions regarding the effect
of the imposed mechanical deformations on the reaction and the
evolving topology. This problem is especially important in the
case of the fragile structures in the vicinity of the gel
point. In addition, extension of this method to on-line
testing is not a straight forward process due to the

complexity of the required equipment (dynamic mechanical spectrometer).

Ultrasonic (US) methods are attractive due to their obvious non destructive nature and relatively simple application to curing systems [17]. Many of the objections raised above regarding in-situ rheological measurements are eliminated in this case. Several workers have used US techniques to study the formation of covalent and reversible networks. Most of the work was related to curing epoxy or tight thermosetting systems [15,16]. Only a very limited number of attempts to apply US to gelation have been reported [14]. The results obtained in these experiments were inconclusive and lacked the accuracy to clearly detect the gel point. An interpretation of these measurements in terms of the evolving molecular structure is impossible since the extent of the chemical reaction in the network has not been measured.

In this work we have demonstrated the ability of US interferometry to determine the gel point of well characterized poly(dimethyl siloxane), PDMS, networks. We suggest that US can be utilized as a non interfering method for the characterization of gelation.

EXPERIMENTAL

Vinyl terminated linear PDMS was crosslinked by means of the hydrosilation reaction with four functional hydride silanes to form an endlinked network. The crosslink density of the network was determined by the size of the initial chains. The PDMS obtained from Petrarch was used in three different molecular weights: 3500, 11000, and 20000 respectively. Oligomers and low molecular contaminants were removed by vacuum stripping. The polymer was characterized by means of SEC and its functionality was checked by NMR analysis and found to be 2.0 within experimental accuracy . The crosslinker purity (tetra kis(dimethyl siloxy) silane from Petrarch) was 98% as determined by GC/MS. It was used as obtained from the manufacturer. The reaction was carried out at 298K in the presence of a platinum based catalyst [10,18]. The extent of reaction was determined by monitoring the disappearance of the silane groups by means of infrared spectroscopy at the 2100 (1/cm) wavelength. The extent of side reactions under these conditions was estimated to be negligible [18].

The network formation reactions took place within the thermostatically controlled sample cell of an acoustic interferometer [19] composed of two X-cut quartz transducers mounted at the ends of two quartz delay rods. The rods were inserted into the two opposing ends of the tube like cell which serves to maintain the alignment of the rods and to hold the liquid reaction mixture. The transducers are used for the transmission and detection of the 10 MHz longitudinal waves within the tested material. The output signal from the sample cell, was subsequently heterodyned with a 40MHz RF signal to improve the signal to noise ratio. After amplification, the sample signal was compared to the original signal and the phase angle changes, linearly proportional to the changes in wave velocity, were monitored.

RESULTS AND DISCUSSION

The changes in the velocity of longitudinal waves traversing a reacting PDMS sample have been obtained as function of extent of reaction for three different crosslink densities: 78, 28, and 17.2 moles of crosslinks/meter3 corresponding to molecular weights of the starting material of 3500, 11000, and 20000 Daltons respectively. The results are shown in Fig. 1 in which the relative changes in wave velocity (velocity change reduced by the velocity in the sample before the onset of the reaction) are plotted against conversion of the reaction.

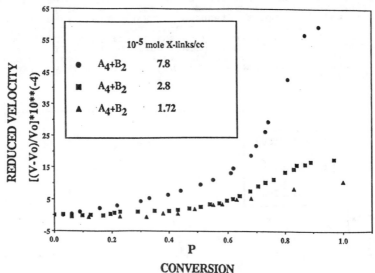

Fig.1   Relative change in longitudinal wave velocity as function of conversion of the crosslinking reaction in endlinked PDMS networks. Three different crosslink densities are shown.

Two distinctive features emerge from this figure: 1. The wave velocity is initially constant and at some conversion it rises quite sharply to attain a new value which is retained till the end of the reaction. 2. The magnitude of change in velocity and subsequently the final velocity value is characteristic of the network structure as reflected by its dependence on the crosslink density. The higher the concentration of crosslinks in the network the larger the change in velocity.

The conversion at which the liquid-gel transition occurs may be predicted from the details of the chemical system by means of a statistical branching theory. We have computed the theoretical gel points for our systems using the Miller-Macosko [20] theory. When the data in Fig. 1 are plotted in terms of the normalized velocity change (the change in velocity at a given conversion, p, relative to the change in velocity for p=1) as function of the relative distance from the computed gel point the three curves collapse into one universal curve as seen in Fig. 2. This universal curve is characterized by a sharp step-like increase in normalized velocity change in the vicinity of the gel point. Furthermore there is a good agreement between the mid-point of the velocity change (normalized velocity = 0.5) and the theoretical gel point (reduced conversion = 0). We consider this as an indication that the gel point of a curing system and the state of cure may be characterized by this type of experiments with the gel point defined by half the total change in wave velocity. Further work is still needed in order to generalize the proposed technique to chemically different curing systems.

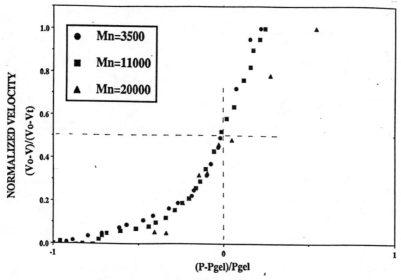

Fig.2 Normalized longitudinal wave velocity change as function of distance from the theoretical gel point for three crosslink densities in endlinked PDMS.

REFERENCES

1. P.J. Flory, Principles of Polymer Chemistry , (Cornell Univ. Press, Ithaca, NY, 1953).

2. L.R.G. Treolar, The Physics of Rubber Elasticity , 3rd ed., (Clarendon Press, Oxford, 1975)

3. J.E. Mark, Adv. Polym. Sci. 44, 1 (1982).

4. B.E. Eichinger, Ann. Rev. Phys. Chem. 34, 359 (1983).

5. S.F. Edwards and T.A. Vilgis, Reports on Prog. Phys., in press (1988).

6. O. Kramer, Biological and Synthetic Networks , (Elsevier, London, 1988).

7. S. Lipshitz and C.W. Macosko, Polym. Eng. Sci. 16, 803 (1976).

8. M. Adam, M. Delsanti, D. Durand, G. Hild and J.P. Munch, Pure Appl. Chem. 53, 1489 (1981).

9. M. Adam, M. Delsanti, and D. Durand, Macromolecules 18, 2285 (1985).

10. E.M. Valles and C.W. Macosko, Macromolecules 12, 521 (1979).

11. R.J. Farris and C. Lee, Polym. Eng. Sci. 23, 586 (1983).

12. H.H. Winter and F. Chambon, J. Rheol. 30, 367 (1986).

13. J.C. Bacri and R. Rajaonarison, J. de Phys. Lett. 40, L-5 (1979) .

14. J. Dumas and J.C. Bacri, J. de Phys. Lett. 41, L-279, L-282 (1980).

15. B. Hartmann and G. Lee, J. Polym. Sci. Polym. Phys. Ed. 20, 1269 (1982).

16. D.L. Hunston, Rev. Prog. Quant. Non Destructive Eval. 2B, 1711 (1983).

17. B. Hartmann, in Methods in Experimental Physics Vol. 16, edited by R.A. Fava (Academic Press, NY, 1980).

18. A. Fisher and M. Gottlieb, Proc. of Networks 86, Elsinore Denmark, 1986.

19. W.P. Mason, Physical Acoustics - Principles and Methods, vol. 1- Part A, (Academic Press, N.Y., 1964).

20. D.R. Miller and C.W. Macosko, Macromolecules 9, 199 (1976); 9, 206 (1976).

APPLICATION OF ULTRASONIC TECHNIQUES TO STUDY THE PROPERTIES OF SILICONE ELASTOMERS IN HIGH PRESSURE GAS MEDIA

BRIAN J. BRISCOE AND SALMAN ZAKARIA
Department of Chemical Engineering and Technology, Imperial College of Science Technology and Medicine, London SW7, England.

ABSTRACT

Polymers absorb large quantities of gas under high pressure. This process is usually accompanied by a significant volumetric strain and also by changes in the mechanical properties of the sample. We will describe how these properties may be studied by using an ultrasonic technique. The sorption and volumetric changes are progressive in nature following the application of an incremental increase or decrease in ambient pneumatic stress. These techniques are also used to follow the time dependent changes in both the linear strain and the mechanical properties of the system. Data will be cited for a range of nitrogen gas pressures up to 27 MPa at 20°C for a virgin Poly-dimethylsiloxane (PDMS) elastomer and for its composite with glass filler to exemplify the experimental and analytical procedures. The final section of the paper shows how the same techniques may be used to monitor the inception and progressive development of internal cracks during certain ambient gas decompression profiles.

The information obtained provides a basis for better material selection and for the specification of the service life of particular materials in hostile environments.

INTRODUCTION

The high elasticity of the elastomers make them suitable materials for many high pressure service environments. When they are exposed to gases at high pressures significant amounts of the gases are imbibed. The effect of absorbed gas on the physical and mechanical properties of elastomers is believed to be strongly dependent upon the type of the gas and the elastomer. Commonly, dilation of elastomers under pressure [1,2] is accompanied by a modulus change. Subsequent rapid release of the ambient gas pressure often has a detrimental effect on the elastomer. The elastomer may develop internal fractures and surface blisters in these conditions [1,2,3,4]. The absorbed gas within the elastomer is at a high pressure and escapes slowly. In such conditions the elastomeric specimen behaves very much like a pressure vessel with a high internal gas pressure which exerts negative hydrostatic stress. The 'trapped gas' expands and the polymer fractures internally and consequently inflates.

In this article we report selected observed changes in the dimensions and mechanical properties of a silicone elastomer and its composite with a glass filler when subjected to high pressure nitrogen gas. The experimental techniques used to study these effects involve the application of ultrasonic signals.

For the sake of brevity, this paper does not provide explicit accounts of the experimental methods and the data analysis nor does it attempt a comprehensive interpretation of the data cited. The purpose of the paper is to illustrate the

Mat. Res. Soc. Symp. Proc. Vol. 142. ©1989 Materials Research Society

experimental method and outline the characteristics of the data
obtained for a particular system

PRINCIPLE OF THE EXPERIMENTAL TECHNIQUES

An ultrasonic pulse echo technique has been used to enable
the measurements of changes in physical and mechanical properties
of elastomers in elevated pneumatic pressure environments. In
this method a short duration stress pulse is transmitted through
a specimen by a piezoelectric transducer attached to the surface.
The signal traverses the length of the specimen, which is in the
form of a short rod, and is reflected back from the opposite face
of the test piece, Figure 1,a. This process continues until the
original signal is completely attenuated. The same transducer
also detects the reflected signals. The lapsed time 'T' between
two consecutive reflections is the time the signal requires to
travel through the specimen length twice [5]. The velocity of
sound is,

$$V = 2 \cdot L \ / \ T \tag{1}$$

where 'V' is the velocity of sound in $m \cdot s^{-1}$, 'L' is the specimen
length in m and 'T' is the lapsed time in sec. The velocity of
sound in the material is related to the specimen density and its
Young's modulus by the following relation,

$$V = \sqrt{(E/\rho)} \tag{2}$$

where 'E' is a Young's modulus in MPa and 'ρ' is the specimen
density in $kg \cdot m^{-3}$. The changes in the transit time of the signal
with pressure have been used to sense 'E' and the linear strain
as a function of pressure.

In addition the rapid release of the ambient pneumatic
pressure results in the formation of internal cracks in the
specimen and the transmitted ultrasonic signal is scattered from
the new interfaces; hence signal attenuation occurs.

The filled systems are naturally more dissipative and it is
difficult to transmit high frequency signals through them. For
such systems an indirect method was used in which the ultrasonic
signal does not pass through the specimen but travels through
distilled water in a container placed over a flat surface of the
specimen, Figure 1,b. This arrangement operates as a displacement
transducer. With the necessary calibrations, it can not only
monitor the dimensional changes of the test piece under gas
pressure but can also quantify the extent of specimen inflation
that accompanies rapid depressurisation. The latter effect could
not be measured when the ultrasonic signal was travelling through
the test piece. The scattering produced by the internal cracks
prevents signal recovery when numerous cracks are formed.

More details covering these experimental techniques and the
data analysis are given elsewhere [6].

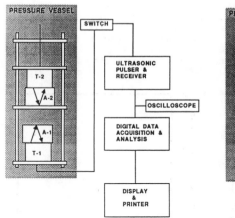 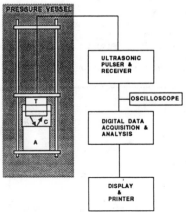

(a)   (b)

Figure 1,a: Experimental arrange-   Figure 1,b: Experimental ar-
ment to determine the physical   rangement to determine the
and mechanical properties of   dimensional changes of
elastomers, T-1 & T-2 are ultra-   specimens over the whole pres-
sonic transducers, A-1 is an un-   sure cycle, T is an ultrasonic
constrained specimen and A-2 is   transducer, C is a liquid
a constrained specimen.   container and A is a specimen.

EXPERIMENTAL METHOD

The silicone elastomer used to prepare the cylindrical
specimens was Dow Corning Sylgard 184. The composite specimens
had 20% by wt glass filler. Similar materials have been used to
study the failure processes in elastomers by Gent et al [7]. The
specimens were housed in a steel pressure vessel and were
pressurised to ca. 27 MPa with oxygen free nitrogen in increments
of 3 - 6 MPa. An equilibrium time of more than 45 min was allowed
after each step for the system to reach a quasi-equilibrium
state. During these time intervals the specimen absorbs
significant amounts of gas and the system also reverts to the
ambient temperature (20°C). Adiabatic heating of the samples was
quite apparent. At the end of each pressure step ultrasonic
signals were transmitted and recorded. The time lag between two
consecutive reflections of the signal from the specimen back wall
was then estimated. Details of the data analysis are given
elsewhere [6].
Depressurisation from the high pressure state to ambient was
completed in ca. 5 min. Ultrasonic signals were continuously
transmitted during this time at intervals of 5 sec.

RESULTS AND DISCUSSION

The changes in the velocity of sound in the pure silicone
elastomer as a function of pressure are given in Figure 2. A
linear increase of ca. 12% was shown for a pressure of ca. 27
MPa. The mutation of the signal velocity in the material was a
combined effect of the changes in the specimen Young's modulus

and density. The gas uptake by the pure elastomer at high pressure leads to an increase in its density in spite of an increase in volume. The absorption of the gas at various pressures was measured using a vibrating reed mass sorption sensor [8]. Figure 3, shows the percentage change in the mass of the specimens at equilibrium conditions. A density increase of ca. 19 % was computed for virgin PDMS for a pressure of ca. 27 MPa.

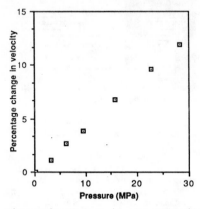

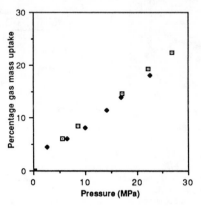

Figure 2: Change in the velocity of sound for virgin PDMS as a function of nitrogen gas pressure, 20°C.

Figure 3: Change in specimen mass on polymer basis as a function of nitrogen gas pressure, 20°C, □ ,virgin PDMS, ● filled PDMS.

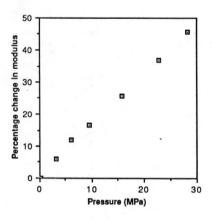

Figure 4: Increase in Young's Modulus of virgin PDMS as a function of nitrogen gas pressure, 20°C.

The modulus changes of the virgin PDMS are given in Figure 4. The increase in modulus of ca. 45% at ca. 27 MPa was considerably larger than the increase in density at the same pressure and thus an overall increase in the velocity of sound in the material was observed.

The dimensional changes for both the virgin PDMS and its composite are given in Figure 5. During initial pressurisation both the systems showed contraction. This could be in response to the contraction of microscopic air bubbles trapped in the bulk of the test piece or due to the readjustment of molecular chains induced by the plasticisation effect of absorbed gas. Further increase in the ambient pneumatic pressure and more absorption of

the gas in the elastomer causes dilation. For a maximum pressure
of 27 MPa the linear dilation in the virgin PDMS was ca. 1.4%,
while it was only 0.4 % for the filled specimen. The filler
particles constrain dilation although the amount of nitrogen gas
absorbed on the basis of elastomer mass fraction was same in both
cases, Figure 3. A similar reinforcement action was observed when
the ambient pressure was rapidly released. The inflation of
virgin PDMS specimen was significantly higher and it also started
at a much greater ambient pressure. This indicates that a lower
pressure gradient was required to inflate the unfilled specimen,
Figure 6. However, for the filled system the inflation was
reduced by more than half relative to the unfilled system (ca.
6.0%), Figure 6. Also the ambient pressure at which the specimen
inflates was portentously low. The presence of fillers in the
elastomeric specimens thus suppresses the extent of gas induced damage.

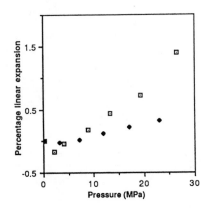

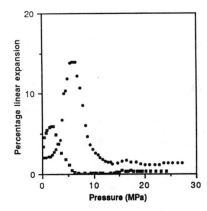

Figure 5: Linear expansion as a
function of nitrogen gas pressure,
20°C, □ virgin PDMS, ◆ filled PDMS.

Figure 6: Linear expansion as
a function of nitrogen gas
depressurisation (see text),
• virgin PDMS, ▪ filled PDMS.

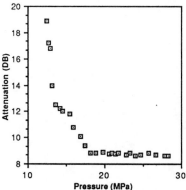

Figure 7: Attenuation of ultrasonic signal in virgin PDMS as a
function of nitrogen gas depressurisation (see text).

During depressurisation the increase in attenuation of the transmitted signal is an indication of the inception and extent of internal fracture in the specimen. It has been noted that the high frequency ultrasonic signal was not attenuated during the pressurisation stage of the experiment. However, the attenuation increase was very apparent in the pressure release stage, Figure 7. When the attenuation change was detectable the volumetric strain in the specimen was still very small ca. 4.5% compared to the strain to break the specimen when uniaxially stressed. The gas induced negative stress was triaxial in nature and the strain to break for elastomeric systems is considerably lower in this type of stress field [9].

CONCLUSIONS

1. An ultrasonic technique in which the signal travels through the specimens provides a means of evaluating unique physical and mechanical properties for an elastomer when subjected to high pneumatic pressure and changes in pneumatic pressure.
2. The modulus of the silicone elastomer increases by ca. 45.0% under a nitrogen gas pressure of ca. 27 MPa.
3. The silicone elastomer dilates linearly by ca. 1.4% in a nitrogen gas ambient at a pressure of ca. 27 MPa. The addition of 20% filler particles reduces the extent of specimen expansion to ca. 0.4% for similar gas pressures.
4. The initiation of the crack damage is a result of the application of negative hydrostatic stress and occurs at relatively low volumetric strains; ca. 4.5%.
5. The inflation of the elastomeric specimen decreases from 13.5% to only 6.0% with the addition of 20% glass filler. Filled systems can thus withstand larger pressure fluctuations in service environment.

ACKNOWLEDGEMENTS

One of the authors S. Zakaria is grateful to the Government of Pakistan and SUPARCO, Pakistan for their continuing financial support during this research work.

REFERENCES

1. Eudes et al., Nucl. Eng., July, 586 (1968).
2. D. H. Ender, Chemtech, January, 52 (1986).
3. B. J. Briscoe and D. Liatsis, "Diffusion in Polymers," P.R.I., 2nd International. conference., paper no. 17, Univ. of Reading, England, (1988).
4. A. N. Gent, D. A. Tompkins, J. Poly. Sci. A-2, 7, 1483, 1969.
5. A. Vary, in 'Research Techniques in Nondestructive Testing', Vol. IV, ed. R.S. Sharrpe, Academic press, London, 1980, pp. 159-204.
6. B. J. Briscoe and S. Zakaria, submitted to J. Phys. E.
7. A. N. Gent and B. Park, J. Mat. Sci., 19, 1947, 1984.
8. B. J. Briscoe and H. Mahgerefteh, J. Phys. E., 17, 1701, 1984.
9. G. H. Lindsey, J. Appl. Phys., 38(12),4843, 1967.

ULTRASONIC MONITORING OF POLYMER PROPERTIES FROM THE MELT TO THE SOLID STATE: INFLUENCE OF CHEMICAL/STRUCTURAL DETAILS

L. PICHÉ, G. LESSARD, F. MASSINES AND A. HAMEL

Industrial Materials Research Institute, National Research Council Canada, 75 De Mortagne Blvd, Boucherville, Québec, Canada J4B 6Y4

ABSTRACT

An ultrasonic technique is described for the simultaneous measurement of specific volume, V, sound velocity, v, and attenuation, a, at frequencies between f = 0.5 and f = 15 MHz, in a wide range of temperature (- 150 to + 400°C) and pressure (up to 2 kbars). The results (V,v,a) are translated into a complex modulus, M* = M' + iM'' and analyzed in terms of the thermodynamic state of the material. Typical results for amorphous and semi-crystalline polymers are presented which show that the technique is a probe of the fundamental features of these materials (glass transition, crystallization, melting, molecular structure) which determine process-ability and end use properties. The method should prove of great interest for quality and process control.

INTRODUCTION

The success story of modern plastics is very much indebted to an increased input from materials science. As materials, a most important feature of polymers is their viscoelastic character. Although most studies on polymer viscoelasticity are performed with low frequency (≈ Hz) rheology techniques, such methods are by no means unique, nor universal, and a different approach could be of interest, especially if its field of appli-cation were broad and if it required minimum effort to implement. In this perspective, ultrasonic techniques should constitute an interesting approach to viscoelastic polymers, as otherwise available reports indicated - for a review see [1,2]. In spite of this, the technique has seldom been used for this application. On the one hand, it can be that the diffi-culties with working on polymers using the traditional methods of ultrasonics [1,3-7] made the approach impracticable. On the other hand, rheology measurements are made at different frequencies under isothermal conditions, whereas ultrasonic experiments are performed at a constant frequency over a range of temperatures. Because polymer properties are both time and temperature dependent, the thermodynamic state of the material must be defined before attempting to correlate the low and high frequency results. This problem was addressed only rarely [6,7], which has lead to uncertainties and ambiguities in the interpretation of the results.

We investigate the adequacy of ultrasonics for the thermomechanical analysis of polymers from solid-like to liquid-like state and study its sensitivity to the morphological, structural and chemical details of the materials. First, we present our technique, that incorporates a simul-taneous measurement of specific volume as a means of characterizing the thermodynamic state of the material, then we describe results on different grades of amorphous polystyrene and semi-crystalline polyethylene.

Mat. Res. Soc. Symp. Proc. Vol. 142. ⊛1989 Materials Research Society

EXPERIMENTAL: Techniques, Samples and Procedure

Basically we used a transmission technique [2,8] whereby the sample (a disk of thickness e $\simeq 0.4$ cm) was held confined between two buffer rods. The assembly allowed axial displacement to compensate thermal expansion. Also, allowance was made for the application of a clamping pressure, controlled to ∓ 1.0 bar in the range to 2 kbars. The displacement was monitored to ∓ 1.0 μm, giving access to thickness, e, and specific volume, V. Finally, a heating/cooling device was attached to the buffers as a means to scan the temperature from -150 to $200°$C and program heating/cooling rates from $50°$C/min to $1°$C/hr, and constant temperature conditions of $\mp 0.1°$C. The setup was under full computer control, and the collection of all data concerning time, t, temperature, T, pressure, p, specific volume, V, sound velocity, v, and attenuation, a, was accomplished every 10 sec.

As examples for amorphous materials, we investigated two atactic polystyrenes of different molecular weights, Mn: PS1 with Mn(PS1) = 130,000 g/mol, and PS2 with Mn(PS2) = 9,000 g/mol. The ratio Mn(PS1)/Mn(PS2) $\simeq 14$ measures the relative chain lengths in the materials and it is recognized [9,10] that the longer chains in PS1 may entangle while Mn(PS2) stands below the critical value, Mc $\simeq 38,000$ g/mol, for these effects. For the case of semi-crystalline polymers, where ordered regions (crystallites) are dispersed in an otherwise amorphous, disordered matrix, we compare results on high and low density polyethylenes (HDPE and LDPE). The higher values of density, ρ(HDPE) $\simeq 0.964$ g/cm^3 compared to ρ(LDPE) $\simeq 0.924$ g/cm^3 (at $20°$, depending on processing), reflect the closer packing of the ordered crystalline phase. In turn, crystallinity is influenced not so much by molecular weight, Mn(HDPE) $\simeq$ Mn (LDPE), as by stereoregularity in the chain. For polyethylene, random short-chain branching correlates with lower crystallinity: typically 1 to 5 branches per 1000 monomers in HDPE and from 15 to 30 per 1000 carbons in LDPE.

The samples were annealed in the apparatus for 30 min at $190°$C. A fixed thermal path was used, so that the results could be reproduced and compared: a pressure, p = 225 bars was applied on the liquid which was cooled to $20°$C at the rate of $25°$C/min for polystyrenes and $2.0°$C/min for polyethylenes. The measurements were carried out at the same pressure with a heating rate dT/dt = $2.0°$C/min, and using longitudinal waves at f = 1.75 MHz for PS1 and PS2, and f = 2.5 MHz for HDPE and LDPE.

Viscoelastic behaviour is usually described in terms of a complex modulus, L*, associated with the storage and loss properties of the material. Writing the equation for the propagation of longitudinal waves in terms of L* = L' + iL", one finds that L' and L" are related to specific volume, V, velocity, v_1, and attenuation, a_1, through:

$$L' \simeq v_1{}^2/V \qquad\qquad (1)$$

$$L" \simeq 2a_1 v_1{}^3/V \qquad\qquad (2)$$

RESULTS: Amorphous Polystyrenes

Specific Volume, Figure 1A

At the lower temperatures, the specific volume, V, and thermal expansion $\alpha = \partial \ln V/\partial T$, are very similar for both glasses. Then the specific volumes suddenly rise much faster. For PS1, this occurs at T_V(PS1) = $105°$C. Up to T_V(PS1) + $35°$C = $140°$C the results depend on thermal

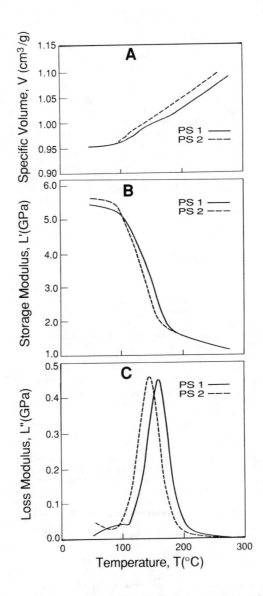

Figure 1: A) Specific volume; B) Storage Modulus; C) Loss Modulus for Polystyrene PS1 with M_n(PS1) = 130,000 and PS2 with M_n(PS2) = 9,000. Measurements were made at constant heating rate, 2.0°C/min and pressure 225 bars, using 1.75 MHz longitudinal waves.

history: structural rearrangement processes lag changes in temperature and it is only above 140°C that the material comes to thermodynamic equilibrium. Near $T_{11} \simeq 175$°C, the "liquid-liquid transition" [11], the entanglement network breaks-up (reptation) and the material becomes free-flowing-liquidlike with constant thermal expansion, α_{11}. For PS2, the increase in specific volume begins at $T_V(PS2) = 96$°C, is more sudden, and already at $T_V(PS2) + 17$°C = 123°C, the expansivity equals that of PS1 above T_{11}.

The behaviour near T_V is typical of the "glass transition phenomenon" which is defined [9] as a kinetic process of volume contraction. Therefore dilatometry is basic for the determination of the glass temperature, Tg. Also, because of the rate aspect of the process, the value of Tg is not unique. However, for operational purposes, Tg is identified by the marked rise of specific volume, hence we associate T_V with Tg.

On the one hand, the larger specific volume, V(PS2), corresponds to excess free volume associated with shorter chains. In turn, this results in Tg depending on molecular weight: Tg(°C) = 100 − 10^5/M, [9]. Here, Tg(PS1) − Tg(PS2) = 9.0°C, in good agreement with published values [12]. On the other hand, the absence of a T_{11} feature in PS2 is in line with the idea that short chain entanglement effects are small.

Ultrasonic Modulus, Figure 1B, 1C

Below Tg, the storage modulus, L_g'(PS1) decreases linearly with temperature: $\partial \ln L'/\partial T \simeq - 12.0 \times 10^{-4}$/°C. As with common solids, this is interpreted through anharmonicity. Here the effect is anomalously strong (x 10), pointing out the highly anharmonic character of the disordered structure in the glass. As temperature exceeds a critical value, $T_L(PS1)$, L'(PS1) suddenly drops, while L"(PS1) begins to increase. In Figure 1, this occurs precisely at the glass transition: $T_L(PS1) = 105$°C = Tg(PS1). The decrease of the storage modulus at Tg reflects the collapse of the structure while the increase in the loss modulus characterizes the irreversibility of the structural rearrangements.

Near Tg, the dominant characteristic times, τ, are long so Tg depends on the time scale for the measurement. Here, the time scale for probing the long relaxation times is given by the rate of heating, dT/dt = 2.0°C/min. The slow molecular movements appear frozen-in to the high frequency sound which measures the instantaneous unrelaxed properties of the material as they evolve with time and temperature.

On approaching equilibrium, 140°C, the regime changes: L' describes a sigmoïd while L" goes through a strong maximum at $T_{max}(PS1) = 163$°C. This is typical of a relaxation process where the time scale is that of the ultrasound, $\omega = 2\pi f$. The description for $L*(\omega) = L'(\omega) + iL''(\omega)$ proceeds from the model for the linear viscoelastic solid [9,10]:

$$L'(\omega) = L_R + (L_U - L_R) \int H(\tau) \, (\omega\tau)^2/[1 + (\omega\tau)^2] d\tau \tag{3}$$

$$L''(\omega) = (L_U - L_R) \int H(\tau)(\omega\tau)/[1 + (\omega\tau)^2] d\tau \tag{4}$$

where $H(\tau)$ is the distribution of relaxation times, τ. At low temperature, $\omega\tau \gg 1$, the material appears an elastic solid with $L'(\omega) = L_U$ (unrelaxed modulus) while at high temperature, $\omega\tau \ll 1$, it behaves as a liquid with $L'(\omega) \simeq L_R$ (relaxed modulus) and viscosity $\eta(\omega) = L''(\omega)/\omega$.

In all evidence, the dynamic modulus is closely linked to specific volume. However, below Tg, V(PS1) = V(PS2) while L'(PS1) $>$ L'(PS2), indicating that configurational details are also important. Above Tg, entanglement coupling of long chains in PS2, impedes molecular mobility, τ(PS1) $>$ τ(PS2), leading to higher stiffness, L'(PS1) $>$L'(PS2). Above T_{ll}, the entanglement network collapses, and L'(PS1) $\simeq$ L'(PS2) while the high molecular weight liquid remains more viscous, L"(PS1) $>$L"(PS2), again pointing to entropic effects. Also, T_{max}(PS2) = 154°C such that the difference T_{max}(PS1) $-$ T_{max}(PS2) = 9°C is the same as for Tg(PS1) $-$ Tg(PS2), suggesting that the high temperature degrees of freedom at T_{max} are in essence similar to the molecular movements at Tg.

RESULTS: SEMI—CRYSTALLINE POLYETHYLENES

Specific Volume, Figure 2A

The specific volume of HDPE increases almost linearly with temperature up to the softening point, T_S(HDPE) $\simeq$ 125°C, where, the rise is more pronounced. At T_M(HDPE) = 147°C, there appears a discontinuity in the V-T curve which identifies [13] the first order thermodynamic transition for melting of the crystallites. Above T_M, the expensivity is constant and typical of free flowing polymer liquids. For LDPE, the behaviour is more gradual and smooth: the softening point is poorly defined T_S(LDPE) $\simeq$ 96°C, also melting occurs at lower temperature, T_M(LDPE) = 118°C. Both these observations are consequences of the higher amorphous content in LDPE compared to HDPE.

Ultrasonic Modulus, Figure 2B, 2C

On increasing temperature the behaviour of L'(HDPE) and L"(HDPE) is reminiscent of the relaxation features in amorphous matter, Figure 1, and may be understood as a characteristic of the amorphous phase in the semi-crystalline composite. At higher temperature, L*(HDPE) tends to level off. It is usually speculated [9,10] that the "amorphous cement" between crystallites looses strength, allowing additional degrees of freedom for the composite structure. Finally, L'(HDPE) goes through a discontinuity at T_M(HDPE) and thereafter decreases linearly with temperature.

The comparatively smaller value of L'(LDPE) is due to the lower content of the stiff crystalline phase component in the composite material. The maximum value of L"(LDPE) is larger than that of L"(HDPE) and occurs at lower temperature, pointing out the higher mobility for the molecules in the more disordered structure. In addition, the plateau region is unnoticed in the case of LDPE, meaning that the interaction between the crystallites is weak and that the material properties are mainly governed by the amorphous matrix. Here also, L'(LDPE) is discontinuous at T_M(LDPE). For the melts, L*(HDPE) = L*(LDPE) which suggests the similarity for the molecular interaction in both liquids.

CONCLUSION

We described an apparatus for measuring the ultrasonic modulus of poly-

250

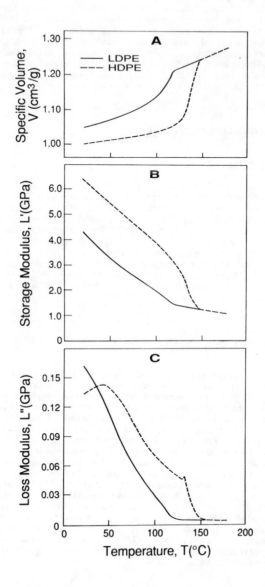

Figure 2: A) Specific volume; B) Storage Modulus; C) Loss Modulus for high and low density polyethylenes (HDPE and LDPE). Measurements were made at constant heating rate, 2.0°C/min, and pressure 225 bars, using 2.5 MHz longitudinal waves.

mers over a wide range of temperature and pressure with simultaneous determination of specific volume for the characterization of the thermodynamic
state of the material. Results on typical amorphous and semi-crystalline
polymers were presented. On the one hand, the method probes the important
thermodynamic and hydrodynamic properties (glass transition, crystallization, melting, long and short relaxation times) and on the other hand
reflects the details of the microscopic structure (chain length, chain
branching, entanglement coupling, crystallinity). For polymers, the above
characteristics have a strong bearing on both processability of the melt
and end use properties of the solid. Therefore, the method which is non
invasive and allows real time monitoring of viscoelasticity should prove
unique as a tool for quality and process control in polymer manufacturing.

REFERENCES
1. B. Hartmann, "Ultrasonic Measurements" in "Methods of Experimental Physics", eds. R.A. Fava (Academic Press, New York 1980), Vol. 16-C, Chap. 12.1, pp. 59-90.
2. L. Piché and F. Massines, "Ultrasonic Investigation of Amorphous Polymers (Polystyrene) in a Wide Range of Temperature and Pressure: I - Introduction, Method, Technique and Results", submitted to J. Polym. Sci.: Polym. Phys. Ed.
3. H.J. McSkimin, "Ultrasonic Methods for Measuring the Mechanical Properties of Liquids and Solids", in "Physical Acoustics", ed. W.P. Mason (Academic Press, New York, 1964), Vo. I-A, Chap. 4, pp. 271-334.
4. Y. Wada and K. Yamamoto, J. Phys. Soc. Jpn, 11, 887 (1956).
5. R. Kono, J. Phys. Soc. Jpn, 15, 718 (1960).
6. E. Morita, R, Kono and H. Yoshizaki, Jpn, J. Appl. Phys. (1968), 7, 451.
7. G.W. Paddison, Proc. IEEE Ultrasonics Symposium, 502 (1979).
8. L. Piché, F. Massines, A. Hamel, C. Néron, "Ultrasonic Characteriza-tion of Polymers Under Simulated Processing Conditions", U.S. Patent No.: 4,754,645, Date: Jul. 5, 1988.
9. J.D. Ferry, "Viscoelastic Properties of Polymers" (John Wiley & Sons, New York, 3rd Edition, 1980).
10. I.M. Ward, "Mechanical Properties of Solid Polymers" (John Wiley & Sons, New York, 1983).
11. R.F. Boyer, "Evidence for $T_{\ell\ell}$ and Related Phenomena for Local Structure in the Amorphous State of Polymers" in "Order in the Amorphous "State" of Polymers", edited by S.K. Keinath, R.L. Miller, J.K. Rieke (Plenum Press, 1987).
12. R.F. Boyer "Styrene Polymers, Physical Properties" in "Encyclopedia of Polymer Science and Technology" (John Wiley and Sons, New York, 1970), Vol. 13, pp 251-326.
13. D.W. Van Krevelen, "Properties of Polymers, Their Estimation and Correlation with Chemical Structure" (Elsevier, New York, 1976).

STUDY OF SPHERULITES IN A SEMI-CRYSTALLINE POLYMER USING ACOUSTIC MICROSCOPY

J.Y. DUQUESNE*, K. YAMANAKA+, C. NERON, C.K. JEN, L. PICHE and G. LESSARD

Industrial Materials Research Institute, National Research Council of Canada, 75 Boul. de Mortagne, Boucherville, Quebec J4B 6Y4, Canada

ABSTRACTS

The local elastic properties of a semi-crystalline polymer (isotactic polypropylene) have been investigated with the use of scanning acoustic microscopy (SAM) techniques. The operating frequency was 775 MHz and the temperatures ranged between 25 and 60°C. A new and effective signal processing method was developed to process the material signature V(z) curve. This method avoids complicated calibration procedures involved with operating the SAM lens at different temperatures. Longitudinal velocities $v_L(\alpha)$ and $v_L(\beta)$ in individual α and β-type spherulites of isothermally crystalized samples have been measured. Acoustic velocity of a quenched sample was also obtained. It was found that $v_L(\alpha)$ and $v_L(\beta)$ were nearly equal but the acoustic impedances of α and β-spherulites were unexpectedly very different.

INTRODUCTION

Mechanical properties of semi-crystalline polymers are of great interest both for fundamental research and industrial applications. Many experiments have been devoted to studying the influence of polymorphism on elastic properties [1,2]. It is well known that isotactic polypropylene (PP) is organized into semi-crystalline structures called spherulites. In a spherulite, lamellae crystallize in either of three polymorphic forms : α (monoclinic), β (pseudohexagonal) and γ (triclinic) [3]. Previous studies on the influence of polymorphism have been performed on bulk samples containing various ratios of α and β spherulites (from 0% to 90% of β type) [1,2]. Those samples were either oriented [1] or contained various amounts of β type nucleant [2].

The aim of this work is to study the elastic properties of different kinds of spherulites in a typical semi-crystalline polymer (isotactic PP) by probing individual spherulites using scanning acoustic microscope (SAM) techniques. It is then possible to compare the elastic properties of different spherulites grown under identical conditions, within the same sample. Because of the local character of the measurement, possible problems arising with spherulites boundaries are avoided. In a previous study [4], different authors obtained qualitative information on PP morphology uisng an acoustic microscope operating at 375 MHz. Here we use a reflection SAM operating at 775 MHz and with a 2 μm resolution. On one hand, the instrument allows one to obtain acoustic images and on the other, it also provides quantitative measurements. The absolute values for the longitudinal velocities of α and β spherulites were obtained within the temperature range between 25 and 60 °C. We have also measured the difference of their acoustic impedances, Z, which is the product of the density, ρ, and velocity, v. Similar experiments are also performed for quenched PP samples.

SAMPLE PREPARATION

Isotactic PP were used as samples. The base polymer was obtained in the form of powder, labelled PRO-FAX 6301 and manufactured by Himont Canada Inc. Thin films were obtained by melting small amounts of powder between two microscope glass slides seperated by 50 μm thick spacers. One group of samples was prepared by quenching the molten material to room temperature and another corresponded to isothermal crystallization (IC) at 135°C. Thin films rather than bulk samples were studied for various reasons. For quenched samples, a high cooling rate was easily obtained. In the case of IC samples the spherulites are flat discs rather than spheres. The radial distribution of fibrils probed by the acoustic microscope can be assumed to be the same from one spherulite to the other. Then, possible contrast between spherulites does not arise primarily from their anisotropy. However, the films must be thick enough to avoid spurious excitation of Lamb's waves for which the velocity depends on the film thickness, and is therefore not intrinsic. These thin films were glued at the sample edges on a microscope slide for the optical or acoustic evaluations.

The films were first examined under a transmission polarized optical microscope (TPOM). Quenched samples were found to be made up of α type spherulites exclusively, having a typical diameter near 50 μm. For the IC samples, the diameter of the α type spherulites was in the range of 200 μm but in this case we observed the presence of a number of β type spherulites. Figures 1(a) an 1(b) show typical TPOM and SAM micrographs respectively of an IC sample. In Fig.1 (a) the contrast between α and β-spherulites has its origin in the changes of optical birefringence while in Fig.1(b) it is due to different elastic properties.

SAM TECHNIQUES

The principles of reflection SAM are well documented elsewhere [4,5]. A focused spherical acoustic wave generated by an acoustic lens in a coupling fluid is reflected from a sample. The reflected beam is collected by the same lens and contains information of the elastic properties of the sample. The depth of acoustic penetration in the sample depends on the acoustic attenuation. In general, lower frequency operation has less attenuation, therefore penetration depth is larger. Figures 2(a) and 2(b) show SAM micrographs taken at 775 and 179 MHz respectively of an identical area of a bulk PP sample quenched at room temperature. The dark zone at the top of the picture was an indentation mark, and this mark together with scratch lines were used for the purpose of position identification. As expected the sizes of spherulites in this quenched sample are smaller compared to those in IC samples shown in Fig.1. It can be seen that Fig.2(a) only probes those spherulites near the surface. Figure 2(b) displays a very similar overlaid structure as those commonly obtained by TPOM. With the help of Figs.2(a) and 2(b) and other optical micrographs we deduced that the acoustic penetration depth in PP samples was less than 20 μm at 775 MHz. Therefore, thin film samples of 50 μm thick can be regarded as semi-infinite substrates for SAM operated at these high frequencies.

By keeping the SAM lens in out-of-focus conditions and moving toward the surface of the sample, leaky acoustic surface waves may be excited efficiently. These waves come to interfere with the specularly reflected waves that propagate in a direction close to the normal of the surface. On changing the position, z, of the lens, there results oscillations (beat pattern) in the amplitude (and phase) of the output signal, $V(z)$, from the piezoelectric transducer in the SAM lens. Proceeding in this manner one obtains $V(z)$ curves [4,6] that contain the quantitative information on the leaky surface wave in the vicinity of the SAM focal point. The distance between the crests in the $V(z)$ curve is given approximately by λ_{sw}^2/λ_w, where λ_{sw} and λ_w are the acoustic wavelength of the leaky acoustic wave and of the longitudinal wave in the coupling liquid respectively. The velocity of the leaky acoustic wave may be computed by Fourier analysis [6] of such $V(z)$ curves. The technique involves surface area whose radius at most extends to roughly $10 \lambda_w$. In the present case, $\lambda_w \sim 2$ μm such that SAM actually constitutes a probe for local acoustic or elastic properties.

Our technique records $V(z)$ curves while scanning an arbitrary x-axis along on the surface of the sample: $V(x,z)$ curves are then obtained. Figure 3(a) shows a zoomed SAM image of the left part of PP sample shown in Fig.1(b), and Fig.3(b) is the $V(x,z)$ curve measured along a horizontal line across the middle of Fig.3(a). In Fig.3(b) the horizontal and the vertical axes correspond to x and -z respectively, and the brightness is proportional to V. At each x value one $V(z)$ curve can be displayed along the vertical, -z, axis direction. Because of acoustic scattering effects by the fibrillar structures of PP samples, in some cases, we found it useful to define a $V(z)$ curve by averaging the $V(x,z)$ curve for a set of x values : $V(z) = \langle V(x,z) \rangle_x$. Figure 4 shows the difference

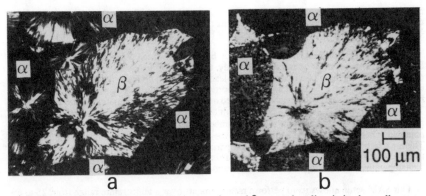

Fig.1 (a) TPOM and (b) reflection SAM image of α and β-type spherulites in isothermally crystallized PP thin films (grown at 135°C).

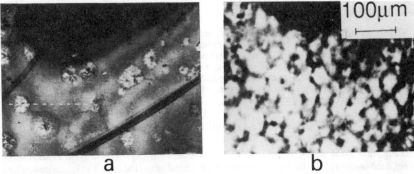

Fig.2 SAM micrograph obtained at (a) 775 and (b) 179 MHz of a bulk PP sample quenched to room temperature.

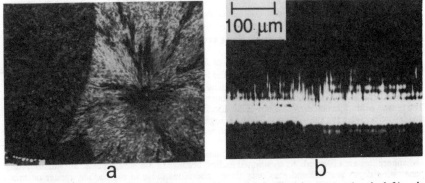

Fig.3 (a) Zoomed SAM micrograph of an interface region in Fig.1 between α (on the left) and β-spherulies (on the right). (b) Measured V(x,z) curve across the horizontal axis at middle height of 3(a), where the horizontal and vertical axes are x and -z axis respectively. The brightness is proportional to the received voltage signal V of the SAM lens.

between a single (upper trace) and an 'averaged' (here 50 x's, middle trace) V(z) curve. The V(z) curves obtained using this averaging approach are more statistically meaningful and then processed by a Fourier analysis technique including a Fast Fourier Transform (FFT) process [6].

Before the FFT process, the low frequency background of the V(z) curve (arising from the SAM lens response) must be removed. Usually, this is performed using a calibration procedure of the lens with the help of a material (usally lead) supporting no leaky surface wave [6]. In our experiments, we used a different and a new approach whereby the low frequency backgound was removed with a high pass numerical filter based on a first order differentiation of the V(z) curves. The bottom trace in Fig.4 shows a typical result of this differentiation. After the FFT process the spatial frequency spectrum versus the wavenumber k (rad/μm) shown in the inset of Fig.4 can be obtained. Then, v_{LSSCW} can be deduced from the wavenumber at the peak of this spectrum [6]. This approach has been cross-checked by comparing the results to those for a well known material, fused quartz, and proved to be quite accurate. This technique also proved efficient when changing the temperature of the sample, since no calibration of the lens versus temperature is required.

Our SAM setup allows us to perform measurements over a temperature range from 25 to 60 °C. The temperature was measured in the coupling fluid (here, water) between the lens and the sample with a miniature thermocouple. The velocity of the longitudinal waves in water versus temperature was obtained from Ref.[7]. It is noted that the temperature limitation came from the adhesive bond between the film and glass substrate, and not from the acoustic attenuation. This bond did not hold at temperatures above 65°C.

The leaky acoustic waves excited at the sample surface may be of various types [6]. After careful examination it was found that leaky surface skimming compressional waves (LSSCW) [6,8] along the surface of spherulites of PP were responsible for the oscillation of the measured V(z) curves. The velocity of LSSCW is close to the velocity of the longitudinal velocity, v_L. To our knowledge, no theoretical investigation has been reported on the deviation of v_{LSSCW} from v_L. However, experimental data showed that this deviation is around a few percent in typical polymers [6]. In the later text v_L is used to represent v_{LSSCW}.

QUALITATIVE RESULTS (ACOUSTIC MICROGRAPHS)

In Fig.3(a) a SAM micrograph of an IC thin sample grown at 135°C is displayed. In particular, it contained one β type (on the right) and one α type spherulite (on the left). The identification for the nature of the spherulites was asserted on the basis of their optical appearance, in agreement with previously published photographs [9,10]. Moreover, the boundaries between the two kinds of spherulites are curved, indicating different growth rates. From the curvature, we can evaluate the difference in growth rates between them to be around 27% in good agreement with published values [11]. In Fig.3(a) clear acoustic contrast is observed between α and β-spherulites. From the measured V(z) curves of these two types of spherulites, it is found that β type has a higher acoustic impedance ($Z = \rho v$) which is responsible for this contrast.

It is of interest to examine the organization of fibrillar structure at the interface between two different spherulites, Fig.3(a), between a α on the left and a β sperulite on the right. The lamellae are clearly identified. From the magnified acoustic micrographs of this interface it was observed that most lamellae stop at the interface, but in some regions they seem to cross over the boundary. The reason behind this crossover is not understood. A higher resolution SAM is planned to study this behavior. We have also obtained SAM micrographs of the fibrillar structures of quenched PP films and they are nearly the same as those reported in [4].

In the SAM image shown in Fig.2(a) of a quenched bulk PP sample, both 'uniform' regions and spherulites are observed. As mentioned earlier, because of the high acoustic attenuation, the penetration depth is small and thus only those spherulites close to the surface appear on this acoustical micrograph. The centers of the spherulites in the center left are located on the surface and could be identified with the optical microscope as being heterogeneous nucleating centers orginated from the dusts on the microscope glass slides used for fabricating these films. The spherulites in the lower center is deeper. The 'uniform' region between the spherulites is attributed

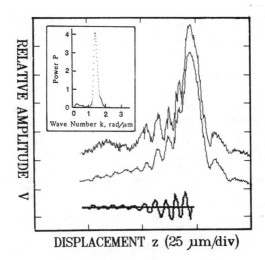

DISPLACEMENT z (25 μm/div)

Fig.4 Upper trace: V(z) curve for a single value of x; middle trace: averaged V(z) curve for 50 x's; Bottom trace: after the numerical differentiation of averaged V(z). Inset: spatial frequency spectrum after FFT of the bottom trace.

Fig.5 A V(x,z) curve obtained along a x scan-line indicated by a dashed line in Fig.2(a) for a quenched bulk PP sample. Unperturbated fringes are from uniform regions and perturbated ones are from surface spherulites.

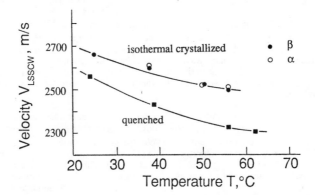

Fig.6 Velocity of the leaky surface skimming compressional waves in α and β-spherulites of an isothermally crystallized film (grown at 135°C) and α spherulites in a quench PP film, versus temperature. SAM was operated at 775 MHz.

to spherulites with a deep inside center which was confirmed by the TPOM images. In the 'uniform', region of Fig.2(a), the acoustic beam probes only the 'top' of the sperulites. The medium appears as an array of oriented fibrils with a narrow angle distribution about the direction normal to the surface.

QUANTITATIVE RESULTS

As mentioned earlier, the velocity of leaky surface skimming compressional wave, $v_{LSSCW} \sim v_L$, can be obtained from the measured $V(z)$ curve of PP sample. In order to allow a direct comparison between velocities of α type, $v_L(\alpha)$, and β type, $v_L(\beta)$, spherulites and eliminate as much as possible common systematic uncertainties in the measurements, $V(x,z)$ curves were recorded simultaneously for two adjacent α and β-spherulites, by scanning an x-line across their boundary. A typical $V(x,z)$ curve was shown in Fig.3(b). The random fluctuations of the brightness can be attributed to the scattering of the acoustic waves propagating along the surface. The scattering centers could possibly be the crystallites since their dimensions typically in range around 2 μm by 200 A° are commensurable with the wavelength, λ_{sw} is about 3 μm. The distribution of the scattering centers may change from point to point, leading to random purturbations of $V(z)$ curves when x is scanned. As mentioned in the previous section, $V(x,z)$ curves were averaged over a range of x to eliminated such fluctuations. It is then possible to measure the v_{LSSCW} with a good accuracy: within a same spherulite the velocity difference is about 0.8% when different areas were selected for averaging.

We also examined the contribution of the sample surface condition to the random fluctuation of $V(x,z)$ curves. Figure 5 is a measured $V(x,z)$ curve along a x-scan line indicated by a dashed line in Fig.2(a) which is a SAM image of a quenched bulk PP sample. The fringes are well defined when x is scanned through a 'uniform' region described in the previous section, however, they are strongly perturbated across a surface spherulite. Therefore it is believed that the random fluctuation of $V(x,z)$ curves comes from the acoustic scattering by crystallites having their radially oriented fibrils lying close to a plane near the surface of the sample. Acoustic contrast could result from phase cancellation.

It is interesting to note that the velocity difference between $v_L(\alpha)$ and $v_L(\beta)$ for α and β spherulites is negligibly small, less than 1%. Figure 6 shows the variation of these velocities versus temperature and also the comparison with that of the quenched thin film. The spherulites in quenched thin films have lower velocity than those in isothermally crystallized thin films.

According to Reference [2], the variations of density, ρ, between α and β-spherulites is 1%. Using $V(z)$ curves we found the difference in their velocities is little. However, the acoustic impedance, Z, of PP sample, derived from the reflection coefficient, Γ, at the water-PP sample interface, indicates a 10% difference between these two types of spherulites. Γ is assumed to be equal to $(Z - Z_w)/(Z + Z_w)$, where Z_w is the acoustic impedance of the coupling liquid (here, water). This formula is only valid for plane waves at normal incidence and may lead to large errors when used to calculate the absolute value of the reflection factor of a spherical wave [12]. Nevertheless, it is expected to give a correct estimate of the variation of that factor versus Z provided that the variation is small.

The origin of the above mentioned 10% variation of acoustic impedance is not clear. The published values for the density of β-spherulites have been obtained by indirect methods [2], therefore the actual difference in density between α and β-spherulites could be possibly larger than 1%. Moreover, the elastic anisotropy of the spherulites may be another contributing factor. The spherulites measured are flat discs rather than spheres. The velocity propagating parallel or normal to the flat surface of spherulites may be different, although the variations are expected to be small. In addition, in our analysis the spherulites are assumed to be homogeneous. In fact, the acoustic wavelength is comparable to the size of the lamellae within the spherulites. Then, scattering effects can be important and may invalidate the classical expression for Γ. These scattering effects can be different from one type to the other since the arrangement of the lamellae is quite different in α (cross-hatched structure) and β type spherulites (parallel stacked lamellae) [10].

CONCLUSIONS

The local elastic properties of a semi-crystalline polymer (isotactic polypropylene, PP) were measured by means of a SAM operated at 775 MHz and at temperatures between 25 to 60 °C. A new and effective method involving numerical filtering was adopted to process the $V(z)$ curve and which avoids the traditional lens calibration procedures at different temperatures. The longitudinal velocities of α, $v_L(\alpha)$, and β-type spherulites, $v_L(\beta)$, prepared by isothermal crystallization at 135°C have been obtained. It was found that $v_L(\alpha)$ is nearly equal to $v_L(\beta)$ in the direction parallel to the plane of thin film spherulites. A large contrast in SAM micrographs between these two spherulites was observed. The origin of this high contrast is not clear. It could reveal unexpected large difference in density or velocity in the direction normal to the plane of thin film spherulites, however the acoustic scattering within the spherulites by the crystalline lamellae should be also considered. In addition, the acoustic velocity of the quenched film was found to be slower than that of isothermal crystallized films.

ACKNOWLEDGEMENT

The authors would express their sincere gratitude to Dr. J. Kushibiki of Tohoku university for providing the scanning acoustic microscope lens and Dr. J.F. Bussiere of IMRI for his strong encouragement.

REFERENCES

[1] J.M. Crissman, J. Polymer Sci., A-2, 7, 389 (1969).

[2] P. Jacoby, B.H. Bersted, W.J. Kissel and C.E. Smith, J. Polymer Sci., B, 24, 461 (1986).

[3] A. Turner-Jones, J.M. Aizlewood and D.R. Beckett, Makromol. Chem., 75, 134 (1964).

[4] P.A. Tucker and R.G. Wilson, J. Polymer Sci., Polymer Lett. Ed., 18, 97 (1980).

[5] R.D. Weglein, Appl. Phy. Lett., 34, 179 (1979).

[6] J. Kushibiki and N. Chubachi, IEEE Trans. Son. and Ultrason., SU-32, 189 (1985).

[7] M. Greenspan and C.E. Tschiegg, J. Res. Nat. Bureau of Standards, 59, 249 (1957).

[8] W.G. Neubauer, J. Appl. Phys., 44, 48 (1972).

[9] F.J. Padden and H.D. Keith, J. Appl. Phys., 30, 1479 (1959).

[10] D.R. Norton and A. Keller, Polymer, 26, 704 (1985).

[11] A.J. Lovinger, J.O. Chua and C.C. Gryte, J. Polymer Sci., Polymer Phys. Ed., 15, 641 (1977).

[12] K.K. Liang, Proc. Ultrasonics Symp., 1141 (1987).

* permanent address: Laboratoire de Physique des Solids, Bat 510, Universite de Paris-Sud, 91405 ORSAY, France.

+ permanent address: Mechanical Engineering Laboratory, 1-2 Namiki, Tsukuba City, Ibaraki-Ken, 305 Japan.

PART V

Ceramics

ULTRASONIC NON DESTRUCTIVE EVALUATION OF MICROSTRUCTURAL CHANGES AND DEGRADATION OF CERAMICS AT HIGH TEMPERATURE

CHRISTIAN GAULT
ENSCI, UA CNRS 320, Avenue Albert Thomas, 87065 LIMOGES, FRANCE

ABSTRACT

This paper deals with the use of pulsed ultrasonic waves to monitor microstructural changes and degradation of ceramics in the field of high temperature. Two types of devices are described. One is a low frequency system for the measurement of Young's modulus at temperatures up to 1800°C. The second is an ultrasonic spectroscopy system for the evaluation of damage in structural ceramics during thermal fatigue experiments. Applications concern phase changes, porosity evolution or microcracking induced by thermal stresses when heating monolithic or composite ceramics.

INTRODUCTION

The physical properties of ceramics, and the associated engineering parameters (i.e. modulus of rupture, toughness, thermal expansion coefficients, thermal conductivity, etc...), strongly depend on the microstructure. Therefore the improvement of ceramic component reliability involves the control of microstructure and of its stability or evolution in service conditions. As ceramic mechanical parts generally are made to be used at high temperatures (> 1000°C), any non destructive method for monitoring microstructural changes or/and degradation in these temperature ranges appears to be as useful for material science researchers as for engineers.

Since the fifties, ultrasonic pulse echo techniques have been widely developed for non destructive testing (NDT) in metallurgical engineering. More recently, the extension to the detection of microscopic critical flaws in non metallic materials such as ceramics, involved the use of very high frequencies (> 100 MHz). This, associated with the development of numerical signal processing facilities, have lead to sophisticated NDT ultrasonic imaging installations such as acoustic microscopes [1].

Another field of use of pulse echo techniques is the measurement of ultrasonic parameters, i.e. attenuation and phase velocities of longitudinal or shear waves in solids. As ultrasonic velocities are related to elastic constants (or elastic moduli in the case of isotropy) and to density which both depend on the material parameters, their measurements can give information about phase distribution, microstructure, microcracking (damage), etc. This is called "non destructive characterization" (NDC) of materials [2]. But, though very accurate measurements have been performed at room or rather low (< 500°C) temperature [3], few results are reported in the high temperature range (up to 2000°C) of interest for thermomechanical ceramic components.

EXPERIMENTAL

Ultrasonic measurement of Young's modulus at high temperature

High temperature ultrasonic measurements require the thermal insulation of the transducer. The principle of the method here presented, is to use magnetostrictive transducers which are able to generate ultrasonic pulses in long thin waveguides. Such an arrangement had been successfully used elsewhere for ultrasonic thermometry up to 2500°C [4].

Fig. 1 shows three types of ultrasonic lines for three different sample geometries. The corresponding ultrasonic patterns are given on the same figure.

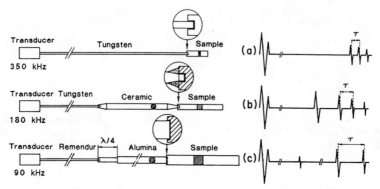

Fig. 1 - Ultrasonic lines used with magnetostrictive transducers, and corresponding echo patterns. τ is the measured time.

These lines are vertically fitted into a vessel with controlled atmosphere and pressure which is put into the isothermal zone of a tubular furnace [5].
    The ultrasonic wave propagates under the "long beam" condition $\lambda \ll s^{1/2}$, λ being the wavelength and s the cross section of the propagation medium. This condition is achieved by the use of suitable low frequencies. The coupling of the different parts of the line is achieved by soldering for metals (remendur/tungsten), or with a refractory alumina cement for the waveguide/sample fitting [6]. When two waveguides are used, an impedance matching is machined in order to minimize the interface echo [7]. The ferromagnetic rod of the transducer (remendur) is placed into a coil excited by a current pulse [8]. The frequency of the broadband ultrasonic pulse depends on the length of the coil.
    Fig. 1 (a) shows a tungsten lead-in (1 mm in diameter, 400 mm in length) soldered to a short ferromagnetic rod and cemented to small beam shaped samples (25 x 2.5 x 1 mm^3). The small dimension of the sample requires the highest frequency (350 kHz) for the transducer. This line works up to 1300°C under inert gas.
    For higher temperatures, the thermal expansion mismatch between tungsten and ceramics causes the coupling to fail. Then, an intermediate ceramic wave-guide, machined as shown on Fig. 1(b), must be used. Samples are rectangular (50 x 2.5 x 2.5 mm^3) beams. This ultrasonic line uses a 180 kHz transducer and the maximum temperature is 1800°C under inert gas.
    The most versatile line which can work under any atmosphere (oxidizing, reducive, inert) and pressure (from 10^{-2} to 1000 mbars) at temperatures up to 1550°C is described on Fig. 1 (c). It requires large volume samples (10 x 10 x 100 mm^3) because of the use of an alumina waveguide of 5 mm in diameter directly coupled to a ferroelectric rod. The respect of the "long beam" propagation condition implies a low frequency (90 kHz). This system is well suited for investigations in composites with large macrostructural scales [9].
The ultrasonic velocity in the sample is determined by an electronic device previously described [5]. It automatically records the round trip time **τ** between two sample echoes. If necessary, a time correction θ is taken into account for the phase delay at the waveguide/sample interface [6]. The velocity is then given by :

$$V_L = \frac{2l}{\tau - \theta} \qquad (1)$$

where l is the sample length.
In the "long beam" mode the velocity of a compressional wave is related to the unidirectional Young's modulus $\bar{E}$ and to the density ρ of the material by [4] :

$$V_L = (\bar{E}/\rho)^{1/2} \qquad (2)$$

For isotropic materials $\bar{E}$ is Young's modulus E, otherwise it is a function of the elastic constants and of the propagation direction cosines.

When structural changes occur in the material during the heat treatment, the mass and the length of the sample may vary. Therefore corrections are to be made for the calculation of E from equations (1) and (2). A HP86B computer monitors the environment parameters (temperature T and pressure P) and calculates E at a temperature T, using equations (1) and (2), from the measured time τ and the previously stored correction parameters [7].

Ultrasonic spectroscopy equipment for damage evaluation in ceramics submitted to thermal fatigue.

Two considerations governed the building of this apparatus.
i - Ultrasonic pulse techniques are well suited for the evaluation of thermal fatigue damage in ceramics [10]. But because measurements are generally made before and after sollicitations, and because samples must be removed between two operations, dispersed results are obtained. As a consequence, the ultrasonic test was studied to directly take place on the thermal fatigue set during thermal cycles.
ii - When inhomogeneities (porosities, cracks, fibers, etc.) are dispersed in the material, the ultrasonic attenuation due to multiple scattering increases. This is particularly obvious in ceramic fiber composites where separated echoes cannot be clearly detected because of multiple reflections. A solution can be Fourier analysis techniques to determine the specimen response (transfer function) to the ultrasonic disturbance versus frequency [11].

Fig. 2. Gives the principle of the apparatus detailed in a previous paper [12].
The samples are disks 30 mm in diameter and 15 mm in thickness. The lower side of the disk is regulated at a temperature T_o (100°C - 300°C) by convective exchange into a silicon oil bath which acts also as propagating medium for ultrasonic waves. The steel tank is cooled by natural convection in air and by a water jacket which maintains the ultrasonic transducer at low temperature. In the vicinity of the sample, the temperature of the oil is also regulated by a thermocoax wire via a platinum probe.
The upper side of the sample is alternately heated by a radiating cavity (up to a temperature T_M), and cooled by convection with compressed air (down to a temperature T_m) (Fig. 3). The highest T_M was chosen to be 1000°C for a low thermal conductivity ceramic (5 $W.m^{-1} K^{-1}$) in order to obtain a transverse thermal gradient of about 5.10^4 °C.m^{-1} with T_o = 300°C. The typical cycle period is less than 100 s.
The compressed air cooling system and the furnace are both on a mobile frame translated in front of the sample by a pneumatic jack driven by a computer through electrosluice gates. The same computer monitors the three control thermocouples and the mobile frame when the temperatures reach predetermined values [12].
A low frequency (1 MHz) has been chosen to avoid dramatic attenuation by microstructure inhomogeneities. As shown on Fig. 4, the frequency spectra of the signals issued from the reflection at the oil-sample interface (reference spectrum), and from the propagation through the sample (sample spectrum) are calculated. Then the transfer function of the sample (which depends on the ultrasonic propagation parameters of the material) is derived from the ratio of the two spectra. A LeCroy storage oscilloscope digitalizes the signals at a rate of 100 MHz and calculates the Fourier spectrum. The data are stored in the floppy disk unit of the computer which calculates the transfer function and displays the results on a printer or an X-Y plotter.
From theoretical considerations the transfer function has been calculated with a model of transmission - reflection of the ultrasonic pulse [13]. It was found that, when reflection is assumed to take place in the backside of the sample

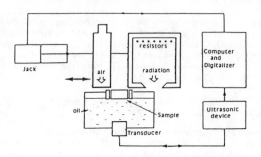

Fig. 2 - General block diagram of the thermal fatigue set.

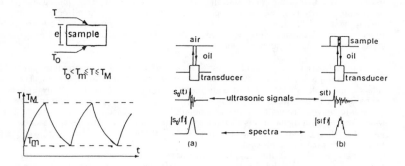

Fig. 3 Typical temperature cycles.

Fig. 4 - Principle of ultrasonic spectroscopy (a)Reference spectrum, (b)Sample spectrum

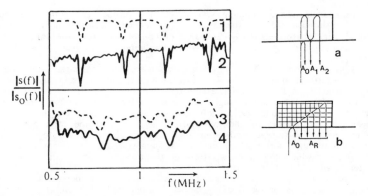

Fig. 5 - Transfert functions for two ceramics (arbitrary units).
ZrO_2 : (1) calculated with model (a), (2) experimental.
2D SiC/SiC composite : (3) calculated with model (b), (4) experimental.

(monolithic ceramic), the transfer function is characterized by narrow lines at given frequencies $f_n = nV/2l$, n being integer, V phase velocity, and l sample length. When reflections take place on scatterers distributed in the sample (composites), the transfer function exhibits widened and attenuated lines [13].

Fig. 5 shows transfer functions in a zirconia sample and in a 2D long fiber SiC - SiC composite. The additional small lines in the experimental curves are attributed to spurious effects due to diffraction of the ultrasonic beam.

ULTRASONIC PROPAGATION AND CERAMIC MICROSTRUCTURES

The propagation of an ultrasonic wave through a solid is characterized by propagation parameters, phase velocity V and attenuation , which all depend on the atomic structure of the material. V is related to density and elastic constants which can be theoretically calculated from interatomic potentials.

This is the case of single crystals which are anisotropic homogeneous materials and of glasses which are isotropic homogeneous solids. But for most industrial ceramics, the processing by sintering powders at high temperature leads to rather complex microstructures : polycrystalline, multiphased and porous. From an ultrasonic point of view, these microstructural features act as inhomogeneities, i.e. domains with different propagation parameters separated by interfaces acting as reflectors. Then, the measured ultrasonic parameters will depend not only on the nature of the constituents, but also upon the geometrical arrangement of the microstructure.

Propagation in polycrystals

It has been widely studied in metals, and theoretical analyses have been made assuming that the polycrystal is a single-phased agglomerate of equiaxed randomly oriented single crystals and behaves as an isotropic medium [14].

The main results are that ultrasonic velocities are independent of grain size, but that there is a strong dependence of attenuation on grain size and frequency due to multiple scattering mechanisms. Quantitative calculations of the multiple scattering parameters are not of a great interest because of the restrictive hypothesis (single-phased and equiaxiality) which are unrealistic for ceramics. Nevertheless a qualitative result is that, for the "long wavelength condition" ($\lambda \gg \overline{D}$, $\overline{D}$ being the average grain size) which is fulfilled for low frequency ultrasonic experiments, ultrasonic attenuation increases with grain size. Its measurement can be used for grain growth control when sintering.

Propagation in multiphased materials

Ceramics are multiphased materials. This is obvious for ceramics made from natural powdered raw materials (for example industrial refractories or concretes), but even when chemically pure synthetic powders are used, small amounts of "sintering aids" are to be added to improve densification. This results in secondary phases often segregated at grain boundaries.

Furthermore a lot of high tech ceramics contain strengthening particles (for example fully or partially stabilized zirconia, or alumina - zirconia compounds) (Fig. 6), or can exhibit crystalline domains embedded in a glassy phase (Fig. 7). As far as ultrasonic velocities are not too much different from one phase to another, multiple reflections can be neglected and the total transit time of the wave can be considered as the summation of the transit times through each phase domain, which gives :

$$\frac{1}{V} = \sum_i \frac{c_i}{V_i} \qquad (3)$$

where V is the ultrasonic velocity in the multiphased material and V_i the velocity in the phase i of volume concentration c_i.

For a two constituent system we get :

$$V = \frac{V_1 V_2}{cV_2 + (1-c)V_1} \qquad (4)$$

c being the volume concentration of phase 1.
Provided V_1 and V_2 are known, the measurement of ultrasonic velocity V can be used to study the phase concentration evolutions (for example crystallisation in an amorphous material).

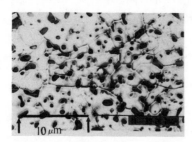

Fig. 6 - Al_2O_3 - ZrO_2 microstructure from ref. [15]. Dark domains are ZrO_2.

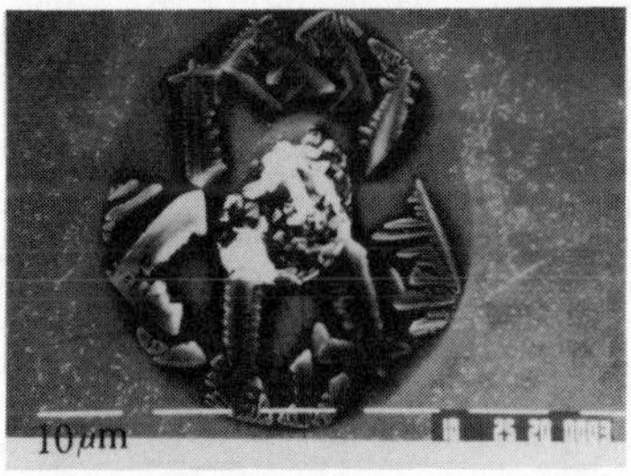

Fig. 7 - Dendritic crystallisation into a SiAlYON glass. After ref. [16].

When the nature and crystalline structure of the two phases are quite different this simple "melting law" model only gives a qualitative information.

If the wavelength is large compared with the phase domain size, it is possible to consider that propagation takes place in an equivalent homogenous material and to evaluate velocity from the average elastic moduli calculated from the individual elastic properties of the constituents. Such calculations have been made by severals authors [17, 18] and they lead to upper and lower bounds for the bulk and shear moduli of the equivalent isotropic material. The best results can be obtained by taking into account the shape of the microstructure [19].

Another interesting approach was carried out by Nielsen et al. [20] who developed a theoretical model using finite element analysis for the prediction of the elastic moduli for two-phased materials. They derived general equations involving geometrical parameters, which can be also extended to transversaly isotropic composites and porous bodies.

All these models predict, as experimentally verified, that the ultrasonic velocity always increases with the concentration of the stiffest phase.

Propagation in porous ceramics

When sintering, porosity cannot be completely eliminated, generally because of concurrent grain growth occuring at high temperature which traps intergranular porosity.

Contrary to the case of grains in polycrystals, both ultrasonic velocities and attenuation are affected by pores : velocity decreases and attenuation increases when the fractionnal volume of pores increases.

Numerous theoretical analyses have been made to describe the dependence of elastic moduli on porosity including bound techniques previously mentionned for two-phased materials (one phase is assumed to be porosity). Nielsen's model [20] is well suited to take into account the shape of the porosity.

The comparison between results from theory (with some adjustable shape parameters !) and from empirical equations, shows that the decrease of elastic moduli is approximatively linear with increasing porosity as long as porosity does not exceed 20 % [21]. Then ultrasonic velocity is found to decrease linearly too.

Therefore, the measurement of ultrasonic velocities is a useful tool to monitor porosity, and to follow the achievement of densification if measurements are performed during high temperature sintering.

Propagation in composite ceramics

The term "composite" involves various classes of ceramics.
- Randomly distributed short fibers or particle reinforced ceramics which are in fact two-phased isotropic materials from an ultrasonic point of view, provided the wavelength remains large with respect to inhomogeneity sizes.
- Whiskers reinforced ceramics because of the shape of the whiskers which leads to preferred orientations when processing, are to be considered as orthotropic or transversely isotropic two-phased ceramics.
- The case of long fiber reinforced ceramics (with 1D, 2D or 3D textures) is particular, because two geometrical scales of inhomogeneities are to be considered. The fiber and the matrix are ceramics with their proper microstructure, and the geometrical arrangement of the fabric has a macrostructural scale of a few millimeter's range. Furthermore a poor bonding between fiber and matrix is often researched to improve mechanical properties. This results in strong multiple scattering effects of ultrasonic waves which involve the use of low frequencies (in the 100 kHz range) and particular techniques, as described before, for ultrasonic testing.

ULTRASONIC MONITORING OF THERMAL EFFECTS IN CERAMICS

For any ultrasonic propagation mode, the corresponding phase velocity is always related to the density and elastic constants of the material which are temperature dependent quantities :
$$V(T) = f[\rho(T), C_{ij}(T)]$$
In the particular case of a longitudinal "long beam" mode the simple relation for ultrasonic velocity is related to Young's modulus by equation (2).
When high temperature measurements are made in "thermally stable" crystalline ceramics, i.e. ceramics which do not encounter any microscopic or macroscopic transformation, a regular decrease of elastic moduli is observed, induced by the decrease of the interatomic bond rigidity. It has been shown that Young's modulus for crystalline oxides can be related to the absolute temperature by a semi-empirical equation [22] :

$$E = E_o - BT \exp(-T_o/T) \qquad (5)$$

where Eo is Young's modulus at 0K, and B and T_o are characteristic constants of the material. This equation gives a linear E (T) decrease at high temperature because exp (- T_o/T) ≠ 1 for T > 600 K [22]. If temperature involves any microstructural or macrostructural changes into the material, they will be accompanied by ultrasonic velocity variations as mentionned before, and a change will be observed on the V (T) or E (T) curves.
Two kinds of mechanisms may be responsible for material modifications in ceramics under the effect of temperature :
- diffusion and transport phenomena involving phase transformations or microstructural evolutions,
- mechanical effects due to thermal stresses which entail microcracking (damage).

Monitoring of microstructural evolutions with temperature

The first application of high temperature ultrasonic measurements can be the monitoring of the processing of ceramic bodies.
Fig. 8 shows the Young's modulus evolutions during firing of a traditionnal china at a rate of 60°C/h. Parallelepipedic samples 50 x 2.5 x 2.5 mm³ were cut out from a dried plate, and measurement was made using the "long beam method"

at a frequency of 180 kHz. Corrections for mass and length variation of the sample were applied using the results of thermogravimetric and dilatometric experiments carried out in samples of the same composition and under similar firing conditions [23]. The low modulus of the unfired body comes from the weak bonding between particles in the green state. Three effects are observed :

- a first increase at 573°C occurs when the $\alpha \longrightarrow \beta$ polymorphic transformation of quartz takes place,
- from 1000°C to 1300°C, the strong increase of E (about 8 E_o at 1300°C) accompanies the densification of the body : crystallisation of mullite, formation of a glassy phase,
- above 1300°C the decrease of E is due to closed porosity development and softening of the glassy phase.

When cooling the E/E_o = f (T) curve becomes regular, and reversible after a second heating, as in a densified stable ceramic.

Such results are very helpful for the calculation of deformations and volume variations when firing a ceramic piece [23].

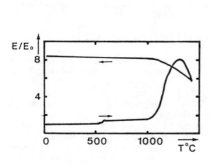

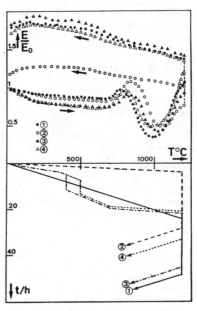

Fig. 8 - Relative variations of Young's modulus during firing of a china sample. E_o=8 GPa. f=180 kHz.

Fig. 9 - (on the right)
Relative variation of Young's modulus in a refractory concrete during four heat treatments corresponding to T=f(t) curves E_o=43 GPa. f=90 kHz.

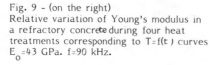

Ultrasonic measurements can also be used to improve heat treatments in order to achieve the best mechanical properties, as reported on Fig. 9. Measurements were performed in a basic chemically bonded refractory concrete used as furnace lining in cement plants [7]. As the green state was characterized by periclase grains and pores as large as 5 mm in diameter, large samples of (100 x 10 x 10 mm³) were used. Before measurement they were dried 48 h at 120°C in air.

By correlation with differential thermal (DTA) and X ray diffraction (XRD) analysis, various stages have been characterized in the E/E_o evolution [7] :

- for T > 700°C the increase of E toward a maximum is linked to the crystallisation of an amorphous phase (2 CaO - Fe_2O_3),
- the following decrease of E denotes the fusion of an eutectic phase in the system SiO_2 - Al_2O_3 - CaO - F_2O_3,

- the strengthening of the concrete occurs above 1000°C by the crystallisation of new phases.

These curves show the role of the heating rate in the achievement of the final mechanical properties of the concrete.

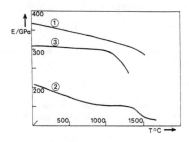

Fig. 10 - Young's modulus vs. temperature for three commercial sintered ceramics :
(1) Al_2O_3, (2) $ZrO_2(MgO)$, (3) SiAlON. f=180 kHz, heating rate 60°C/h.

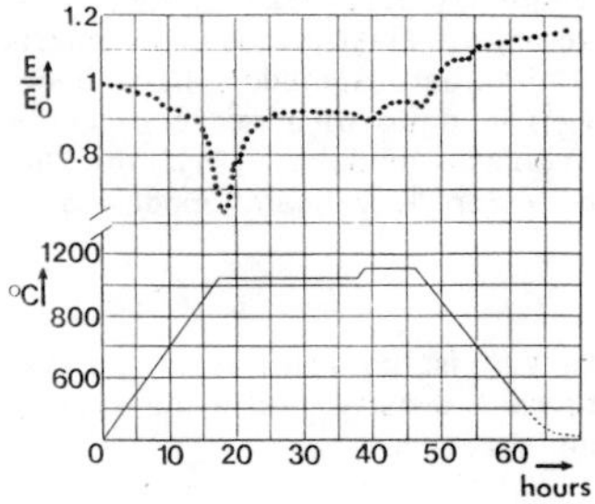

Fig. 11 - E/E_o vs. time in a SiAlYON glass (3wt.% N) during the heat treatments reported on the same figure E_o=150 GPa, f=350 kHz.

Moreover, ultrasonic measurement of Young's modulus with temperature in engineering ceramics is a helpful tool to determine the temperature range where phase transformations may occur.

The differences in behaviour between three commercial sintered ceramics $(Al_2O_3$, stabilized ZrO_2 and SiAlON) are clearly pointed out on Fig. 10. As reported before, the regular and reversible variation E (T) for Al_2O_3 denotes the microstructural stability in the range 20°C - 1500°C. The slight departure of linearity above 1300°C is due to sliding at grain boundaries [6].

SiAlON encounters a drastic decrease of E after 1000°C because of the softening of the intergranular glassy phase.

Zirconia has been stabilized by MgO additives in the tetragonal form which transforms into the cubic form (with a lower Young's modulus) in the 1400°C - 1500°C range. This transformation is reversible when thermal cycles are made.

Fig. 11 illustrates the sensitivity of ultrasonic measurements to crystallisation phenomena in glasses. The material is a SiAlYON glass with 3 wt. % nitrogen which is at first heated at 60°C/h up to the temperature where crystallisation has been found to occur by XRD analysis (1050°C). Then, the ultrasonic velocity is measured during ageing at this temperature [16]. It can be seen on the figure that, when heating, a strong decrease of Young's modulus is observed after 900°C which characterizes the softening of the glass. During ageing at 1050°C, the nucleation and growth of crystallites (Fig. 7) involve an increase of about 30 % in E. The maximum of crystallized phase is reached after 12 h at 1050°C. Then an augmentation of the ageing temperature leads to a new increase of about 5 % denoting the activation of the crystallization process with temperature. A quantitative interpretation of these curves has been undertaken using correlations with XRD results.

Ultrasonic monitoring of damage due to thermal stresses

Most thermomechanical engineering ceramics are used because they are refractory materials with high mechanical properties. But their rather low thermal conductivity, their high elastic moduli and their brittleness make them sensitive to thermal stresses which involve microcraking when the local stress field becomes superior to the ultimate strength. The origin of microcracking may be dynamical temperature gradients (thermal shocks and thermal fatigue), thermal expansion mismatch in anisotropic large grained ceramics or in fiber composites, phase transformation accompanied by a volume change in a rigid matrix, etc ...

The effects of microcracking on the propagation parameters are a strong increase of attenuation (multiple scattering, provided the wavelength is larger than the geometry parameter of the crack array), and a decrease in velocity due to the weakening of elastic moduli by microcracks.

Numerous models have been proposed for the calculation of the average Young's modulus of a microcracked isotropic material. From Hasselman's equation for Young's modulus, [24], the longitudinal wave velocity in a microcracked material for "long beam" mode may by written [13] :

$$V_L = V_{Lo} \left(1 + \frac{16 Nb^3}{9} \right)^{-1/2} \tag{6}$$

where V_{Lo} is the velocity in the uncracked body assumed to be isotropic, N is the microcrack density, and b the average diameter of cracks which are assumed to be "penny shaped".

A first example is given by polycrystalline ceramics with a non cubic crystalline structure which are affected by microcracking when they encounter temperature variations. If temperature cycles are made, crack opening and crack closure phenomenon are observed, which lead to hysteretic effects on the curves E(T) [25]. The condition for grain-boundary cracking can be theoretically analysed on the basis of an energy criterion from which a critical grain size is determined which is proportionnal to the ratio $\gamma_f/(\Delta\alpha)$max, γ_f being the fracture surface energy and $(\Delta\alpha)$max the maximum anisotropy of the thermal expansion coefficient of a grain. For alumina, this critical grain size is about 70 µm, thus this phenomenon is only observed in large grained refractories [6]. On the contrary for strongly anisotropic ceramics such as aluminium titanate $TiAl_2O_5$, the critical grain size is very small (about 2 µm) [26]. On Fig. 12 an important Young's modulus defect (70 %) due to microcracking is observed in a $TiAl_2O_5$ sample cycled between 20°C and 800°C at 60°C/h.

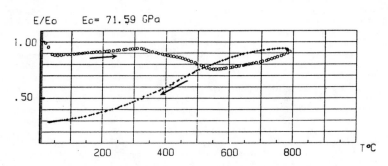

Fig. 12 - E/E_o vs. temperature for Al_2TiO_5.
Heating and cooling rate : 60°C/h. f=350 kHz.

When heating, two decreasing zones are observed on the curve $E/E_o = f(T)$. The first between 20°C - 50°C, the second between 300°C - 550°C. The lower temperature effect may be attributed to the relaxation of residual intergranular stresses which involve an extension of microcracks. The second effect denotes microcracking due to thermal expansion anisotropy. When cooling the regular decrease is due to crack opening.

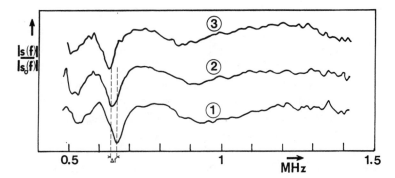

Fig. 13 - Ultrasonic spectroscopy results in a SiC - SiO$_2$ composite submitted to 300°C - 1000°C thermal fatigue cycles.
The thermal cycle period was 60 s.
(1) Transfer function before thermal cycling
(2) Transfer function after 1 cycle
(3) Transfer function after 100 cycles

Severe thermal cyclings with superimposed thermal gradients are among the most important sources of damage by thermal fatigue in structural ceramics. Fig. 13 shows the results of tests on a SiC - SiO$_2$ long fiber composite supplied by the Aerospatiale society [9]. This composite is made from a 3D fabric of SiC fibers (Nicalon) filled with SiO$_2$ by a sol gel impregnation process and densified at rather low temperature (< 1000°C). The SiO$_2$ matrix is amorphous. Samples were disks (30 mm in diameter, 15 mm in thickness) with their axis parallel to one of the three fiber directions. Periodic 300°C - 1000°C cycles were applied on the upper face while the lower face was regulated at 200°C. Ultrasonic spectroscopy analysis was performed and the transfer function of the sample plotted after each cycle. This transfer function is characteristic of a composite macrostructure with widened lines due to multiple scattering at the fibre/matrix interfaces [13]. After the first cycle the frequency of the higher amplitude line is lowered from 0.655 MHz toward 0.627 MHz and then remains stable up to 100 cycles.

The interpretation was made using a model, based on optical microscopic observations, which allowed the calculation of the transfer function of a damaged sample [13]. It confirmed that the modifications observed after the first cycle were due to intense cracking of the matrix through a thin layer of the heated surface. After one cycle, this layer acts as a protection against the propagation of damage inside the material which is not further affected after 100 cycles.

CONCLUSION

As they are sensitive to microstructural changes and to microcracking, the measurement of elastic moduli by suitable ultrasonic pulse methods are helpful tools for the monitoring of the high temperature behaviour of ceramic bodies.
The field of interest may concern :
- the improvement of ceramic processing by the determination of structural changes and densification stages during heat treatments,
- for material science research, the investigation of the kinetic parameters of phase changes (for example crystallisation of glasses at high temperature),
- the study of damage in ceramics submitted to thermal stresses.
This last application is very important for the determination of optimised service conditions for ceramic mechanical components. Nevertheless, care must be taken when using results obtained from experiment carried in samples with rather simple shapes, for the prediction of reliability of real components.

REFERENCES

1 - L.W. Kessler and T.M. Gasiel, Adv. Ceram. Mater. 2, 4, 107 (1987).

2 - M.C. Bhardwaj, Adv. Ceram. Mater. 3, 3A (1987).

3 - E.P. Papadakis, J. Acoust. Soc. Amer. 42, 5, 1046 (1967).

4 - E.P. Papadakis, K.A. Fowler, L.C. Lynnworth, A. Robertson, and E.D. Zysk, Jour. of Appl. Phys., 45, 6, 2409 (1974).

5 - C. Gault, P. Lamidieu, F. Platon, at 6th International Conference on Non Destructive Testing Methods, Strasbourg, France, CGA - Alcatel Publ. (1986), p. 85.

6 - C. Gault, F. Platon and D. Le Bras, Mater. Sci. Engn, 74, 105 (1985).

7 - P. Lamidieu and C. Gault, Mater. Sci. and Engn., 77, L 11 (1986).

8 - L.C. Lynnworth, IEEE Trans. Sonics and Ultrasonics, Su-22, 2, 71 (1975).

9 - P. Lamidieu and C. Gault, Rev. Phys. Appl. 23, 201 (1988).

10 - P. Boch, J.F. Coudert, C. Gault, in Ceramic Components for Engines, KTK Scientific Publishers, TOKYO (1984) p. 682.

11 - R.A. Kline, J. Acouts. Soc. Amer. 76, 2, 498 (1984).

12 - P. Lamidieu, D. Jacques, C. Gault, J.F. Coudert, in Ceramic Materials and Components for Engines, edited by W. Bunk and H. Hausner (Published by Verlag Deutsche Keramusche Gesellschaft, 1986), p. 869.

13 - P. Lamidieu, Dr. Thesis, Limoges University, France (1987).

14 - E.P. Papadakis, in Physical Acoustics, vol IV, part B, chap. 15, edited by W.P. Mason (Academic Press, New York and London, 1968) p. 269.

15 - P. Boch and J.P. Giry, High Technology Ceramics, edited by P. Vincenzini (Elsevier, 1987).

16 - T. Rouxel and J.L. Besson, ENSCI Limoges, France (Private Com.).

17 - Z. Hashin and S. Shtrikman, J. Appl. Phys. 33, 10 (1962), 3125.

18 - R. Hill, J. Mech. Phys. Solids 11 (1963), 357.

19 - G. Ondracek, Mater. Chem. and Phys., 15, 281 (1986).

20 - L.F. Nielsen, Mater. Sci. and Engn. 52, 39 (1982).

21 - K.K. Phani, S.K. Niyogi, Jour. of Mater. Sci. Let. 5, 427 (1986).

22 - J.B. Wachtman Jr., W.E. Tefft, D.G. Lam and C.S. Apstein, Phys. Rev. 122, 6, 1754 (1961).

23 - J.M. Gaillard, E.N.S.C.I. Limoges, France (Private Communication).

24 - D.P.H. Hasselman, J.P. Suigh, Am. Ceram. Soc. Bull. 58, 9, 856 (1979).

25 - E.D. Case, J.R. Smyth and O. Hunter, Mater. Sci. Engin. 51, 175 (1981).

26 - Y. Ohya, Z.E. Nakagawa, K. Hamano, J. Am. Ceram. Soc. 70, 8, C-184 (1987).

ELASTIC PROPERTIES OF POROUS CERAMICS

H.M. Ledbetter[*], M. Lei[*], and S.K. Datta[**]
[*]Institute for Materials Science and Engineering, National Institute of Standards and Technology, Boulder, Colorado 80303
[**]Department of Mechanical Engineering and CIRES, University of Colorado, Boulder, Colorado 80309

ABSTRACT

Using theoretical models, we consider the sound velocities and elastic constants of ceramics containing pores. As an example, we consider alumina. However, the approach applies to all ceramics. As a point of departure, we consider spherical pores. For all the usual elastic constants--Young modulus, shear modulus, bulk modulus, Poisson ratio--we give relationships for both the forward and inverse cases: predicting the porous ceramic properties and estimating the pore-free ceramic properties. Following a suggestion by Hasselman and Fulrath that sintering or hot pressing can produce cylindrical pores, we derive a relationship for the elastic constants of a distribution of randomly oriented long cylinders ($c/a = \infty$, the prolate-spheroid limit). This model predicts elastic constants lower than for spherical pores, but well above measurement. We obtain agreement with observation by assuming the pores are oblate spheroids. For alumina, the necessary aspect ratio equals one-ninth. Besides pore aspect ratio, the model requires only the pore-free alumina elastic constants. It contains no adjustable parameters.

INTRODUCTION

Despite many experimental and theoretical studies, for ceramics there remains a major unsolved problem: how does porosity affect physical properties, especially the elastic constants? When Wachtman [1] reviewed this topic in 1969, he gave twenty references and concluded that nonspherical pores present the principal obstacle to theory.

Since 1969, many more studies appeared; Phani and Niyogi [2] gave many references. Notable studies include those by Wang [3,4] and by Sayers and Smith [5,6]. Considering uniform-sized spherical particles packed in a simple-cubic array in a porous material, Wang derived a theoretical relationship between porosity and Young modulus and showed the exact solution graphically. He also proposed an approximate solution that agreed with the exact solution throughout the porosity range 0 to 0.38.

Using a self-consistent theory, Sayers and Smith [5,6] calculated the longitudinal-wave velocity for a porous material in the long-wavelength limit. Their calculation applies to the entire range of pore volume fraction, but, it applies only to spherical pores.

The present study focuses on pore shape. Using methods developed previously and for composite materials, we consider the well-understood case of spherical pores. And we consider two further: (1) an ensemble of randomly oriented rod-shaped (prolate) pores, a geometry suggested by Hasselman and Fulrath [7], but developed incompletely; (2) an ensemble of randomly oriented disc-shaped (oblate) pores. For the present study, we treat a ceramic as a composite containing pores (particles with zero elastic resistance).

MODELING

We consider pores as "particles" distributed in a homogeneous isotropic matrix. Using a multiple-scattering plane-wave approach, Ledbetter and Datta [8] studied the elastic properties of a composite material containing randomly distributed and randomly oriented ellipsoid-shape particles. They calculated the effective wave speeds of plane waves, both longitudinal and shear, in the long-wavelength limit and gave relationships for the bulk and shear moduli:

$$\frac{B - B_m}{B_p - B_m} = \frac{\frac{1}{3} \bar{C} \sum\limits_{i,j}^{3} T_{iijj}}{1 - \dfrac{\bar{C}(B_p - B_m)}{3B_m + 4G_m} \sum\limits_{i,j=1}^{3} T_{iijj}} \tag{1}$$

$$\frac{G - G_m}{G_p - G_m} = \frac{\frac{1}{5} \bar{C} G_m \left(\sum\limits_{i,j=1}^{3} T_{ijij} - \frac{1}{3} \sum\limits_{i,j=1}^{3} T_{iijj} \right)}{G_m - \dfrac{2\bar{C}}{75} \dfrac{(G_p - G_m)(3\lambda_m + 8G_m)}{\lambda_m + 2G_m} \left(\sum\limits_{i,j=1}^{3} T_{ijij} - \frac{1}{3} \sum\limits_{i,j=1}^{3} T_{iijj} \right)} \tag{2}$$

Here, B and $\bar{C}$ denote the bulk modulus and the volume fraction of particles; λ and G the Lamé constants; subscripts m and p matrix and particle. Also,

$$T_{ijij} = T_{jiji} = \frac{1}{2(1 + 2US_{ijij})} \tag{3}$$

$$[T_{iijj}] = \begin{bmatrix} 1+US_{1111}+VS_1 & US_{1122}+VS_1 & US_{1133}+VS_1 \\ US_{2211}+VS_2 & 1+US_{2222}+US_2 & US_{2233}+VS_2 \\ US_{3311}+VS_3 & US_{3322}+VS_3 & 1+US_{3333}+VS_3 \end{bmatrix}^{-1} \tag{4}$$

$$U = \frac{G_p}{G_m} - 1 \tag{5}$$

$$V = \frac{1}{3}\left(\frac{B_p}{B_m} - \frac{G_p}{G_m} \right) \tag{6}$$

$$S_i = \sum\limits_{j=1}^{3} S_{iijj} \tag{7}$$

The S_{ijkl}, which relate constrained and stress-free strains, and which contain the particle radii a, b, and c, were defined by Eshelby [9].

Since for pores the elastic stiffnesses B and G equal zero, in Eqs. (1) and (2) we put $B_p = G_p = 0$ and obtain

$$B = B_m \left[1 - \frac{\frac{1}{3} \bar{C} \sum\limits_{i,j=1}^{3} T_{iijj}}{1 + \dfrac{\bar{C}B_m}{3B_m + 4G_m} \sum\limits_{i,j=1}^{3} T_{iijj}} \right] \tag{8}$$

$$G = G_m \left[1 - \frac{\frac{1}{5} \overline{C} \, G_m \left(\sum\limits_{i,j=1}^{3} T_{ijij} - \frac{1}{3} \sum\limits_{i,j=1}^{3} T_{iijj} \right)}{G_m + \frac{2\overline{C}G_m}{75} \frac{3\lambda_m + 8G_m}{\lambda_m + 2G_m} \left(\sum\limits_{i,j=1}^{3} T_{ijij} - \frac{1}{3} \sum\limits_{i,j=1}^{3} T_{iijj} \right)} \right] \tag{9}$$

Because $U = -1$ and $V = 0$ when $B_p = G_p = 0$, then

$$T_{ijij} = T_{ijij} = \frac{1}{2(1 - 2S_{ijij})} \tag{10}$$

$$[T_{iijj}] = \begin{bmatrix} 1-S_{1111} & -S_{1122} & -S_{1133} \\ -S_{2211} & 1-S_{2222} & -S_{2233} \\ -S_{3311} & -S_{3322} & 1-S_{3333} \end{bmatrix}^{-1} \tag{11}$$

Model for spherical pores

First, we consider pores as spherical "particles." In this case, $a = b = c$, we obtain

$$\sum\limits_{i,j=1}^{3} T_{iijj} = \frac{9 \, (1 - \nu_m)}{2(1 - 2\nu_m)} \tag{12}$$

$$\sum\limits_{i,j=1}^{3} T_{iijj} = \sum\limits_{i,j=1}^{3} T_{ijij} = \frac{75(1 - \nu_m)}{7 - 5\nu_m} \tag{13}$$

Here, ν denotes the Poisson ratio. Substituting Eqs. (12) and (13) into Eqs. (8) and (9), we obtain expressions for the bulk and shear moduli for a material containing spherical pores:

$$B = B_m \frac{4(1 - \overline{C})G_m}{4G_m + 3\overline{C}B_m} \tag{14}$$

or

$$B = B_m \left[1 + \frac{A\overline{C}}{1 - (1 + A)\overline{C}} \right] \tag{14a}$$

$$G = G_m \frac{(1 - \overline{C})(9B_m + 8 \, G_m)}{(9 + 6\overline{C})B_m + (8 + 12\overline{C})G_m} \tag{15}$$

or

$$G = G_m \left[1 + \frac{D\overline{C}}{1 - (1 + D)\overline{C}} \right] \tag{15a}$$

Here,

$$A = -\frac{3B_m + 4G_m}{4G_m} \tag{16}$$

$$D = -\frac{5(3B_m + 4G_m)}{9B_m + 8B_m} \tag{17}$$

Equations (14) and (15) are identical with Hashin's [10] composite-sphere-model equations for the bulk and shear moduli. We invoke the well-known isotropic-material relationships for Young modulus and Poisson ratio:

$$E = \frac{9BG}{3B + G} \tag{18}$$

$$\nu = \frac{1}{2} \frac{3B - 2G}{3B + G} \tag{19}$$

And, for the elastic contants E and ν of a porous material, we obtain:

$$E = E_m \frac{4(1 - \overline{C})(3B_m + G_m)(9B_m + 8G_m)}{108B_m^2 + 132B_mG_m + 32G_m^2 + \overline{C}(99B_m^2 + 168B_mG_m)} \tag{20}$$

or

$$E = E_m \left[1 + \frac{F\overline{C}}{1 - (1 + F)\overline{C}} \right] \tag{20a}$$

$$\nu = \nu_m \frac{(3B_m + G_m)[36B_m^2 + 8G_mB_m - \frac{64}{3}G_m^2 + \overline{C}(6B_m^2 + 32B_mG_m)]}{(36B_m - 2B_m)[36B_m^2 + 44G_mB_m + \frac{32}{3}G_m^2 + \overline{C}(33B_m^2 + 56B_mG_m)]} \tag{21}$$

or

$$\nu = \nu_m \left[1 + \frac{H\overline{C}}{1 - (1 + A)\overline{C}} \right] \tag{21a}$$

Here,

$$F = - \frac{207B_m^2 + 300B_mG_m + 32G_m^2}{108B_m^2 + 132B_mG_m + 32G_m^2} \tag{22}$$

$$H = - \frac{27B_m(3B_m + 4G_m)(3B_m - 4G_m)}{324B_m^3 + 180B_m^2G_m - 168B_mG_m^2 - 64G_m^3} \tag{23}$$

From Eqs. (14), (15), (20), and (21) we obtain the inverse relationships:

$$G_m = \frac{-R + \sqrt{(R^2 - 4QS)}}{2Q} \tag{24}$$

$$B_m = \frac{4G_mB}{4(1 - \bar{C})G_m - 3\bar{C}B} \tag{25}$$

$$E_m = \frac{36G_mB}{4(3B + G_m) - \bar{C}(4G_m + 3B)} \tag{26}$$

$$\nu_m = \frac{2(3B - 2G_m) + \bar{C}(4G_m + 3B)}{4(3B + G_m) - \bar{C}(4G_m + 3B)} \tag{27}$$

Here,

$$Q = \frac{8}{3}(1 - \bar{C}) \tag{28}$$

$$R = (3 - 2\bar{C})B - (\frac{8}{3} + 4\bar{C})G \tag{29}$$

$$S = -3(1 + \bar{C})BG \tag{30}$$

We can use Eqs. (24)-(27) to estimate the pore-free material properties.

As an example, we choose porous alumina. Table I shows the physical properties of pore-free alumina obtained from monocrystal elastic constants [11] by a Voigt-Reuss-Hill averaging method.

TABLE I. Properties of nonporous alumina.

ρ (g/cm^3)	E (GPa)	B (GPa)	G (GPa)	λ (GPa)	ν
3.986	402.9	252.1	163.3	143.2	0.2336

For the Young modulus of porous alumina, we use the Phani-Niyogi [2] empirical fitting results to Knudsen's [12] measurements. For the Young modulus, Fig. 1 shows a comparison between measurements and a spherical-pore model. Calculated results for the other elastic constants will appear elsewhere.

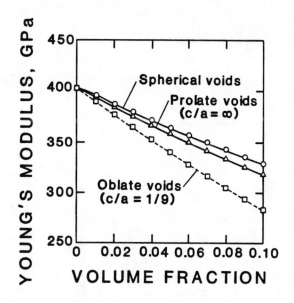

Fig. 1. For Young modulus versus pore volume fraction, this figure compares three sets of modeling results--spheres, prolate-limit cylinders, discs (c/a - 1/9)--with observation. Because we fit the disc case to observation, these two curves coincide. Symbols represent calculations. Dashed curve represents measurements.

Model for nonspherical pores

Second, we consider pores as ellipsoidal "particles".

(a) Prolate limit

For a prolate ellipsoid, which approximates a rod, a - b ≪ c. If we take the limit c/a → ∞, then we get

$$\sum_{i,j=1}^{3} T_{iijj} = \frac{5 - 4\nu_m}{1 - 2\nu_m} \tag{31}$$

$$\sum_{i,j=1}^{3} T_{ijij} - \frac{1}{3} \sum_{i,j=1}^{3} T_{iijj} = \frac{8(5 - 3\nu_m)}{3} \tag{32}$$

Substituting Eqs. (31) and (32) these two equations into Eqs. (8) and (9), we obtain expressions for the bulk and shear moduli in the prolate limit:

$$B = B_m \left[1 - \frac{\overline{C}(5 - 4\nu_m)}{3 - 6\nu_m + \dfrac{3\overline{C} B_m(5 - 4\nu_m)}{(3B_m + 4G_m)}} \right] \tag{33}$$

or

$$B = B_m \frac{(3B_m + 4G_m)G_m - 4\overline{C}G_m(B_m + G_m)}{(3B_m + 4G_m)G_m + 3\overline{C}B_m(B_m + G_m)} \tag{33a}$$

$$G = G_m \left[1 - \frac{8\overline{C}(5 - 3\nu_m)}{15 + \dfrac{16\overline{C}}{15} \dfrac{(5 - 3\nu_m)(3\lambda_m + 8G_m)}{\lambda_m + 2G_m}} \right] \tag{34}$$

Similarly, we can obtain relationships for the Young modulus and the Poisson ratio by using Eqs. (18) and (19). For the Young modulus, Fig. 1 shows the prolate-limit results.

(b) Oblate limit

Because the prolate-limit-pore-shape approach gives results only slightly below the spherical-pore-shape model and well above measurement, we consider oblate-ellipsoid-shape pores (a = b > c). The oblate limit (c/a = 0) is singular, and we plan to discuss elsewhere some analytical expressions for the elastic constants. Here, we give numerical results. Figure 1 shows the effect of changing aspect ratio from unity (sphere: c/a = 1) toward the oblate limit.

DISCUSSION

In considering the elastic properties of a porous composite, we take a different approach than used recently [2-4]. We ignore empirical parameters, particle shape, and particle packing. Instead, we focus on pore geometry. Figure 1 shows that the elastic properties depend strongly on the pore shape. We confirm that oblate pores, more than prolate pores, reduce elastic stiffness, depending on pore aspect ratio. Also, for alumina, we see that both spherical and prolate-limit pores fail to agree with observation.

To obtain the best model fit to the observed Young modulus, we used oblate pores with an aspect ratio of one-ninth. With an aspect ratio less than one third, oblate pores predict lower bulk and shear moduli and Poisson ratio than spherical and prolate-limit pores. To check model predictions, for elastic constants other than the Young modulus, we need measurements on the corresponding properties, measurements generally unavailable.

The theoretical approach described here applies to any material containing voids, not only to porous ceramics. We can also use the model inversely to study pore geometry. Thus, the model provides a possibly valuable nondestructive-evaluation tool. Because the present model fails to allow for pore-pore overlap and because it breaks down for nonspherical particles at high volume concentration [13], we must apply the model to lower volume fractions. We believe the model can be extended to higher volume fractions. The model ignores the occurrence of microcracks [14].

CONCLUSIONS

From this study, we reach the following five conclusions:

(1) From measurements of the elastic constants of a material containing spherical pores, we can estimate the pore-free material properties.

(2) Physical properties of a porous material depend strongly on the pore shape. Discs produce larger effects than rods.

(3) A model calculation shows that randomly oriented cylinder-shaped pores (even in the prolate limit) predict a too-high elastic stiffness, only slightly lower than for spherical pores.

(4) Theoretical calculations suggest that porous alumina contains oblate-ellipsoid-shaped pores, not cylindrical pores suggested by other studies.

(5) Further experimental studies should include elastic constants other than the Young modulus: for example, Poisson ratio and bulk modulus. These other elastic constants would provide invaluable checks of our predictions.

References

1. J.B. Wachtman, in Mechanical and Thermal Properties of Ceramics. Spec. Publ. 303 (Nat. Bur. Stands., Washington, 1969), p. 139.

2. K.K. Phani and S.K. Niyogi, J. Mater. Sci. 22, 257 (1987).

3. J.C. Wang, J. Mater. Sci. 19, 801 (1984).

4. J.C. Wang, J. Mater. Sci. 19, 809 (1984).

5. C.M. Sayers, J. Phys. D: Appl. Phys. 14, 413 (1981).

6. C.M. Sayers and R.L. Smith, Ultrasonics 20, 201 (1982).

7. D.P.H. Hasselman and R.M. Fulrath, J. Am. Ceram. Soc. 48, 545 (1965).

8. H.M. Ledbetter and S.K. Datta, J. Acoust. Soc. Am. 79, 239 (1986).

9. J.D. Eshelby, Proc. Roy. Soc. Lond. A241, 376 (1957).

10. Z. Hashin, J. Appl. Mech. 29, 143 (1962).

11. W.E. Tefft, J. Res. Nat. Bur. Stand. 70A, 277 (1966).

12. F.P. Knudsen, J. Am. Ceram. Soc. 45, 94 (1962).

13. H.M. Ledbetter, S.K. Datta, and R.D. Kriz, Acta Metall. 32, 2225 (1984).

14. B. Budiansky and R.J. O'Connell, Int. J. Solids Struct. 12, 81 (1976).

THERMAL WAVE CHARACTERIZATION OF GLASS-BONDED ALUMINA

LORRETTA J. INGLEHART[*], RICHARD A. HABER[**], and JOHN B. WACHTMAN, Jr.[**]
[*]The Johns Hopkins University, Center for Nondestructive Evaluation, Materials Science and Engineering, Baltimore, MD 21218
[**]Rutgers University, Center for Ceramics Research, Piscataway, NJ 08854

ABSTRACT

We present the results of a study of glass-bonded Al_2O_3 with controlled composition and particle size using photoacoustic thermal wave nondestructive evaluation. Experimental results are related to particle size, density, porosity, and thermal properties of the materials.

INTRODUCTION

We are presently investigating ceramic materials with varied processing and material parameters with the use of thermal wave nondestructive evaluation [1]. Previous results for Al_2O_3 samples subjected to varying sintering times with no composition or particle size control gave a clear indication of the sensisitivity of the thermal wave method to the presence of porosity with processing [2]. In this work, we investigate the model system of hot-pressed glass-bonded Al_2O_3 to determine effects of composition and particle size. The samples investigated are alumina with two different silica glass fractions, which have well controlled volume fractions of alumina and varying alumina particle sizes. Many commercial aluminas, as well as other ceramics, are glass bonded. All sample surfaces were studied in the as-prepared state. The sample parameters are presented in Table 1.

Table 1. Sample parameters: Sample and firing schedule for 1 hour, particle size, volume percent of alumina, bulk density, interparticle spacing(MFP), and calculated thermal diffusivity.

Sample	Al_2O_3 Particle Size (μm)	Vol.% Al_2O_3	Bulk ρ	MFP (μm)	α^* (cm^2/s)
F_1 (1100)	5	60	3.26	4.4	0.0521
F_2 (1050)	5	40	2.90	10.0	0.0368
5A (1050)	12	40	2.67	24.0	0.0368
F_3 (1150)	20	60	3.42	17.8	0.0521
F_7 (1050)	20	40	-----	40.0	0.0368
F_6 (1050)	40	40	-----	80.0	0.0368

The mean free path (MFP) or interparticle spacing, was determined according to the method presented by Fullman [3] for the case of spherical particles. The thermal diffusivity, α^*, was calculated according to

$$\alpha^* = \alpha_{Al_2O_3} (\text{Vol \% } Al_2O_3) + \alpha_{SiO_2} (\text{Vol \% } SiO_2), \tag{1}$$

where we have used room temperature values for polycrystalline alumina and fused silica according to [4]. We are presently measuring the thermal diffusivity for these samples and will present the results in a future paper.

Thermal wave nondestructive evaluation is characterized by the periodic diffusion of heat in the form of a critically damped, localized heat wave, a "thermal wave", which propagates in the heated material [1, 2]. The interaction of a thermal wave with features in a solid depends on local variations in the thermal properties of the material such as thermal conductivity (κ), density (ρ), specific heat (c), thermal diffusivity (α, $\alpha = \kappa/\rho c$), thermal effusivity (e, $e = \sqrt{\kappa \rho c}$), and the presence of defects such as cracks, pores, or voids. As the modulation frequency, ω, of the heat source is varied, the effective penetration depth of the thermal wave in the material is varied. This effective range of penetration is given by the thermal diffusion length in the solid, [μ_s, $\mu_s = (\alpha/\pi f)^{1/2}$]. The effective range of penetration used in these experiments ranges from 25 μm to 500 μm.

For ease of data analysis, the photoacoustic thermal wave detection system was chosen for this study. We use a focused CO_2 laser which is absorbed by both the alumina and glass, with a spot size about 200 μm. The intensity of the laser is modulated with a mechanical chopper. Photoacoustic detection involves placing the sample in a closed cell which contains a sensitive microphone (see Fig.1). A ZnSe window transmits the modulated CO_2 beam to the sample surface where it is absorbed by the material. The periodic heating of the sample produces a periodic temperature increase in the cell which is detected by the microphone as a periodic increase in pressure. The pressure variation is detected by the microphone as a low frequency "sound" wave. The photoacoustic measurement is proportional to the temperature averaged over the surface of the sample, which can be related to the pressure variation, and finally to the thermal and optical properties of the material [2]. The voltage of the microphone is detected by a vector lock-in amplifier which we use to monitor the magnitude and phase of the photoacoustic signal. For the samples studied in this paper, we monitor magnitude and phase of the photoacoustic signal as the power of the laser is increased (at a fixed frequency), and the magnitude and phase of the signal as the frequency is varied (at a fixed power). We assume that the samples are optically opaque and thermally thick. Then the real part of the photoacoustic signal is given approximately by [1]

$$|S_p| = C\,P\,(1-R)\,\frac{\mu_g\,\mu_s}{l_g\,\kappa_s} \qquad (2)$$

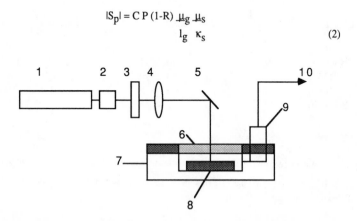

Figure 1. Block diagram of thermal wave photoacoustic detection system. (1) CO_2 laser, (2) Polarizer-attenuator, (3) chopper, (4) focusing lens, (5) mirror, (6) ZnSe window, (7) Photoacoustic cell, (8) sample, (9) microphone (10) signal to lock-in amplifier. The data are stored in a microcomputer.

where $|S_p|$ is the real part of the ac signal magnitude due to the pressure in the cell, P is the incident laser power, R is the sample optical reflectivity, $\mu_{s,g}$ is the thermal diffusion length in the solid or gas, l_g is the gas length in the cell, κ_s is the thermal conductivity of the solid, and C is a constant. The terms which vary from sample to sample are those involving R, l_g, μ_s and κ_s.

RESULTS AND DISCUSSION

Previous studies using the photoacoustic method to monitor alumina samples with varying densities showed that the power dependence of the signal could be directly related to the amount of porosity remaining as the processing of the sample was varied [2]. The optical properties as measured by FTIR were also found to vary with processing, but had no effect on the photoacoustic magnitude or phase measurements. The samples we now examine have had little or no variation in procesing, but have varying particle size and volume fraction of alumina, and are reported to have little or no porosity (92%-99% of theoretical density). The samples were provided in the shape of more-bars, each having approximate dimensions of 3mm x 4mm x 7mm.

A modulation frequency of 100 Hz was chosen for the power dependence measurements, which gave a thermal diffusion length of 108 µm for the 40% alumina samples and 129 µm for the 60% alumina samples. The intensity of the CO_2 laser incident on the sample surface was varied from 0-300 mW. Higher powers were possible, but we did not want to overheat the samples. For these materials, a power of 300 mW could correspond to an increase in ac surface temperature of 75-100 °C. The results of the power dependence for each composition are shown in Figs. 2 and 3.

The slope of the photoacoustic signal plotted as a function of power absorbed by the sample is related to the thermal properties of the sample as shown by Eq. (2). The sample

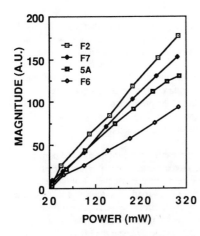

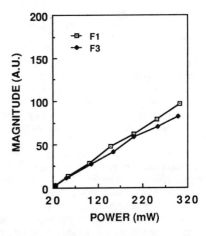

Figure 2. Magnitude of photoacoustic signal vs. power for samples with 40% Al_2O_3. Lines drawn to aid the eye.

Figure 3. Magnitude of photoacoustic signal vs. power for 60% Al_2O_3 samples. Lines drawn to aid the eye.

properties are contained in μ_s/κ_s, which can be rewritten as the thermal effusivity of the solid. In this case, the photoacoustic signal magnitude due to the sample is proportional to $(\sqrt{\kappa\rho c})^{-1}$ with a frequency dependence of ω^{-1}. The variation in the slopes of the curves in Figs. 2 and 3 can be understood physically as follows: for samples with low values of effusivity, as more power is deposited in the sample, the surface temperature increases faster. This had been observed in an earlier work [2], but the effect of particle size was not then known. Here we also see the direct relation of the effusivity with particle size. The slope of the photoacoustic signal increases with decreasing particle size with the exception of the 12 μm particle size sample (5A). We also note that the higher diffusivity materials have a smaller slope in general, while lower diffusivity materials have higher slopes, which can be seen by comparing Fig. 2 with Fig. 3. We also point out that the effusivity increases as the particle size increases, ie., larger particle sizes are conducting heat away at a faster rate than small particle sizes. We note that the slopes also scale with the interparticle spacing (MFP), even though the thermal diffusion lengths are much larger than the MFP. By comparing Figs. 2 and 3 in terms of density, we see that there is no direct correspondence with density.

The effect of periodic surface heating on composition (same particle size) can be observed by comparing the slopes of F3 with F7, or F1 with F2 in Figs. 2 and 3. The compositions with the smaller volume percentage of alumina have distinctly higher slopes, indicating that they reach higher surface temperatures more quickly, due to the increased amount of glass phase.

Thermal effusivity is the relevant thermophysical parameter when considering time dependent surface heating and cooling processes. It determines the surface temperature of the material being heated, and is a measure of the heat energy stored in a solid per degree of temperature rise after the beginning of a surface heating process. Changes in the surface of a solid introduced by machining, heat treatments, roughness, or porosity, can contribute to changes in the effective effusivity at the solid surface, which will affect the heating and cooling processes across the solid surface.

When we also consider effects due to porosity, then we see that the thermal effusivity is a most sensitive measurement of porosity. This can be understood by considering how thermal conductivity, heat capacity and density are affected by porosity [5]. Each of these physical parameters become reduced by the presence of porosity, which reduces the thermal effusivity and therefore increases the surface temperature. It may be just this effect that we are observing: for a given particle size sample the increase in surface temperature (photoacoustic signal) scales inversely with the thermal effusivity, which is sensitive to the presence of porosity. The photoacoustic measurement provides us with a straightforward way to determine relative effusivities and near surface porosities, including both open and closed porosities.

We have also conducted studies to determine if the effects of particle size and porosity could be observed by monitoring the behavior of the photoacoustic signal as a function of frequency, for a given power (100 mW). By varying the modulation frequency of the heat source, the effective range of the thermal wave in the solid is varied. The results of these studies showed that for identical compositions with different particle sizes, no measurable difference in the frequency behavior could be observed. We also note that the range of thermal diffusion lengths for these experiments is always greater than the particle size, and the MFP.

Measurements of the slope of the thermal wave signal magnitude as a function of frequency provide information about the thermal diffusivity of the solid. However, we have found that the effects of porosity are only weakly observed through the thermal diffusivity since this property is a ratio of conductivity, density, and heat capactiy and the effects of porosity effectively cancel. We do observe, however, in Figs. 4 and 5, that the frequency dependence of the phase is sensitive to the volume percentage of alumina, for a given particle size. This result is consistent with the effect of composition on the thermal diffusivity. At low modulation frequencies, the thermal diffusion length in the two materials is much larger than the particle size or the MFP. As the frequency increases and the thermal diffusion length decreases, we observe that the phase-shift increases faster for the smaller particle size samples than it does for the samples having larger particle size, independent of composition. It is evident that the composition has an even larger effect on the phase-shift as the frequency increases.

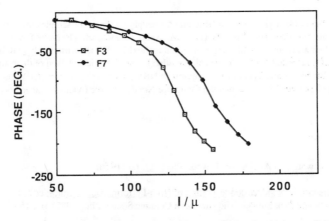

Figure 4. Phase shift of the photoacoustic signal corresponding to the thickness of the sample, l, normalized by the thermal diffusion length, μ, comparing two samples with 20 μm particle size and different volume % alumina.

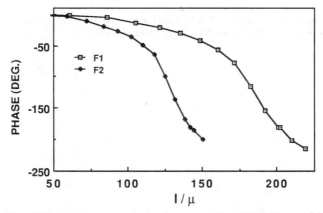

Figure 5. Phase shift of the photoacoustic signal corresponding to the thickness of the sample, l, normalized by the thermal diffusion length, μ, comparing two samples with 5 μm particle size and different volume % alumina.

CONCLUSION

We have investigated the model ceramic insulating system of glass-bonded alumina with controlled particle size and volume percentage of alumina with photoacoustic thermal wave nondestructive evaluation and found that samples having the same composition but different particle sizes ranging from 5 μm to 40 μm will have ac surface temperatures which scale with the particle size. This effect is determined to be related to the thermal effusivity of the material, which is a sensitive measure of surface porosity.

ACKNOWLEDGEMENTS

We thank Ms. Lynn Kowzan (Rutgers) for supplying the set of samples used in this investigation. We also thank Ms. Xia Teng, Mr. Surya Chadda, and Dr. Ivette F.C. Oppenheim (JHU) for their assistance in collecting and graphing the data. We are particularly grateful for the use of the CO_2 laser and photoacoustic cell provided by the Ceramics Divison of the National Institute of Standards and Technology (previously NBS). One of the authors (LJI) acknowledges the support of Battelle Memorial and the Center for Nondestructive Evaluation at Johns Hopkins University.

REFERENCES

1. A. Rosencwaig and A. Gersho, J. Appl. Phys. 47, 64 (1976).

2. L.J. Inglehart, in Nondestructive Testing of High-Performance Ceramics, edited by Alex Vary and Jack Snyder (The American Ceramic Society, Inc. 1987), pp. 163-176.

3. R.L. Fullman, Trans. of AIME, Vol. 197, No. 3, 447-52 (1953).

4. Y.S. Touloukian, R.W. Powell, C.Y. Ho, and M.C. Nicolasu, Thermal Diffusivity, (Plenum, New York, 1973).

5. W.D. Kingery, H.K. Bowen, D.R. Uhlmann, Introduction to Ceramics, 2nd Ed. (Wiley-Interscience Publication, New York, 1976), pgs. 518-590.

THERMAL PROPERTIES OF NON-METALLIC FILMS BY MEANS OF
THERMAL WAVE TECHNIQUES*

H.P.R. FREDERIKSE, X.T. YING, AND A. FELDMAN
NATIONAL INSTITUTE OF STANDARDS AND TECHNOLOGY, GAITHERSBURG, MD.

ABSTRACT

   The propagation of a thermal wave into a thin film or coating depends
on the thermal properties of the material. Consequently, thermal wave
generation and detection can be used to obtain the heat conductivity of
the material. The method is also useful because thermal wave propagation
is sensitive to inhomogeneity, porosity, inclusions, voids, and
delaminations. The results of two specific applications of the thermal
wave technique are presented, the heat resistance of oxide coatings and of
diamond films.

INTRODUCTION

   Thermal wave techniques hold considerable promise for the
nondestructive characterization of thermal and structural properties[1,2] of
solid materials. Because the attenuation of thermal waves is large, the
method is particularly well suited for thin films and coatings, typically
several μm to several mm thick. This implies a broad spectrum of
applications: protective coatings, corrosion resistant films, wear
resistant films, thermal barrier layers, etc. In some applications
thermal insulation is important while in other applications high thermal
conductance is important; however, structural integrity is usually
essential. In this paper the emphasis is on measuring the heat resistance
of a coating, although the technique can also be used to inspect for
inhomogeneity, porosity, inclusions, voids, and delaminations.
   Coatings and films produced by various deposition techniques[3],
usually have properties that deviate from the corresponding bulk
materials: the density is nearly always smaller; the refractive index is
not the same; and, the thermal conductivity is different. Furthermore, it
would be useful if these parameters could be determined at the time of
deposition, making it possible to change the processing conditions when
the measured properties are not within the desired range.
   In this paper we will describe the principles of the thermal wave
method and report on two different applications, the determination of the
thermal resistance of plasma sprayed oxide layers, and the measurement of
the heat conductivity of diamond films produced by chemical vapor
deposition.

THERMAL WAVES

   A representation of the thermal wave technique is shown in Fig. 1. A
laser beam, modulated at frequency f, produces a periodically varying
temperature on the surface of a film; it is desirable that the laser light
is absorbed within several hundred nanometers from the surface in order to
allow for the use of a one-dimensional analysis of the heat transport. A
thermal wave of frequency f will propagate into the film; at the
interface, part of the wave is reflected while the remainder continues
into the substrate. The reflected wave will interfere with the heat
source at the surface, resulting in a surface temperature variation that
differs in phase from the heating source.
   The different versions of the thermal wave method use different means
to determine the modulated surface temperature. The version shown in
Fig. 1, which is the method we are using, is known as photothermal
radiometry (PTR). In this method, the temperature of the spot heated by

the laser is measured with an infrared detector. Our particular
experimental set-up uses a liquid nitrogen cooled InSb photodiode with a

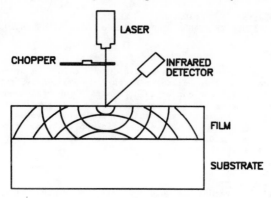

Fig 1. Principle of photothermal radiometry (PTR)

maximum sensitivity at $\lambda=5\mu m$, just below the cutoff wavelength of the
detector.

In the photoacoustic technique (PA), the specimen is enclosed in a
small chamber; the modulated surface temperature causes oscillations in
the chamber pressure that are detected with a microphone. In the optical
beam deflection method (OBD), a second laser probe-beam skims the specimen
parallel to the surface; the periodic variations in the refractive index
of the gas near the surface, induced by the periodic variations of the
surface temperature, produces a periodic deflection of the probe beam
("the mirage effect"[6,7]) that is detected by a position sensitive
detector. Other approaches to measuring the temperature modulation
include a transducer[8] in contact with the sample and a surface deformation
indicator[9] sensitive to thermal expansion.

The temporal and spatial variation in temperature depends on the
thermal diffusivity α of the film, the heat reflection coefficient R at
the film-substrate interface, and the modulation frequency f of the laser
beam. Consequently, by measuring either the magnitude of the detector
signal $|S|$ or the phase difference $\Delta\phi$ between the detector signal and the
incident laser beam modulation, one can determine the thermal diffusivity
of the film or coating. If we solve the thermal diffusion equation,
taking into account the boundary conditions at the film-ambient interface
and at the film-substrate interface, we find the phase difference to
be[5,10]

$$\Delta\phi = \phi - \phi_{ref} = \tan^{-1}\left[\frac{2R\sin(2x)}{\exp(2x)-R^2\exp(-2x)}\right] \qquad (1)$$

where the terms within eq.(1) are defined by $x=L/\mu_T=F\sqrt{f}$, $R=(1-b)/(1+b)$,
$F=L/\sqrt{(\pi/\alpha)}$, $e=\sqrt{(\kappa\rho C)}$, $b=e_s/e_c$, $\mu_T=\sqrt{[\alpha/(\pi f)]}$, $\alpha=\kappa/(\rho C)$, and where, L=film
thickness, , κ=thermal conductivity, ρ=density, C=specific heat,
μ_T=thermal diffusion length, e=effusivity, ϕ_{ref}=phase offset correction
term, and the subscripts s and c refer the substrate and coating,
respectively. ϕ_{ref} is measured as a function of f,

Table I.   THERMAL DIFFUSION LENGTH μ_T AS A FUNCTION
OF MODULATION FREQUENCY

	α	μ_T (μm)		
	cm^2/sec	f=10Hz	f=1kHz	f=1MHz
diamond	3.83	3431	343	10.9
aluminum	0.82	1614	161	5.1
stainless steel	0.037	348	35	1.1
chromia	0.029	304	30	0.96
crown glass	6x10^{-3}	139	14	0.44
polyethylene	1x10^{-3}	57	5.7	0.18

on a thermally thick sample (L>>μ_T). Equation (1) shows that $\Delta\phi$ is a
function of x and consequently of $\sqrt{f}$. By fitting the experimental data to
eq.(1) by the method of least squares, the parameters ϕ_{ref}, R and α can be
determined.

The method described above applies to a single spot on the surface of
the coating or film. Moving the spot to different locations on the
surface yields information about the lateral variation of the thermal
diffusivity. This approach can also be used to detect defects like voids,
cracks, inclusions of other phases or dissimilar materials, regions of a
different crystal orientation, in short, anything which produces a
discontinuity in the depth profile of the thermal diffusivity in the
sample. Because simultaneous motion of the laser beam and the detector,
both aimed at the same point, is cumbersome, one usually moves the sample
parallel to its surface. Defects at depths ranging from several μm to
several mm can be observed by varying the modulation frequency of the
laser beam, and, hence, the thermal diffusion length. Thus, distances
below the surface that can be probed easily will depend on the thermal
characteristics of the material under investigation, as shown in Table I.

OXIDES

Oxide coatings and films are found both as naturally occurring layers
on metals (due to oxidation or corrosion), and as man-made deposits for
prevention of oxidation, wear, heat loss, or thermal degradation. For
example, the low thermal conductivity (diffusivity) of thermally
protective oxide coatings plays an essential role in devices such as heat
engines, combustors, gas turbines, space vehicles, nuclear reactors,
cutting tools, etc.

We have applied the PTR technique to determine α and κ of three high-
temperature oxides, chromia, zirconia and alumina, at temperatures between
295 and 1173 K. (22 and 900°C). These materials were plasma sprayed to
thicknesses between 50 and 100 μm on 1.58 mm thick stainless steel
substrates. The experimental details have been described in a recent
paper[11] .
Results of these measurements are shown in Table 2. For comparison,

handbook values for bulk ceramics are also presented. The values of κ for
the coatings are smaller than the values for the bulk materials,
suggesting a higher degree of porosity in the coating. In addition, the
temperature dependence of κ in the coatings is not the 1/T proportionality
expected by theory.

Table II.   THERMAL DIFFUSIVITY α AND THERMAL CONDUCTIVITY κ OF
THREE DIFFERENT OXIDE COATINGS BETWEEN 20 AND 900°C [†]

Chromia on Steel CS2			Zirconia on Steel ZS3			Alumina on Steel AS3		
L=55μm			L=92μm			L=73.5μm		
T °C	α cm^2/sec	κ W/cm.K	T °C	α cm^2/sec	κ W/cm.K	T °C	α cm^2/sec	κ W/cm.K
22	.0059	.0163	22	.0051	.0103	22	.0085	.0068
252	.0050	.0138	251	.0042	.0060	280	.0055	.0054
500	.0051	.0141	544	.0034	.0070	500	.0060	.0054
594	.0055	.0175	605	.0036	.0054	595	.0059	.0053
688	.0059	.0186	696	.0040	.0033	699	.0056	.0066
793	.0066	.0205	800	.0045	.0036	799	.0055	.0087
900	.0072	.0200	898	.0044	.0026	899	.0042	.0100
20	-	.10-.33[‡]	20	.0044[‡]	.018-.022[‡]	127°C	.08[‡]	.29[‡]

[†] Samples obtained from Accumetrix Corp., Arlington, Virginia.
[‡] Ceramics Source 86 (American Ceramics Society, Westerville, OH, 1986),
    pp. 350-351

DIAMOND

In recent years there has been a rapidly growing interest in diamond
films made by chemical vapor deposition. These films are nearly
indistinguishable from bulk diamond and their properties approximate the
fascinating and potentially useful characteristics of diamond crystals,
including great hardness, high heat conductivity, and optical transparency
over wide wavelength ranges[12]. It has been shown by several workers[13,14]
that a chemical reaction at about 800°C involving hydrocarbons, such as
methane (or even C-H-O compounds) and atomic hydrogen leads to the
deposition of diamond. Films and coatings up to a mm thick have been
deposited on various substrates. A variety of properties of these films
have been measured which confirm that they are diamond. These include
crystal structure, lattice constant[13], hardness[15], Raman spectrum and
several other optical properties[13].

Two groups of researchers have determined the thermal conductivity of
diamond films. The first group[16] used a conventional steady state method
to measure the thermal conductivity of two films, 13 μm and 18 μm thick,
at temperatures from 10 to 300K. Below room temperature the heat
conductivity of the films appeared to be up to two orders of magnitude
smaller than the single crystal values reported by Berman et al.[17], but at
room temperature the film results begin to approach the Berman data. The
small heat conductivities at low temperatures probably are caused by grain
boundary scattering because the diamond films consisted of very small
crystallites.

The other group [18] used a quasi-static method in which a free
standing film was clamped between two heated posts in vacuum. A
thermograph measured the temperature distribution along the film, which

exchanged heat with its surroundings by radiation. The thermal
conductivity was deduced from the balance between the emission of heat
from the film and the heat conduction along the length of the film.
Measurements were performed on a series of films made from methane-
hydrogen mixtures covering a range of CH_4/H_2 ratios by volume from 0.1 to
3%. The thermal conductivity was strongly dependent on the fraction of

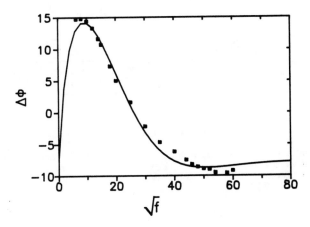

Fig 2. Relative phase vs square root of the modulcation frequency for a thick diamond film (sample GE-1)

CH$_4$ used in the film deposition; for a very small CH$_4$ fraction (0.1%) the thermal conductivity of the film exceeded that of Type I and Type IIb diamond crystals.

The thermal wave method is an attractive way to measure the diffusion of heat in diamond films. However, the method is practical only if the thermal diffusion length, which depends on f, can be varied from about 0.3L to 2L. Table 1 shows the diffusion lengths of different materials, including diamond, at several modulation frequencies. In the case of pure diamond films several micrometers thick, a dimension typical of CVD diamond films, f in excess of one MHz would be required, making the measurements exceedingly difficult to perform. However, two specimens of CVD diamond, 0.45mm and 0.25mm thick, designated GE-1 and GE-2 respectively were made available to us[19]. This permitted us to perform measurements in the frequency range 60 to 4000 Hz.

An Argon laser beam was focused on the sample surfaces which were blackened by a very thin ink layer. Blackening was necessary in order to cause the heating to occur at the specimen surface; diamond has a small absorption coefficient over nearly the entire visible and infrared spectral range, except for a small band between 2 and 6.5 μm. Because the specimens were unsupported, the "substrate" was air. This would imply that the reflection coefficient was close to 1; however, rough surfaces, present in the specimens under measurement, can cause considerable scattering of the thermal wave and lower the effective reflection coefficient.

A plot of the relative phase angle vs √f for sample GE-1 is shown in Fig. 2. A least squares fitting program yielded values for ϕ_{ref}, R and the thermal diffusivity α. Using handbook data for the C and for ρ of diamond at room temperature[21], we have calculated κ. A similar procedure was used for the other sample, GE-2. Results of these calculations are presented in Table 3. It appears that the thermal properties of the two films are not the same. These differences are probably due to differences in the deposition conditions during preparation, which are expected to strongly affect the thermal properties. Furthermore, because thermal diffusivity is a tensor property, and deposition is a layer upon layer process leading to a possible material anisotropy, α perpendicular and α parallel to the surface may differ.

Table III. THERMAL DIFFUSIVITY OF CVD DIAMOND

SAMPLE	THICKNESS	METHOD	THERMAL DIFFUSIVITY	THERMAL CONDUCTIVITY
GE-1	0.45 mm	PTR	$4.8\ cm^2/s$	8.7 W/cm/K
GE-2	0.25 mm	PTR	$0.3\ cm^2/s$	0.5 W/cm/K

SUMMARY

    The thermal wave technique provides a way to test the thermal
properties of a wide variety of coatings and films. This technique is a
non-destructive, non-contact method. Several versions, in particular PTR
and OBD, have the advantage that one can probe the surface point-by-point
or line-by-line and thus gain information about the lateral uniformity of
the sample.

REFERENCES

*Supported by the Office of Non-Destructive Evaluation of the National
    Institute of Standards and Technology.
1.   A. Rosencwaig and A. Gersho J. Appl. Phys 47, 64 (1976).
2.   A. Rosencwaig in "International Advances in Non-destructive Testing"
     Vol. 11, pp. 105-174, Ed. : W. McGonnagle, (Gordon & Breach, London,
     1985.)
3.   Deposition Techniques for Films and Coatings, R.F. Bunshah, editor,
     (Noyes Publications, Park Ridge, NJ, 1982.)
4.   P.E. Nordal and S.O. Kanstad, Physica Scripta 20, 609-662 (1979).
5.   H. Frederikse and A. Feldman in "Nondestructive Testing of High
     Performance Ceramics", Eds: A.Vary and J. Snyder (American Ceramic
     Soc., Westerville, OH, 1987), pp. 177-182.
6.   A. Boccara, D. Fournier and J. Badoz, Appl. Phys. Lett. 36, 130
     (1980).
7.   K.R. Grice, L.J. Inglehart, L.D. Favro, P.K. Kuo and R.L. Thomas,
     J. Appl. Phys. 54, 6245 (1983).
8.   C.E. Yeach, R.L. Melcher and S.S. Jha, J. Appl. Phys., 53, 3947
     (1982).
9.   M.A. Olmstead, N.M. Amer, S.Kohn, D. Fournier and A.C. Boccara,
     Appl. Phys. A32, 141 (1983).
10. A. Lachaine, J. Appl. Phys., 57, 5075 (1985).
11. H.P.R. Frederikse and X.T. Ying, Applied Optics, in press.
12. J.E. Field, The Properties of Diamond [Academic Press, London, 1979].
13. S. Matsumoto, Y. Sato, M. Kamo and N. Setaka, Jap. J. Appl. Phys. 21,
     L183 (1983).
14. S. Ashok, K. Srikanth, A. Badzian, T. Badzian and R. Messier,
     Appl. Phys. Lett., 50, 763 (1987).
15. A. Sawabe and T. Inuzuka, Appl. Phys. Lett. 46, 146 (1985).
16. D.T. Morelli, C.P. Beetz and T.A. Perry, J. Appl. Phys., 64, 3063
     (1988).
17. R. Berman, E.L. Foster and J.M. Ziman, Proc. Roy. Soc., London, Ser.
     A237, 344 (1956).
18. A. Ono, T. Baba, H. Funamoto and A. Nishikawa, Jap. J. Appl. Phys.
     25, L808 (1986).

19. We thank Thomas Anthony of the General Electric Research Center for
     supplying these samples.
20. American Institute of Physics Handbook, 3rd Edition McGraw Hill, New
     York (1972), p.4-106.

ULTRASONIC CHARACTERIZATION OF
SOL/GEL PROCESSING

S. Chiou and H. T. Hahn

Department of Engineering Science and Mechanics
The Pennsylvania State University, University Park, PA 16802

Abstract

The sol/gel process has recently received attention because it enables
the production of pure and homogeneous glass without external heating,
eliminating the high processing temperatures needed in conventional
methods. A disadvantage of the process is that if the rate of evaporation
during syneresis is too high, cracking due to mechanical shrinkage can be
extensive. A slower evaporation rate produces less cracking, but results
in a longer processing time. The key to decreasing processing time while
simultaneously eliminating shrinkage cracks lies in monitoring the change
of the mechanical properties during the evaporation process. Ultrasonic
techniques are ideal for this purpose.

To monitor the process, an ultrasonic signal was passed through the
material. Spectrum changes were measured and then converted to changes in
modulus and viscosity, while shrinkage and density changes were monitored
using an air-type transducer and an electronic balance. The results show
no observable change up to the beginning of the gelation process, followed
by rapid increase in the modulus and viscosity after the onset of
shrinkage. This finding was confirmed using a viscometer and laser light
scattering measurements. Because of the effectiveness of this ultrasonic
technique in detecting changes in the material, such as the onset of
cracking, it may be used in controlling and improving the process.

Introduction

In recent years, the sol/gel process, based on hydrolysis and
polycondensation of metal alkoxides, has been widely used for the
preparation of glasses and ceramics. It has not yet become a reliable
technique, due to lack of knowledge regarding the formation of residual
stresses during processing. These stresses can cause considerable cracking
in the final product.

A considerable amount of research has been done in the effort to
prevent the formation of these stresses and their subsequent cracking.
Hench [1,2] added DCCA to promote more uniform pore size in the material
and consequently reduce the capillary stress. Scherer [3,4] modelled the
gelation process as contraction due to the reduction of surface energy, and
light scattering techniques and SAXS [5,6] have been used to monitor the
process. So far, however, no reliable method has been found to decrease
both the amount of cracking present and the processing time.

One of the principal difficulties is to find a technique which can
monitor the material property changes continuously. Sol/gel processing has
been used most frequently in film drawing and substrate coating. In these
applications, it is essential to find the change in rheological properties

and especially to identify the transition region of gel point. Sheu [7] has used a viscometer and DMA to find the gel point, but with either of these an external force is applied which affects the chemical process and breaks the weak agglomorate; shear thinning is an example of this behavior. A technique is needed which allows continuous observation of the gellation process without disturbing the reaction.

In order to achieve this goal, the authors employed a non-destructive ultrasonic technique to measure the change in properties during the process, using a 4 MHz wave to monitor the process continuously. Changes in the velocity and attenuation of the input signal were recorded and converted to complex modulus and viscosity using a simple spring and dash-pot viscoelastic model. Using these values, it is possible not only to determine the change in the properties but to find the gel transition point as well. Wide angle light scattering and a low strain rate viscometer are used to confirm the results. So far, however, no attempt has been made to quantitatively compare the ultrasonic results with those from other instruments.

Experimental Procedure

The sol used to make pyrex glass (Corning 7740) was selected for testing in this study; details for preparation of the sol are given in Fig 1. Immediately after mixing the sol it, was poured into a plastic casting dish and covered. A longitudinal wide-band transducer with a central frequency of 4 MHz was used to monitor the gellation process, and an air-type transducer was placed on the top center of the plastic dish to measure the shrinkage. An electronic weight balance was put underneath the dish to record the weight loss, and temperature and humidity were recorded continuously during the process. The details of the set-up are illustrated in the Fig. 2. Before the testing, the system was calibrated using water.

As an alternate means of determining the change in wave speed, wide angle light scattering was used. With a Helium-Cadmium laser (wavelength = 442 nm) as a source, the intensity of the scattered light was recorded continuously. To verify viscosity measurements, a Broofield cone and plate viscometer was used. Measurements were made with about 0.5 ml of sol and a 3.75 (1/second) stain rate while continuously monitoring the shear viscosity.

Analysis

During the processing, the gel particles agglomerate together and expel the excess liquid, causing volume shrinkage. This thickness shrinkage was measured using an air-type transducer. Knowing the time for a sound wave to travel to and from the top of the solution, its thickness $S(t)$ may easily be determined as:

$$S(t) = v_s \frac{\left(t_o - t(t)\right)}{2}$$

(1)

where t_o is the traveling time of the empty container and v_s is the speed of sound in air (347 m/s).

PROCESSING SEQUENCE

STEP 1:

```
+  30.36 ml  EtOH
+  30.36 ml  TEOS (Si(OC2H5)4)
+  3    ml  H2O
+  0.3  ml  HCL
```

STEP 2: HEAT UP TO 70 C

```
+  2.1  ml Al-sec-BUTOXIDE
+  0.5  ml H2O
```

STEP 3:

```
+  4.3  ml TRIMETHYL BORATE
+  2.5  ml H2O
```

STEP 4:

```
+  2    ml  HOAc
+  6.45 ml  2M NaOAc
+  7    ml  H2O
```

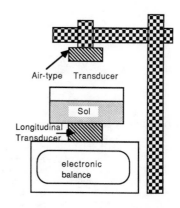

Figure 1   PROCESSING SEQUENCE   **Figure 2   EXPERIMENTAL SET-UP**

The areal shrinkage was measured using image processing. Since both the weight and the thickness of the solution are known, the density $D(t)$ as a function of time may be determined as:

$$D(t) = \frac{W(t)}{A(t)\ S(t)}\qquad(2)$$

where $W(t)$ is the weight and $A(t)$ is the area. Since the pulse–echo method was used during the measurement, the ultrasonic velocity $V(t)$ was found by dividing the solution thickness by half of the time of flight $u(t)$:

$$V(t) = \frac{S(t)}{u(t)/2}\qquad(3)$$

The attenuation was calculated by using the following equation together with the same amount of the water as the reference.

$$\alpha(t) = \frac{20}{S(t)/2}\ \log\left(\frac{A(t)\ T_0}{A_o\ T(t)}\right) + 1\qquad(4)$$

where Ao and A(t) are the amplitude of the received waves for water and sol, respectively. The term T_0, T(t) are the transmission coefficient through the container-water and container-sol and may be calculated using:

$$T(t) = \left(\frac{2\sqrt{D_p \, V_p \, D(t) \, V(t)}}{\left(D_p \, V_p + D(t) \, V(t)\right)^2}\right)^2 \tag{5}$$

$$T_0 = \frac{\left(2\sqrt{D_p \, V_p \, D_0 \, V_0}\right)^2}{\left(D_p \, V_p + D_0 \, V_0\right)^2} \tag{6}$$

where D_p, D_0 are the density of the container and water; V_p and V_0 are the sound speed of container and water.

Linear viscoelastic behavior was assumed during the processing, based on the results of viscosity measurements by Sheu [7]. By using Hahn's model [8,9,10], this linear viscoelatic behavior can be analyzed with a simple spring and dash-pot combination (Fig. 3). If a plane wave is travelling through a material, its complex modulus and viscosity can be calculated by knowing the wave speed and attenuation and using the relations

$$\overline{W}(t) = \frac{W}{\alpha(t) \, V(t)} \tag{7}$$

$$M'(t) = D(t) \, V^2(t) \frac{\overline{W}^{-2}(t) \, (\overline{W}^{-2}(t) - 1)}{(\overline{W}^{-2}(t) + 1)^2} \tag{8}$$

$$M''(t) = 2 \, D(t) \, V^2(t) \frac{\overline{W}^{-3}(t)}{(\overline{W}^{-2}(t) + 1)^2} \tag{9}$$

$$M(t) = M'(t) + \frac{M''(t)}{(M'(t) - M_0)} \tag{10}$$

$$\eta(t) = \frac{(M(t) - M_0)}{(M(t) - M'')} \frac{M''(t)}{W} \tag{11}$$

Where W is the center frequency of the sound wave. M'(t), M"(t) are storage and loss moduli of sol and M_0 is bulk modulus of the water. $\eta(t)$ is the viscosity of the sol.

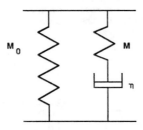

Figure 3   VISCOELASTIC MODEL BY HAHN [9]

Results and Discussions

The thickness and density data is shown graphically in Figs. 4 and 5. Both parameters show no change during the first 17 hours. After 17 hours, shrinkage begins and the density increases gradually. The wave speed and attenuation show a similar initial trend, but increase significantly after 17 hours (Figs. 6, 7). After 30 days (40,000 minutes), some white particles began to appear on the gel surface. It was suspected that B segregation occurred at the surface of the gel, causing the sharp increase in attenuation. The experiment was stopped at this point. The complex moduli and viscosity were obtained by using Eqs. 8, 9 and 11. The storage modulus (Fig. 8) changed after 17 hours from an initial value of 1.7 Gpa to a final value of 8.2 Gpa; an increase of approximately 1 order of magnitude. The loss modulus was initially very small (83 MPa) and then increased rapidly to 8 GPa about 30 days later. The viscosity results using Eq. 11 are shown in Fig. 9. After 30 days, the data also increased sharply from 0 to 25000 cps. If these results are compared with those from the viscometer (Fig. 10), we find the ultrasonic values are much smaller than those from the viscometer. The results from light scattering (Fig. 11) also prove that the change only occurred at the later stages of the gelation process.

Conclusions

An ultrasonic technique was used to monitor the sol/gel processing in this paper. The result of the complex modulus and viscosity measurements clearly show the transition gel point. Confirmation was provided by light scattering and viscometer measurements. However, the results also indicate that this gelation behavior is quite different from the curing of polymers. In polymerization, the complex moduli and viscosity increase and reach a maximum value. However, in sol/gel processing, the moduli and viscosity of the gelled material continue increasing with further aging or drying, as observed previously [8 ,9, 11]. This behavior is most likely attributable to the effects of syneresis and viscosity sintering. Because of these

300

unusual characteristics, ultrasonic techniques are probably the only suitable technology for monitoring the transformation from liquid to solid and beyond. Furthermore, by using a suitable sol, this technique should be able to differentiate between syneresis and drying. By being able to make this distinction, improvements in the process may be made possible which will produce a more refined product.

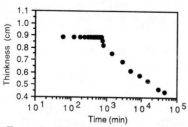

Figure 4   Thinkness shrinkage v.s. gel time of the sol at 25 C, 20% humidity lever.

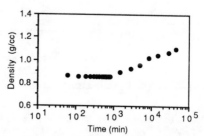

Figure 5   Density change v.s. gel time of the sol at 25 C, 20% humidity lever.

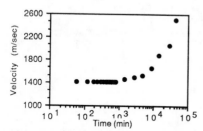

Figure 6   Ultrasonic speed v.s. gel time of the sol at 25 C, 20 %humidity lever.

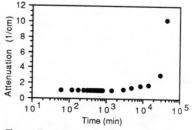

Figure 7   Ultrasonic attenuation v.s. gel time of the sol at 25 C , 20% humidity lever.

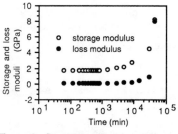

Figure 8 Storage and loss moduli v.s. gel time of the sol.

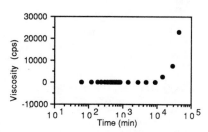

Figure 9 Ultrasonic viscosity v.s. gel time of the sol.

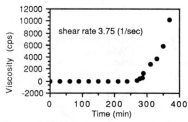

Figure 10 Shear viscosity v.s. gel time by using cone and plate shear viscometer.

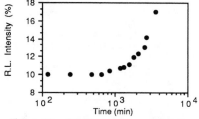

Figure 11 Relative light scattering Intensity v.s. gel time of the sol, light scattering is 90 degree.

Acknowledgment

This paper is based on the work supported by Center for Adanced Materials at Penn State University. An appreciation is extend to Mr. D. Q. Xi and Dr. C. G. Pantano for the preparation of the silicate sol.

Referernces

1. M. Prassas and L. L. Hench " Physical Chemical Factor in Sol-Gel Processing " in Ultrastructure Processing of Ceramics, Glass and Composite, L.L. Hench and D.R. Ulrich eds., John Wiley and Sons, N,Y,

1984, PP 100-125.

2. L. L. Hench, S. H. Wang, and S. C. Park " SiO2 Gel Glasses", SPIE'S 28th Annual International Technical Sympoium on optical and Eletro-optics Aug. ,pp 19-24, 1984, San Diego, California.

3. G. W. Scherer " Drying Gel 1. General Theory ", Journal of Non-Crystalline Solids, 87, 1986, PP 199-225.

4. G. W. Scherer " Drying Gel 2. Film and Plate ", Journal of Non-Crystalline solids, 89, 1987, pp 217-238.

5. C. J. Brinker, K. D. Keefer, D. W. Schaefer, R. A. Assink, B. D. Kay and C. S. Ashley, " Sol-Gel Transition in Simple silicates II", Journal of Non-Crystalline solids, 63, 1984, pp 45-59.

6. A. J. Hunt and P Berdahl " Structure Data from Light Scattering studies of aerogel " in Better Ceramics Through Chemsity ,C. J. Brinker, D. E. Clark, D. R. Ulrich eds, 1984, pp275-280.

7. Michael D. Sacks and Rong-Shenq Sheu " Rhelogical Characterization during Sol/Gel Transition",in Science of Ceramic Chemical Processing, L. L. Hemch and D. R. Ulrich eds, 1986, pp 100-107.

8. H. T. Hahn, " Application of Ultrasonic Technique to Cure Characterization of Epoxies", in Nondestructive Method for Material Property Determination, C. O. Ruud and R. E. Green, Jr., 1984, pp 315-326.

9. E. J. Tuegel and H. T. Hahn "Ultrasonic Cure Characterization of Epoxy Resins: Constitutive Modeling", October, 1986, report for Office of Naval Research.

10. T. A. Litovitz and C. M. David " Structural and Shear Relaxationin Liquid" in Physical Acoustic IIA by Warren P. Mason, 1969, pp 281-349.

11. A. M. Lindrose, " Ultrasonic Wave and Moduli Chgange in a Curing Epoxy Resin", Experimental Mechanics, Vol 18, 1978, pp 227-232.

ULTRASONIC REAL-TIME MONITORING OF SLIP CAST CAKE THICKNESS

DAVID JARMAN', ADAM J. GESING', BAHRAM FARAHBAKHSH', GENE BURGER',
DAVID HUCHINS'', AND H.D. MAIR''
'Alcan International Ltd., Kingston Research and Development Centre, Kingston, Ontario, K7L 5L9.
''Queen's University, Dept. of Physics, Kingston, Ontario, K7L 3N6.

ABSTRACT

An ultrasonic method was developed for real-time monitoring of cake thickness during the casting of an alumina slip. The technique uses pulse-echo ultrasound to measure the time of flight and hence the thickness of the cast layer. Variations on the method are applicable to both production process control and to the fundamental studies of slip casting process kinetics.

INTRODUCTION

One of the main factors limiting the application of slip casting to the production of advanced technical ceramics is the difficulty in maintaining tight wall thickness tolerances. In the slip casting process, typically, a ceramic powder is dispersed in water to form a stable suspension called a slip. The slip is poured into a porous mold, typically made of plaster of Paris or porous urethane. The capillary suction, sometimes assisted by applied pressure or vacuum, draws the water from the slip, and a green ceramic filter cake is deposited at the mold-slip interface. Once sufficient cake thickness has been deposited, the excess slip is drained.

Cake formation kinetics have been recently reanalyzed by Schilling and Aksay [1],[3] and by Tiller and Tsai [2] confirming the well known dependence of the filter cake thickness on the square root of the casting time. Commercial practice depends on that relation to predict and control the wall thickness of the cast piece. This task, however, is complicated by changes in the water content of the mold from cast to cast as well as by slip and mold aging.

Some attempts have been made to monitor the slip casting process by non-destructive techniques. In particular, radiation measurement based on Gamma-ray attenuation is a reliable and sensitive method of detecting density variation. It has been applied to the monitoring of casting thickness with a resolution of +/- 0.05 mm by Deacon and Miskin[4], and was used to study the structure of slip-cast deposits by Schilling and Aksay et. al. [5], [6]. Recently Hayashi et. al. [7] used nuclear magnetic resonance (NMR) imaging, based on the spin-lattice relaxation time, to obtain a contrast between the cake and the slip, and to follow cake growth.

In this study, a new, ultrasonics-based approach is reported. The use of ultrasonics permits the real-time measurement of the cake growth using the time of flight of ultrasonic waves in the slip and the cake. This is a direct and precise measurement of the slip and the cake characteristics and the kinetics of the process.

EXPERIMENTAL TECHNIQUE

Apparatus

The equipment used in this study is shown in figure 1. The probe was an immersion piezoelectric longitudinal transducer (5 Mhz, 12.5 mm diameter Harrisonic SIJ-5.0) attached to a 25-cm extension. The extension is held by a fixture which allows fine rotational adjustments of the transducer so that the amplitude of the echo signal is maximized. This assembly is then firmly attached to the vertical axis of an adjustable stand. The vertical axis also has a moveable platform on which the plaster mold is placed. The position of this platform is measured by a linear variable differential transformer (LVDT), the signal from which is sent to a host microcomputer via an analog to digital convertor (LeCroy TR8828C). The transducer is connected to a pulser-receiver (Panametric 5052PR). The echo signal from the transducer is then captured and digitized using a waveform recorder (LeCroy 8013A) via its amplifier (LeCroy 6103) and its digitizer (LeCroy TR8828C).

304

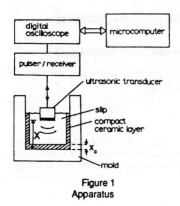

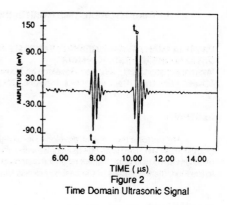

Figure 1	Figure 2
Apparatus	Time Domain Ultrasonic Signal

Method

Precise thickness gauging by ultrasonic methods is well known, but this technology has not been applied in the past to concentrated ceramic dispersions, which attenuate and scatter ultrasonic signals. In order to obtain the ultrasonic signal upon which the method is based, the 5-MHz transducer is placed into the slip, as shown in figure 1. The distance between the transducer and the mold is set at about 1 cm. Ultrasonic pulses emitted by the transducer are partially reflected by the top surface of the growing ceramic cake because of the acoustic impedance mismatch between the slip and the cake. This reflected pulse is detected by the transducer at a time t_a. Part of the emitted pulse is transmitted into the ceramic cake and subsequently reflected from the interface between the cake and the mold. This pulse is detected by the transducer at time t_b.

Because slips contain a high concentration of fine suspended particles, they are highly attenuating to ultrasound. Consequently, for the A-16 slip used in this study, a precise orientation of the transducer with respect to the bottom of the mold is critical for achieving sufficient echo-signal strength. Correct adjustment of the transducer, however, provides ultrasonic signals with a good signal-to-noise ratio as illustrated in figure 2.

A key requirement for a practical monitoring system is that it must be able to calculate the thickness of the growing cake at intervals of a few seconds. For this reason, signal processing and calculation of cake thickness is performed in real-time by a microcomputer providing an updated measurement every 15 s.

The measured parameters are: 1) the transducer location, $(X(T)-X(0))$, measured by a LVDT with a resolution of +/- 2 microns; and 2) the ultrasonic signal amplitude record, consisting of 4096 points measured at a sampling rate of 100 MHz.

Data Reduction

The parameters calculated are: 1) t_a and t_b determined via a peak detection algorithm, and 2) the time delay between the two peaks ($\Delta t = t_b - t_a$), calculated by cross correlation.

One can write an expression for the cake thickness at time T, $x_c(T)$:

$$x_c(T) = V_s \, (t_b - t_a) \, / \, 2 \tag{1}$$

V_s, however, depends on the slip characteristics, and both t_a and t_b depend on transducer location, X and slip casting time, T. Therefore,

$$x_c(T) = V_s \, [t_b(X,T) + 2/V_s \, (X(T)-X(0)) - t_a(X,T)]/2 \tag{2}$$

or,

$$x_c(T) = V_s \, [t_b(X,T) + 2/V_s \, (X(T)-X(0)) - t_b(X,T) + \Delta t(T)]/2 \tag{3}$$

Cake thickness can also be expressed in terms of Δt and the acoustic velocity in the cake, V_c:

$$x_c(T) = V_c(T) \, \Delta t(T) \, / \, 2 \tag{4}$$

The use of equation (4) to measure X_c requires that $V_c(T)$ be known. If for a given type of mold and slip, $V_c(T)$ was a function of cake thickness, then cakes of a fixed thickness could be produced by stopping the slip cast when Δt reaches a preset value. There may be, however, an uncertainty in ultrasonically controlled cake thickness if V_c varies from cast to cast. This is discussed in more detail below.

Post Processing of the Data

Careful analysis of the data was carried out to verify the validity of the Δt - x_c empirical correlation which depends on the constant value of V_c, and to gauge the precision of the measurements. By multiple linear regression analysis on t_b, it was determined that the statistically significant terms are:

$$t_b(X,T) = t_b(0,0) + 2/V_s \, (X(T)-X(0)) + C_{t_b} \, T^{1/2} \tag{5}$$

Where X is the location of the transducer during casting and $t_b(0,0)$, $2/V_s$ and C_{t_b} are the regression coefficients. The interaction term $(T)(X)$ is not statistically significant. This indicates that the acoustic velocity in the slip, V_s, is independent of casting time, T, and can be derived from the regression coefficient of the X term.

The statistically significant terms in the multiple linear regression on Δt are:

$$\Delta t(T) = C_{\Delta t} \, T^{1/2} + D_{\Delta t} \, T^{1/3} + E_{\Delta t} \, T \tag{6}$$

The substitution of (5) and (6) into (3) yields a predicted empirical form of the equation describing the cake growth kinetics:

$$x_c(T) = V_s/2 \, ((C_{\Delta t} - C_{t_b}) \, T^{1/2} + D_{\Delta t} \, T^{1/3} + E_{\Delta t} \, T) \tag{7}$$

The statistical significance of the T and $T^{1/3}$ terms is surprising since the most current theoretical treatments of slip-casting kinetics [2] predict only the $T^{1/2}$ dependence, even in the case when the cake density varies through the thickness due to cake compressibility. Figure 3 shows the time-of-flight values. The step changes in the curves are a result of the transducer movement which allows an instantaneous calculation of V_s. Figure 4 shows the fit of the experimental data calculated from the individual $t_b(T)$ and $\Delta t(T)$ points to the trend lines estimated from the regression coefficients for t_b and Δt. The deviation from the $T^{1/2}$ trend is statistically significant as illustrated in figure 5.

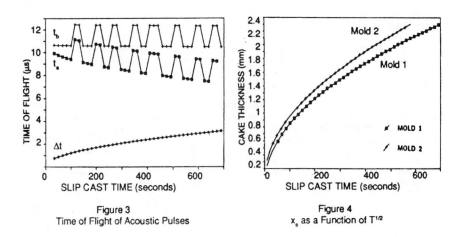

Figure 3
Time of Flight of Acoustic Pulses

Figure 4
x_c as a Function of $T^{1/2}$

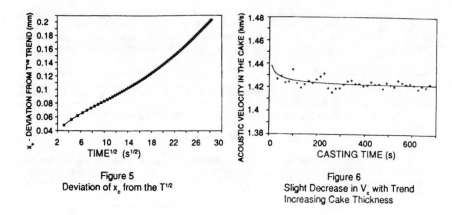

Figure 5
Deviation of x_c from the $T^{1/2}$

Figure 6
Slight Decrease in V_c with Trend
Increasing Cake Thickness

By measuring $\Delta t(T)$ and calculating the $x_c(T)$, we estimate the acoustic velocity in the cake averaged over its thickness, $V_c(T)$:

$$V_c(T) = 2\, x_c(T) / \Delta t(T) \qquad (8)$$

The V_c trend line shown in figure 6 has a non-zero slope at low T values. However, this slope is not significantly different from zero when compared to an optimistic estimate of the 95% confidence interval for the trend line. This suggests that it may be possible to obtain reasonable accuracy by using equation (4) with a constant value of V_c.

Application of the Method to the A-16 Alumina Slip

The validity and the precision of the technique were evaluated by assessing the repeatability of the cake thickness obtained in successive casts of alumina slip into two plaster molds.

Materials

The slip was prepared from an alumina (Alcoa A-16) at a solids concentration of 71 wt% in aqueous medium. The dispersant used was Darvan C (0.5 wt% on a solids basis). Dispersion was accomplished

Table 1
Slip Characteristics

Particle Size Distribution		Slip Viscosity	
Particle size (µm)	% (by mass) under size	RPM	Apparent Viscosity (cP)
0.180	0.0	5	220
0.240	0.8	10	170
0.310	3.0	20	125
0.400	24.1	50	82
0.520	63.9	100	70
0.670	88.2		
0.870	95.0		
1.130	98.7		
1.460	100.0		
Mean particle size = 0.477 +/- 0.212 µm			

by ball milling for four hours using high purity alumina grinding media, with a ball-to-charge ratio of 2:1. The particle size distribution in the slip was determined after milling using a Malvern Autosizer (model IIc). This analysis, given in table 1, suggests that the alumina in the slip was well dispersed. Slip viscosity measurements were determined using a Brookfield viscometer (model RVT), with a no. 4 spindle. These results are also given in table 1.

Two molds were prepared from the same batch of plaster of Paris (standard potter's plaster), using a plaster-to-water ration of 5:4. These were oven-dried at 50°C, and then kept at ambient temperature and atmosphere for several days prior to use.

Procedure

Two molds were used in this experiment and two castings were made in each mold for five successive days. Prior to each casting run, the molds were preconditioned by wetting them with water and then were weighed before the slip was added to the mold. During casting, the ultrasonic measurement system recorded the t_a, t_b, Δt, and the height of the transducer. The casting procedure was terminated when ultrasonic timing measurements indicated that Δt had reached a value of 3.22 μs. (This value was chosen arbitrarily.) After the casting procedure was complete, the thickness of the cake produced was measured with a micrometer.

RESULTS

Cake Thickness

The significant trends of the results, which are summarized in table 2, are as follows. There is no effect of either run number, mold water content or the measured cake thickness for casts drained when Δt reached 3.22 μs. This is also true for the estimate of cake thickness calculated from the ultrasonic measurements at $\Delta t = 3.22$ μs. There is, however, a statistically significant constant difference of 0.19 +/- 0.008 mm between the measured and the calculated values. This difference is attributed to a boundary layer at the slip-cake interface in which the rheology of the slip changes so that the plane from which the slip drains may be displaced from the one reflecting the acoustic pulses. This is in agreement with the increase in thickness of 0.25 mm due to draining reported by Deacon and Miskin [4].

As figures 7 and 8 demonstrate, the coefficient of variation of the cake thicknesses measured at a given Δt is much smaller than that of the casting time or the cake thickness measured at a given casting point in time. The latter two depend on both the casting run and the mold. For the thickness measured at $\Delta t = 3.22$, the mold-to-mold variation is significantly smaller than the cast-to-cast variation. This suggests that a bank of several molds filled and drained simultaneously could be controlled by measurements on a single test mold. This significantly simplifies practical process control applications.

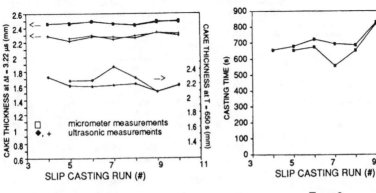

Figure 7
Comparison of Physically Measured
and Ultrasonically Determined x_c

Figure 8
Variation in the Casting Time is
Much Bigger than in x_c in Figure 7

Acoustic Velocity in Cake and Slip

Within an individual casting run, V_s, V_c and $V_c(x)$ all remain constant as a function of casting time, T. From cast to cast, however, a significant variation in both V_c and V_s is observed. This variation accounts for the observation that the precision of the cake thickness measurement from a given cast is better than the observed between-cast thickness variation.

The variations in V_s and V_c are closely correlated and both velocities increase slightly with the slip and mold age (cast #). High acoustic velocities correlate with long casting times (slow casting rates). Because the variation in V_c is translated directly into error in the cake thickness, the correlation of V_c with V_s indicates that V_s may be a useful parameter for characterizing the slip properties that affect slip casting.

Slip Casting Kinetics

The empirical expression for cake thickness as a function of casting time is given by equation (7). The coefficients C, D, and E are listed in table 2. High values of the coefficients imply fast casting rates.

Measurement Precision

The range of moisture content in the molds is not sufficient for a statistically significant study of that parameter. Table 3, therefore, concentrates on the measurement precision within a cast and the variation between the casts and between molds. The measurement precision within a casting run is good, and there is no mold effect on final cake thickness. On the other hand, casting time varied widely from cast to cast and from mold to mold. The variations are attributed to the correlated variation in the V_s and V_c.

Table 2
Experimental Results

MOLD #1							
CAST # CASTING TIME s	4 656	5 679	6 722	7 693	8 686	9 829	10 739
x_c at Δt=3.22 (measured) mm	2.460	2.471	2.488	2.458	2.438	2.480	2.501
x_c at Δt=3.22 (ultrasonic) mm	2.288	2.219	2.277	2.275	2.258	2.335	2.318
x_c at T=650 s (ultrasonic) mm	2.306	2.184	2.177	2.192	2.207	2.096	2.183
V_s m/s	1333.0	1300.4	1319.1	1313.1	1302.5	1354.2	1351.9
V_c m/s	1423.0	1379.3	1417.9	1421.6	1409.9	1475.4	1445.8
x_c reg. coeff's.							
C ($T^{1/2}$) mm/(s$^{1/2}$)	623.66	660.05	660.91	1036.35	648.87	627.72	640.02
D ($T^{1/3}$) mm/(s$^{1/3}$)	563.00	383.71	393.71	-394.90	454.30	370.88	422.33
E (T) mm/s	3.51	2.60	2.31	-1.10	2.45	2.68	2.85
MOLD #2							
CAST # CASTING TIME s		5 655	6 675	7 558	8 654	9 820	10 719
x_c at Δt=3.22 (measured) mm		2.459	2.493	2.458	2.443	2.496	2.487
x_c at Δt=3.22 (ultrasonic) mm		2.255	2.296	2.242	2.291	2.332	2.293
x_c at T=650 s (ultrasonic) mm		2.255	2.260	2.448	2.291	2.098	2.192
V_s m/s		1298.6	1318.9	1296.1	1322.6	1349.2	1334.6
V_c m/s		1403.1	1428.1	1396.6	1426.8	1459.0	1431.3
x_c reg. coeff's.							
C ($T^{1/2}$) mm/(s$^{1/2}$)		724.58	657.47	711.87	748.60	604.58	651.54
D ($T^{1/3}$) mm/(s$^{1/3}$)		317.35	412.12	431.34	280.13	462.14	420.82
E (T) mm/s		2.04	3.48	3.98	2.14	2.41	2.56

Table 3
Measurement Precision

	COEFFICIENT OF VARIATION (%)		
Response　　　　　　　Effect	Within a Cast	Between Casts	Between Molds
Casting time		13.63	3.52
x_c measured	0.19	1.13	0.06
x_c ultrasonic at $\Delta t = 3.22\ \mu s$	0.23	2.01	0.11
x_c ultrasonic at $T = 650\ s$		5.39	2.60
V_s	0.08	2.10	0.26
V_c	0.24	2.41	0.03

DISCUSSION

The process control method based on draining the mold when the ultrasonic peak separation reaches a preset value eliminates most of the errors associated with control based simply on casting time. The remaining errors are due to the variation of acoustic velocities with the slip consistency. The method can be further improved by controlling the slip consistency to achieve a constant acoustic velocity in the slip.

In considering this technique as a laboratory tool for studying process mechanisms, the following observations have been made. Equations (4), (5), and (6) predict that when deviation from the $T^{1/2}$ trend is significant, there should be a dependence of V_c on time and cake thickness. The data set, obtained on a well-behaved slip casting grade of alumina, suggests this behavior. It is expected that in particle dispersions more prone to segregation or exhibiting a more pronounced dependence of packing density on pressure, the present method will detect statistically significant velocity gradients in the cake.

In slip casting, it is well known that the degree of dispersion of the slip affects both the cake density and the casting rate. Well dispersed slips give dense cakes and slow filtration rates. This is in accordance with the correlations between the casting time, V_s, and V_c observed here.

SUMMARY

Ultrasonic real-time measurements were found to give a precise estimate of slip-cast cake thickness. A simplified version of the method is useful as a process control tool. Movement of the transducer during casting combined with post processing of the data gives a very powerful method for precise studies of the slip casting process.

REFERENCES

1. I.A. Aksay and C.H. Schilling. In Ultrastructure Processing of Ceramics, Glasses, and Composites. Edited by L.L. Hench and D.R. Ulrich. New York: John Wiley and Sons, 1984, 439-447.
2. F.M. Tiller and C.-D. Tsai. J. Am. Ceram. Soc., **69**, 12 (1986): 882-887.
3. C.H. Schilling. MSc Thesis, University of California at Los Angeles, Los Angeles, California, 1983.
4. R.F. Deacon and S.F.A. Miskin. Trans. Brit. Ceram. Soc., **63**, (1964): 473-486.
5. C.H. Schilling, et. al. In Proceedings of the 23rd University Conference on Ceramic Science, Materials Research Society, Seattle, Washington, 30 Aug - 2 Sept 1987.
6. C.H. Schilling and I.A. Aksay. The First International Conference on Ceramic Powder Processing Science, Orlando, Florida, 1-4 Nov 1987.
7. K. Hayashi, et al. J. Phys. D: Appl. Phys. **21** (1988): 1037-1039.

NEUTRON DIFFRACTION AS A TOOL FOR
NON-DESTRUCTIVE EVALUATION OF CERAMICS

JOHN H. ROOT AND JAMES D. SULLIVAN
Atomic Energy of Canada Limited, Chalk River, Ontario, KOJ 1JO

ABSTRACT

Neutron diffraction is a powerful probe of the properties of condensed matter. In recent years neutron diffraction has been applied to the non-destructive evaluation of mechanical characteristics of engineering components. This paper presents examples of applications to ceramic composites including the measurement of position dependence of residual strain, grain size and minority phase concentration. In addition, an example of volume-averaged crystallographic texture is presented.

INTRODUCTION

Neutron scattering is perhaps the most versatile single probe of condensed matter properties. It can be applied to the study of atomic and magnetic structure and dynamics in liquids, glasses, polycrystalline aggregates and single crystals. In the field of ceramics research the dominant technique is wide-angle neutron diffraction. Most samples are polycrystalline aggregates formed by the sintering of powders. They may contain more than one chemical phase or species, and more than one structural phase of a given species.

Neutron diffraction is a particularly suitable non-destructive probe of the mechanical properties of ceramic components. The high penetrating power of neutrons through most materials enables the illumination of the bulk of a sample. For instance, whereas X-rays probe to a depth of 40 microns in zirconia before 90% of beam intensity is lost, neutrons reach a depth of 5.2 cm. Neutron diffraction thus permits bulk averaging which reduces the statistical fluctuations associated with large grain size, and eliminates the need for sophisticated sample surface preparation. The penetrability also enables the study of ceramics in extreme environments, where the neutron beam must pass through the walls of furnaces, cryostats and pressure cells. Finally, by defining incident and diffracted beams with narrow slits, one can investigate the variation of mechanical properties versus position inside the volume of intact ceramic components. Equally important for the study of ceramic materials is the relatively narrow range of scattering amplitude for a wide range of atomic mass. Because X-rays are more sensitive to heavy atoms than light atoms, the ratio of scattered intensity of alumina to zirconia is about 1:20. The more favourable 1:1 ratio in a neutron diffraction experiment means that one can observe effects in both phases of an alumina/zirconia composite with relative ease.

In neutron diffraction experiments at Chalk River an incident beam of neutrons of fixed wavelength (tuneable from 1 to 4 Angstroms) is diffracted by the sample to a single ^{3}He detector with cross section 5 cm x 5 cm or to a multidetector array spanning 2° in steps of 0.3°. Figure 1a shows the experimental geometry for a profile measurement of residual strains perpendicular to the axis of a cylindrical pellet viewed from above. The incident and diffracted beams, defined by masks made of absorbing cadmium, intersect in a small region denoted the gauge volume, which can be as small as a few mm^3. The sample is positioned by a computer-controlled XY translator to locate the gauge volume at selected positions to a precision of about 0.05 mm.

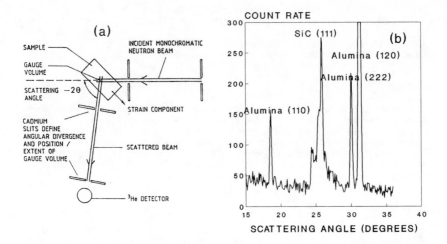

Fig. 1 a) Geometry (plan view) for profile studies. b) Part of neutron diffraction pattern from Al_2O_3/SiC tool bit.

Figure 1b shows a diffraction pattern obtained from an Alumina/SiC composite over a restricted region of scattering angle. The peaks occur at scattering angles, 2Θ, related to the neutron wavelength, λ, and interplanar spacings, d_{hkl}, by Bragg's law:

$$\lambda = 2\ d_{hkl} \sin(\Theta_{hkl}) \qquad (1)$$

The relative intensity of the two phases (Fig. 1b) indicates about 15% of the composite is silicon carbide. Comparison of lattice spacings, d, with those of unstressed component powders, do, yields strain through the relation:

$$\varepsilon_{hkl} = 1 - (d/do)_{hkl} \qquad (2)$$

Monitoring the variation of peak intensity with sample orientation indicates the degree of preferred orientation of grains (ie texture) within the sample. Finally, the shape of the peaks yields information about the distribution of strains over the measured volume and the presence of very small grains, defects (< 100 nm) or short range order in one dimension. This latter phenomenon is observed in Figure 1b. The broad asymmetric feature at the base of the SiC (111) peak arises from stacking faults occurring in SiC. Each of the above properties has a direct bearing on the mechanical behaviour of ceramic components.

RESIDUAL STRAIN IN FRACTURE-TOUGHENED ALUMINA

　　　In an effort to toughen alumina mullite has been introduced as a
second phase by an infiltration technique [1]. The depth of penetration of
ethyl silicate into the green alumina is varied by time of immersion.   The
ethyl silicate combines with alumina upon firing to form the mullite phase.
On cooling, the central volume of pure alumina contracts more than the
mullite-infiltrated outer shell.   This introduces a compressive residual
stress in the plane of the sample surface, enhancing the surface toughness,
and gives rise to the measurable quantities, residual strains.   Strain
variations with position along the axis of an untreated cylindrical alumina
pellet of diameter 20 mm and depth 10 mm are presented in figure 2a.
Strains perpendicular to the cylinder axis are obtained with the geometry
in figure 1a while strains parallel to the axis require a pellet rotation
of 90° in the scattering plane. The strain state of the untreated pellet
is less than 0.01%, as determined by measurement of shifts in the
position of the alumina (201) Bragg peak. By comparison (figure 2b) there
is a definite residual strain distribution in the infiltrated pellet, with
compressive perpendicular strains (i.e. in the plane of the surface), and
tensile parallel strains at the pellet surfaces.

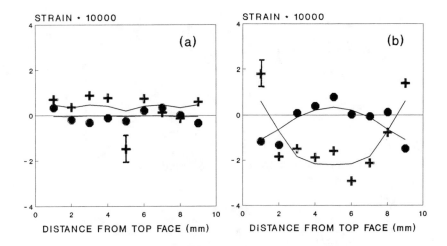

Fig. 2 Residual strain in sintered alumina a) untreated and b) infiltrated
with mullite in an outer shell. (+) Strain parallel to the cylinder axis.
(·) Strain perpendicular to the axis.   (−) Symmetric through pellet centre.

LARGE GRAIN DETECTION IN ALUMINA/MULLITE

　　　Defects such as cracks, voids, and large grains initiate fracture in
ceramic components.   While radiography and tomography are preferable for
the detection of holes in a matrix, these methods are insensitive to the
size of grains present. However, in diffraction, a large grain oriented
such that the (hkl) plane normal is in the scattering plane produces an
intensity increase beyond the general level due to a collection of randomly

314

oriented fine grains. More usually, a large grain is misoriented with
respect to the scattering plane, and constitutes a fraction of the gauge
volume from which no Bragg scattering arises, so that the intensity
decreases.   In a fully infiltrated alumina/mullite pellet, with uniformly
low grain size, mullite extends from the pellet surface through to the
centre with a monotonically decreasing concentration. The intensity of the
alumina (201) peak correspondingly increases smoothly and monotonically
towards the centre of the pellet (figure 3a).   In a partially infiltrated
pellet there is a central region of pure alumina and an outer shell of
mullite-infiltrated alumina. The thickness of the shell is about 1 mm, and
enhanced grain growth is observed in the pure alumina for a distance of
about 1.5 mm beneath the shell. This effect can be observed destructively
by cutting the sample open for visual inspection, or non-destructively by a
loss of Bragg peak intensity (figure 3b).

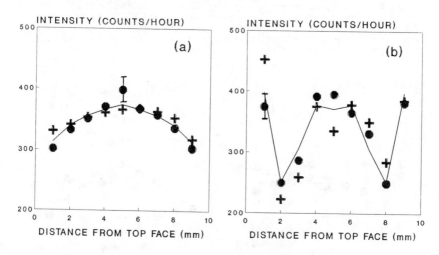

Fig. 3 Intensity of the alumina (201) peak versus position in a sintered
pellet a) infiltrated with mullite throughout volume b) infiltrated by
mullite to a depth of 1 mm. (+) Pellet oriented for strains parallel to
axis. (·) Pellet oriented for strains perpendicular to axis.   (−)
Symmetric through centre.

MINORITY PHASE CONCENTRATION PROFILE IN ZIRCONIA/ALUMINA

     Aluminum nitrate nonahydrate infiltrates green zirconia, and forms
alumina particulate inclusions upon firing. The aim of this treatment is
to enhance the fracture resistance of zirconia by production of compressive
residual stresses in the plane of the component surface [2].   It is seen in
figure 4a that the characteristic lines of alumina and zirconia are readily
resolved. Measurement of the variation of intensities of alumina (123) and
zirconia (112) peaks is a straightforward measure of the fraction of
alumina present in the zirconia matrix at each position. Figure 4b shows
that the mole percentage of alumina decreases with depth in composite
pellets of thickness 10 mm, and increases with infiltration time.

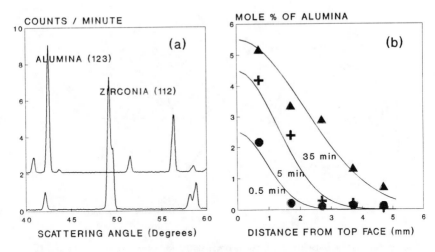

Fig. 4 a) Portion of the diffraction patterns of alumina (upper line) and zirconia (lower line) indicating the diffraction peaks employed to obtain the alumina concentration profile. b) Alumina concentration versus depth for infiltration times 0.5, 5, and 35 minutes.

PREFERRED ORIENTATION OF SiC WHISKERS IN AN ALUMINA TOOL BIT

A promising mechanism for toughening of alumina is the inclusion in the matrix of SiC whiskers which must pull out of their surroundings for cracks to continue propogation. Whiskers of β-SiC are approximately single crystals with [111] direction along the whisker axis, [2$\overline{2}$0] perpendicular to the whisker and a roughly triangular cross sectional shape [3]. Alumina/SiC is employed in the fabrication of machine tool bits which are rectangular prisms with a face of area $(12 \text{ mm})^2$ and thickness 3mm. During fabrication, the whiskers assume a preferred orientation with their long axes randomly directed perpendicular to the tool face normal. Surprisingly, there is also a preferred orientation of the whisker about its long axis such that the flat surfaces of the triangular cross section are parallel to the tool face. These results were obtained by a neutron diffraction measurement of crystallographic texture. Figures 5a,b show pole figures which are projections of Bragg peak intensity versus angles χ and η onto a plane. The tilt, χ, is the angle of the bisector of incident and diffracted neutron beams with the tool face normal, denoted Z. It is presented as the radial distance from the centre of the pole figure and ranges from $0°$ to $90°$. The azimuth, η, is the angle of the bisector from one of the tool edge directions, denoted X. Intensity is denoted by contours with dashed lines representing intensity less than that of a random distribution of grain orientations, and solid lines representing intensity greater than or equal to that of a random distribution. Because of the preferred orientation of whiskers in the tool investigated, there is an enhancement of toughening against cracks with planes perpendicular to the tool face.

316

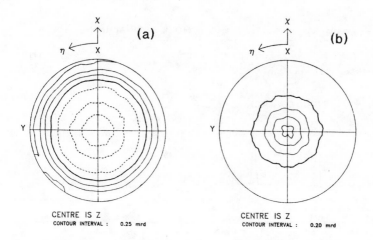

Fig. 5 Intensity pole figures for alumina/SiC whisker composite tool bit.
a) SiC (111) intensity b) SiC (220) intensity with contours in units of
multiples of random density (mrd).

SUMMARY

In this paper we have shown a few examples of the neutron diffraction
techniques applicable to the non-destructive evaluation of ceramic
materials and components. These included the through-pellet variation of
strain in alumina for mullite-infiltrated and uninfiltrated samples, the
detection of subsurface regions of enhanced grain growth in the
mullite-infiltrated sample, the measurement of an alumina concentration
profile in a zirconia/alumina composite, and the degree of preferred
orientation of SiC whiskers in an alumina/SiC(w) cutting tool. Other
neutron diffraction techniques may be employed to determine quantities such
as the aspect ratio of the grains, phase transformation aging, and
temperature gradients, all non-destructively.

References

1. B. R. Marple and D. J. Green, J. Am. Ceram. Soc. 71,C471-C473 (1988).
2. J. D. Sullivan, J. H. Root, B. D. Sawicka, and S. J. Glass, AECL #9728,
    Atomic Energy of Canada Ltd., Chalk River, Ontario (1988).
3. G. A. Bootsma, W. F. Knippenberg, G. Verspui, J. Cryst. Growth 11,
    297-309 (1971).

MICROFOCUS RADIOGRAPHIC CHARACTERIZATION OF MATERIALS AND COMPONENTS FOR HIGH PERFORMANCE APPLICATIONS

D.J. Cotter, W.D. Koenigsberg, E.M. Dunn, and M. Abdollahian,
GTE Laboratories Incorporated, 40 Sylvan Road, Waltham, MA 02254

ABSTRACT

Results are presented from several studies where microfocus radiography has been applied to provide feedback necessary to improve reliability and performance of materials and components. Improving the reliability of advanced silicon nitride ceramics through the use of NDE was studied by exploring the relationship between process-related defects, radiography results, and fracture of test samples. Research in the areas of structural and electronic ceramic joining has been aided by monitoring the effects of process modifications with real-time microfocus radiography and computer-based image processing.

Microfocus projection radiography was used to nondestructively examine a large quantity of silicon nitride modulus of rupture test bars for internal defects. Failure stress prediction was attempted using a fracture mechanics model and quantitative NDE data, and compared to actual failure stress.

In-process NDE of silicon nitride ceramic to Incoloy 909 metal brazed test samples was performed. Correlation between NDE results, optical microscopy, and destructive mechanical strength levels of samples is discussed.

Microfocus x-ray imaging was used to monitor the process of bonding microwave power transistors to metallized BeO ceramic substrates. Nonuniform distribution of the eutectic bond was readily detectable, thereby providing feedback for process improvement. Comparison is made between x-ray, optical, and infrared images.

INTRODUCTION

Nondestructive evaluation (NDE) is improving the reliability of materials and components by providing feedback for process modification and control. This evolution represents a shift in emphasis on NDE solely as a final inspection tool. In his book on quality [1], W. Edward Deming wrote, "Routine 100 percent inspection to improve quality is equivalent to planning for defects ...Inspection to improve quality is too late, ineffective, costly. Characterization of materials and components through NDE should be utilized to improve the production process. In most cases, this requires knowledge of the relationship among process-related defects, NDE results, and performance of the sample under test.

For many materials applications requiring maximum reliability, such as heat engine ceramics, critical defects simply cannot be tolerated, and 100% inspection of finished parts may be mandatory. In the following, examples of NDE by microfocus projection radiography are given, highlighting the efficacy of this technique for both in-process monitoring and final inspection.

Microfocus projection radiography provides magnified two-dimensional images of an object and its internal structure. The microfocus technique is nondestructive, capable of detecting internal defects, applicable to complex-shaped objects, and produces real-time images which are the basis for automation in a production environment. Microfocus radiography has been identified as one of the most effective methods for inspection of ceramics [2–4]. Research has been performed to optimize defect detectability by establishing microfocus imaging parameters (accelerating potential, magnification, etc.) for a wide range of ceramic components [5].

NDE FOR STRUCTURAL CERAMICS

A nondestructive evaluation capability has been established to complement ongoing efforts in silicon nitride materials research [6–8]. It is essential to link mechanical properties of the ceramic, failure origin, and the relevant process parameters to improve reliability. A necessary element is to determine the relationship between NDE results and the mechanical behavior of the ceramic component.

Mat. Res. Soc. Symp. Proc. Vol. 142. ©1989 Materials Research Society

Mechanical Behavior of Ceramics

Most polycrystalline ceramics fail at a fraction of their theoretical strength because of small prevalent flaws. The variety of possible flaw geometries leads to considerable uncertainty in the mechanical strength. According to the modified Griffith equation [9–11], the influence of flaws on the strength of brittle materials can be described by

$$\sigma_f = \frac{Z}{Y} \left(\frac{2E\gamma_i}{c} \right)^{1/2} , \tag{1}$$

where σ_f is the fracture stress, E is Young's modulus, γ_i the effective surface energy for fracture initiation, and c the depth of the surface flaw (or half the flaw size for an internal flaw). Y is a dimensionless term that depends on the flaw depth and test geometry; Z is also dimensionless and depends on the flaw configuration. For an internal flaw that is less than one-tenth of the thickness of the cross section under tensile loading, $Y = 1.77$ [7]. For a surface flaw that is much less than one-tenth of the thickness of the cross section under bend loading, Y approaches 2.0 [7]. Typically, Z varies between 1.0 and 2.0 depending on the ratio c/l (flaw depth to flaw width). A plot of formulated values of Z for varying ratios of c/l is given by Evans and Tappin [11].

Failure Stress Prediction Based on Quantitative NDE[1]

Microfocus radiography has been applied to a large quantity of isopressed and sintered silicon nitride modulus-of-rupture test bars (1500 bars).[2] Failure stress was calculated for bars containing defects by combining Eq. (1) and flaw geometry data obtained from microfocus film radiography. In total, 76 bars were rejected. These contained major naturally occurring flaws between the tensile surface and the neutral axis. All of the flaws characterized were voids; many were roughly spherical in shape. An accelerating potential of 65 kV was employed, and 5X magnification images of the bars were obtained at two mutually orthogonal orientations on high-contrast x-ray film. Flaw data were measured from the radiographs using an image-processing system.

For this study, fracture toughness (K_{IC}) which is equivalent to $(2E\gamma_i)^{1/2}$ was substituted into Eq. (1) [13]. A controlled flaw technique was used to determine fracture toughness of the material [14]. The average fracture toughness measured was 6.1 MPa·m$^{1/2}$, with a standard deviation of 0.2.

An analysis was performed of the modulus-of-rupture test bar to determine the stress distribution present when stressed to failure in four-point bending. For simplicity, only bars with voids located between the tensile surface and neutral axis, and within the inner load span, were analyzed. Within the elastic limit of the material, stress varied directly as the distance from the neutral axis.

Flaw data measured by NDE were used to predict failure with Eq. (1). A plot of predicted versus experimental failure stress is shown in Figure 1.[3] Results show a strong relation between predictions and actual fracture. Ideally, the points would fall along a straight line, drawn for reference, and the correlation coefficient would be unity. Uncertainty in the data is more pronounced as the stress becomes greater, and this behavior can be attributed to several factors. In general, predicted failure stress is higher because flaw size is smaller; therefore, the error in measurement of flaw dimensions is more significant. Furthermore, the microfocus x-ray system has a small but finite size source of x-rays (x-ray focal spot), which leads to geometrical unsharpness at higher magnifications. For instance, a 50-μm void at a 5X magnification imaged using a 10-μm focal spot (nominal spot size for the system used in this study) results in 15% unsharpness around the edge of the void. This effect limits the measurement resolution. The resolution of small flaws at 5X magnification is also compromised by the digital-imaging system. In this case, a 480×640 array of picture elements was used. The smaller the void, the closer in size it is to inherent flaws (residual porosity in this case) so that coalescence of these flaws can occur, resulting in lower failure strength. Finally, void size is measured from radiographs made in only two orientations so there is some ambiguity in alignment with the largest dimension of the voids. Fortunately, where the scatter in prediction is greatest, the voids are less detrimental in reducing strength.

Data are displayed for bars that definitely failed at the flaw detected by NDE. For these 34 bars, the stress at the void location was sufficient to initiate fracture. In total, 39 bars did not break

---

[1]This work was part of an in-depth study [12] on improving the reliability of ceramics through NDE.
[2]Bars made from SNW1000 silicon nitride, GTE Electrical Products, Belmont, California.
[3]The method of display was influenced by Munz et al. [15] from their work on assessment of flaws in ceramic material using NDE.

at the NDE-detected voids because the stress at the void location was not great enough to cause failure. This result was consistent with the predictions. Finally, three bars that did not break at detected voids failed at stresses higher than their predicted failure stresses. In NDE terms, these rejections would be false. In this case, false rejection of the three bars from the total population would only account for a 0.2% reduction in yield. The Weibull modulus of the NDE-rejected bars was 3.9 compared to 13.0 for accepted samples (51 bars were selected randomly from the remaining population to estimate Weibull modulus), demonstrating that NDE can significantly influence reliability.

Fractography was performed on all of the broken test bars to determine failure origins. Flaw size, shape, and location were measured from the fracture surface for failure calculations with Eq. (1). Results are similar to those obtained using NDE and are shown in Figure 2 for comparison. In some cases, voids were actually larger than indicated by fractography because the entire void was not revealed by fracture, for example when the fracture surface intersected the top of an elliptical void rather than its major diameter. This phenomenon was corroborated using radiography. In these instances, the NDE prediction of failure strength was superior to calculations based on fractography. Again, the scatter in the predictions became greater as the flaws became smaller.

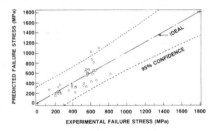

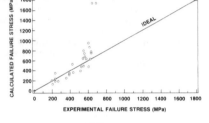

Figure 1.

Predicted failure stress using NDE. Flaw data measured from microfocus radiographs were used to predict failure stress. Ideally, the points would fall along a straight line and the correlation coefficient would be unity. Two other lines are drawn to represent the confidence limits for individual predicted values at the 95th percentile. Where the scatter in prediction is greatest, the voids are generally smaller and consequently less detrimental to strength.

Figure 2.

Calculated failure stress using fractography. Flaw data were obtained from fracture surfaces of broken test bars. In some cases, voids were actually larger than determined by fractography because the entire void was not revealed by fracture, for example, when the fracture surface intersected the top of an elliptical void rather than transversing its major diameter. This behavior was corroborated using radiographs.

NDE FOR CERAMICS JOINING

The utilization of advanced ceramics in high performance applications such as in heat engines requires the development of suitable joining methods. Ceramic-ceramic and ceramic-metal joints are needed in a variety of circumstances, for example coupling ceramic turbine rotors to metal shafts in hybrid automotive engines and attachment of ceramic coupons in critical wear areas or local high temperature zones. Silicon nitride is a prime candidate for heat engine applications because of its high temperature strength, oxidation resistance, and excellent thermal shock resistance.

Silicon nitride has a low thermal expansion coefficient relative to the structural materials to which it is commonly joined. This mismatch results in expansion differences at elevated temperatures, which can cause high local stresses at the interface, compromising the structural integrity of the joint. Brazing is the preferred method for ceramic-metal joining because the ductile braze alloys employed are able to accommodate the thermal expansion mismatch. In ceramic-ceramic joining, brazing is favored because flow of the molten braze alloy allows less stringent joint tolerances than methods such as diffusion bonding. A number of braze materials for joining ceramics are now commercially available [16], and research to develop new alloys is ongoing.

Silicon nitride components for heat engines must generally experience fewer failures than one part per million to satisfy reliability requirements in engine production. Ceramic joining reliability must be commensurate. Nondestructive evaluation plays a key role in ceramic joining, both by serving as a materials research tool for process development and for inspection of the final product.

Influence of the Braze on NDE Signals

Wetting and bonding are the crucial criteria for an effective braze [16]. The ability to monitor the distribution of braze material is advantageous because the degree of wetting and potential for bonding is revealed. Microfocus radiography can be used to measure the presence and distribution of braze material at the interface between joined components because x-ray attenuation is sensitive to small changes in the composition and thickness of the braze material. The factor which governs x-ray attenuation is the atomic number of the constituent elements. Fortuitously, the braze materials contain elements with high atomic numbers compared to the materials being joined. This results in x-ray images with high contrast between the braze and surrounding area, easing radiographic interpretation.

A comparison of linear attenuation coefficients is given in Figure 3 for silicon nitride, Incoloy™[4] 909 metal, and Nioro™[5] braze material (82 w/o Au, 18 w/o Ni) as a function of x-ray energy level. The calculated linear attenuation coefficients are based on tabulated values of mass attenuation coefficients of the constituent elements [17]. These calculated values are approximate because they are based on monochromatic radiation and the microfocus x-ray system has a polychromatic radiation source. The graph exhibits large differences in x-ray attenuation of the materials. These results favor x-ray imaging because the contrast in the image is directly related to differences in linear attenuation coefficient. Decreasing the accelerating potential (kilovoltage) of the x-ray system increases the minimum wavelength of the penetrating radiation and typically improves contrast.[6]

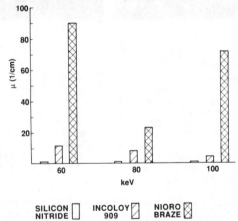

Figure 3. Comparison of linear attenuation coefficients. A comparison of linear attenuation coefficients for silicon nitride, Incoloy 909 metal, and Nioro braze material (82 w/o Au, 18 w/o Ni). Large differences in x-ray attenuation coefficients result in high contrast images, which ease radiographic interpretation.

Microfocus x-ray imaging was used in the present study to examine silicon nitride-silicon nitride and silicon nitride-Incoloy brazed with Nioro. It was expected that images of the ceramic-ceramic braze could be used to detect varying degrees of the distribution of the braze material; however, the similar detectability that occurred in the ceramic-metal braze images was unexpected.

---

[4]Incoloy is a trademark of Huntingon Alloys, Inc., Huntington, West Virginia.

[5]Nioro is a trademark of GTE Electrical Products, Belmont, California.

[6]An exception occurs in the graph shown because of a K edge for Au at 80.9 keV.

Fortunately, imaging was done at low accelerating potentials (60 to 120 kV), and the large difference between linear x-ray attenuation coefficients of the ceramic, metal, and braze material provided sufficient contrast. In particular, the 82% Au component of the Nioro braze material dominated the effect on contrast.

Microfocus Radiography of Ceramic-Ceramic Brazed Samples

The microfocus x-ray images shown in Figure 4 are representative of NDE of one ceramic-ceramic brazing operation. In this case, rectangular silicon nitride bars with chamfered edges are viewed after Nioro brazing. The pictures are of processed real-time x-ray images made at approximately 10X magnification. The only image processing performed was frame averaging to reduce random noise. The accelerating potential was 65 kV. The bars were positioned with a slight rotation and tilt with respect to the image plane. In this orientation, a uniform distribution of braze material would be imaged as a complete rectangle of consistent brightness. The resultant high contrast image reveals varying degrees of distribution of the braze material. Such variations are a measure of the degree of wetting at temperature and potential for bonding.

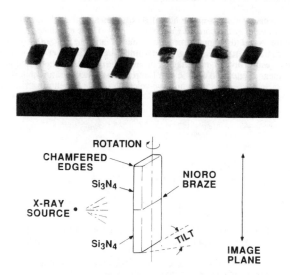

Figure 4. X-ray image of ceramic-ceramic brazed samples. Silicon nitride bars with chamfered edges joined with the Nioro braze are viewed. The bars are positioned with a slight rotation and tilt with respect to the x-ray source and image plane. Varying degrees of distribution of the braze material are revealed.

The microfocus radiographs shown in Figure 5 are representative of cylindrical and square brazed ceramic-ceramic samples that were used to study the effect of brazing parameters and substrate coatings on wettability. Samples were imaged in transmission, with the braze plane perpendicular to the direction of primary radiation. The total thickness of silicon nitride was approximately 20 mm. Variation in the braze material was still detectable.

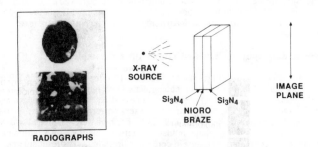

X-RAY
SOURCE
Si_3N_4
Si_3N_4
NIORO
BRAZE
IMAGE
PLANE
RADIOGRAPHS

Figure 5. Through transmission radiographs of ceramic-ceramic brazed samples. Samples were imaged with the braze plane perpendicular to the direction of primary radiation. Nonuniform distribution of the braze material is still discernible.

Microfocus Radiography of Ceramic-Metal Brazed Samples

An immediate requirement is the attachment of ceramic turbine rotors to metal shafts in automotive engines. A simplified illustration of a prototype ceramic-metal joint situated for imaging is shown in Figure 6. The joint can be viewed as a ring of metal brazed to a cylindrical ceramic rod. In this study, silicon nitride (SNW1000) was joined to Incoloy 909 with a Nioro braze.

CERAMIC
BRAZE
METAL
X-RAY
SOURCE
FILM

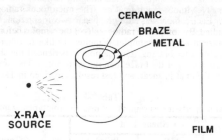

Figure 6. Prototype ceramic-metal joint. The simplified illustration of a prototype ceramic-metal joint can be viewed as a ring of metal brazed to a cylindrical ceramic rod.

The in-process NDE test matrix is shown in Figure 7 for ceramic-metal joining process development. This approach was applied to samples under a variety of braze conditions (e.g., braze cycle time, quantity of braze material, etc.). The flow diagram illustrates the concept of in-process NDE, where x-ray images are obtained at each process step. In this way, the final condition of the specimen can be traced back to intermediate process steps. For instance, if a crack is detected in the ceramic after brazing, was it a result of brazing, or was the crack present in the sample before brazing? Perhaps the most important feature of in-process NDE is that it simplifies interpretation of the final image.

In-process NDE was used to inspect four mechanical shear test samples joined at different braze cycle times. Microfocus film radiography was used to take advantage of the superior contrast of the film compared to the real-time image intensifier. Gamma radiography with a radioactive isotope source was also performed. For comparison, representative radiographs are shown in Figure 8.

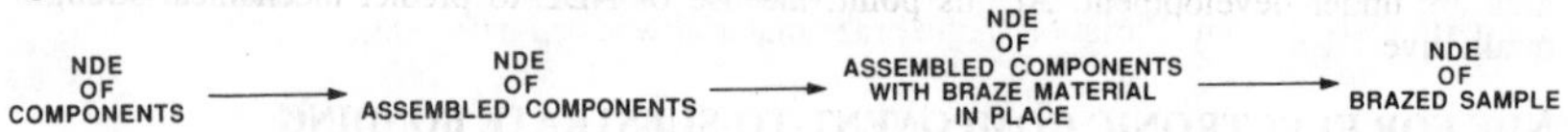

NDE
OF
COMPONENTS
NDE
OF
ASSEMBLED COMPONENTS
NDE
OF
ASSEMBLED COMPONENTS
WITH BRAZE MATERIAL
IN PLACE
NDE
OF
BRAZED SAMPLE

Figure 7. In-process NDE test matrix. In-process NDE allows tracing the final condition of the sample to intermediate processing steps, simplifying interpretation of the image.

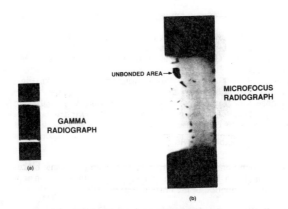

Figure 8.   Radiographs of a braze specimen (a) from gamma rays and (b) from microfocus x-rays about 6X magnification. The microfocus radiographs showed superior contrast and resolution. The gamma radiographs had wider imaging latitude, and multiple samples could be viewed simultaneously.

The gamma radiographs exhibited wide imaging latitude, and several samples could be examined simultaneously without magnification. The microfocus radiographs showed superior contrast and resolution. Each sample was radiographed in two orientations (rotated 90°) so the entire braze area could be studied. By comparing radiographs of the samples before and after joining, it was determined that one braze insert had incompletely melted. Unlike the other samples, the structure of the insert was still visible after brazing. This sample experienced the shortest braze cycle time. Varying degrees of distribution of the braze material were noted. The samples were mechanically tested by placing the braze joint in shear, and test results are given in Table I.

Table I
Mechanical property measurements of brazed prototype ceramic-metal joint

Sample	Shear Test		Comments
	MPa	lb/in²	
A	231	33,500	Substantial unbonded area
B	252*	36,500*	Unbonded area
C	252*	36,500*	No unbonded area visible Ceramic cracks
D	237	34,400	Incomplete melting

* Denotes sample did not break. Test discontinued because of load cell limitation.

Correlation of the NDE results with mechanical performance was only partially possible. For example, the substantial unbonded area in Sample A, compared with the complete bonding in Sample C, was consistent with their relative mechanical strength. Also, the incompletely melted braze insert resulted in visible unbonded area in Sample D and lower strength than Sample C. Comparisons between samples with significant unbonded areas such as A and B could not be used to accurately predict relative strength. Image processing algorithms capable of quantifying the braze distribution area are under development. At this point, the use of NDE to predict mechanical strength is qualitative.

NDE FOR ELECTRONIC COMPONENT-TO-SUBSTRATE BONDING

Development of NDE methods in a materials research environment has led to wider application of techniques like microfocus radiography than anticipated. For example, NDE techniques developed for structural ceramic joining are directly applicable to electronic component/substrate joining.

An example is high power microwave transistor bonding, where the need for increased power dissipation is placing stringent requirements on attachment of the silicon die to the ceramic substrate. Microfocus x-ray imaging is being used to monitor the process of bonding microwave power transistors to metallized BeO substrates, thereby providing feedback for process improvement.

An illustration of the typical die bond components is shown in Figure 9. The components are placed in contact and heated above the liquidus temperature of the Au-Si eutectic ($\approx 400°C$). Then, the die is scrubbed (mechanically vibrated) to ensure proper wetting. Incomplete wetting causes most of the problems in die bonding, e.g., poor mechanical attachment or high electrical resistivity [18]. Microfocus radiography can be used to evaluate the presence and distribution of the Au-Si eutectic because the attenuation of x-rays is largely governed by the Au component compared to the Si die and BeO substrate.

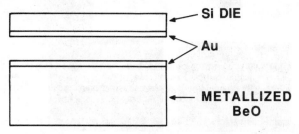

Figure 9. Typical die bond components. The components are placed in contact and heated above the liquidus temperature of the Au-Si eutectic. The die is then scrubbed (mechanically vibrated) to ensure proper wetting.

A microfocus x-ray image of a transistor quad section is shown in Figure 10. Four Si chips bonded to a metallized BeO substrate are shown. A magnification of approximately 15X and accelerating potential of 45 kV ensured high resolution and adequate image contrast. Variations in the distribution of the eutectic are revealed. An infrared image of this section in operation is shown for comparison. The temperature is scaled in pseudo color, with the low end corresponding to ambient. The chip that is operating at excessive temperatures displays an x-ray image which shows no apparent eutectic formation. Infrared imaging is utilized to macroscopically measure the effects of poor bonding; however, detail of the structure of the interface is not discernible. The microfocus x-ray images are complementary.

Figure 10. Microfocus x-ray and infrared images. Infrared imaging is utilized to macroscopically measure the effects of poor bonding; however, detail of the structure of the interface is not discernible. Complementary microfocus images reveal varying degrees of the presence and distribution of the eutectic.

Conventional metallographic techniques facilitated interpretation of the x-ray images. A sample die bonded to Al_2O_3 was sectioned and examined optically. An Al_2O_3 substrate was used rather than BeO to avoid health-related problems in sectioning and polishing. Examination of the sample revealed large unbonded areas which corresponded to regions noted in the radiographs. Figure 11 illustrates both techniques.

The images shown in Figure 12 are from a scrubbing time test where five bonds were formed, with scrubbing time intervals from shortest (1) to longest (5). As expected, the amount of eutectic formed increased with increased scrubbing time; however, the eutectic formation was not balanced across the interface. Subsequent examination of the process revealed that the scrubbing mechanism applied a load preferentially to one side of the Si die.

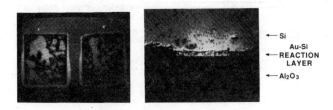

Figure 11.　Metallography corroborates x-ray findings. Microfocus x-ray image of power transistors bonded to alumina and an optical micrograph of sectioned sample #2, 500X. Examination of the sample revealed large unbonded areas which corresponded to regions noted in microfocus x-ray images.

Figure 12.　Feedback for process improvement. Microfocus x-ray images revealed that eutectic formation was not balanced across the interface. Subsequent examination of the process revealed that the scrubbing mechanism applied a load preferentially to one side of the Si die.

SUMMARY AND CONCLUSIONS

Results were presented from several studies where microfocus radiography was employed to provide feedback for improving reliability and performance of advanced materials and components. Quantitative microfocus radiography of structural ceramics has been utilized to explore the relationship between NDE results and the mechanical behavior of components. Failure stress was predicted for a set of test bars known to contain process-related defects with a fracture-mechanics model and flaw data obtained from radiographs. The results showed a strong relation between predictions and actual fracture stress. Data exhibited higher scatter at smaller flaw sizes.

Improvements in failure stress prediction should be realized with future advancements in NDE and fracture mechanics, especially the development of higher resolution imaging systems for more effective characterization of flaw geometry because this factor is critical in determining fracture strength.

Studying the effects of process modifications with microfocus radiography has aided understanding of structural ceramic joining. Detection of varying degrees of the distribution of the braze material was possible for both ceramic-ceramic and ceramic-metal joints. Correlation of the NDE results with mechanical performance was only partially possible. Samples with substantial differences in distribution of braze material had relative strengths consistent with NDE results; however, more subtle comparisons could not be made. The correlation of NDE results with mechanical performance should improve with advances in quantitative image analysis and expansion of the database on relevant testing.

The process of bonding high power transistors to ceramic substrates has been evaluated with microfocus radiography. X-ray images revealed details of the structure of the die bond interface, and led to identification of areas needing process improvement.

REFERENCES

[1] W. Edward Deming, *Out of the Crisis,* Massachusetts Institute of Technology Publishers, Cambridge, MA, p. 28 (1986).
[2] R.W. Parish, AGARD Lecture Series No. 103, pp. 3.1–3.29 (1979).
[3] R.A. Roberts, W.A. Ellingson, and M.W. Vannier, "A Comparison of X-ray Computer Tomography Through Transmission Ultrasound and Low-kV X-ray Imaging for Characterization of Green-State Ceramics," Proc. of the 15th Symp. on Nondestructive Evaluation, pp. 118–124 (1985).
[4] D.W. Richerson, *Modern Ceramic Engineering,* Marcel Dekker, New York, pp. 69–96 (1982).
[5] D.J. Cotter and W.D. Koenigsberg, "Microfocus Radiography of High Performance Silicon Nitride Ceramics," Proc. Conf. on Nondestructive Testing of High Performance Ceramics, Boston, MA, Am. Ceram. Soc., pp. 233–253 (1987).
[6] C.L. Quackenbush, J.T. Neil, and J.T. Smith, "Sintering, Microstructure, and Properties of Si_3N_4 and SiC Based Structural Ceramics," ASME Paper 81-GT-220, pp. 1–9 (March 1981).
[7] A.E. Pasto, J.T. Neil, and C.L. Quackenbush, "Microstructural Effects Influencing Strength of Silicon Nitride," Proc. Int. Conf. on Ultrastructure Processing of Ceramics, Glasses and Composites, pp. 476–479 (1983).
[8] G. Bandyopadhyay and K.W. French, "Fabrication of Near Net Shape Silicon Nitride Parts for Engine Application," ASME Paper No. 86-GT-11, 31st Int. Gas Turbine Conference and Exhibit, Dusseldorf, Fed. Repub. of Germany, pp. 1–4 (June 1986).
[9] A.A. Griffith, "The Phenomenon of Rupture and Flow in Solids," *Philos. Trans. R. Soc. London, Ser. A. 221 [4], pp. 163–198 (1920).*
[10] R.W. Davidge and A.G. Evans, *Mater. Sci. Eng. 6,* pp. 281–298 (1970).
[11] A.G. Evans and G. Tappin, *Proc. Br. Ceram. Soc. 20,* pp. 275–297 (1972).
[12] D.J. Cotter, W.D. Koenigsberg, A.E. Pasto, and L.J. Bowen, "Improving the Reliability of High-Performance Ceramics Using Nondestructive Evaluation," 12th Conf. on Composites and Advanced Ceramics, Cocoa Beach, FL (January 1988).
[13] R.W. Davidge, "Effects of Microstructure on the Mechanical Properties of Ceramics," *Fracture Mechanics of Ceramics, Vol. 2,* Plenum Press, New York, pp. 447–468 (1974).
[14] J.J. Petrovic and M.G. Mendiratta, "Fracture from Controlled Surface Flaws, Fracture Mechanics Applied to Brittle Materials," ASTM STP 678, ed. S.W. Frieman (American Society for Testing and Materials, Philadelphia, PA, pp. 83–102 (1979).

[15] D. Munz and O. Rosenfelder, "Assessment of Flaws in Ceramic Materials on the Basis of Nondestructive Evaluation," *Fracture Mechanics of Ceramics*, Vol. 7, Plenum Press, New York, pp. 265–283 (1986).

[16] R.E. Loehman and A.P. Tomsia, *Ceram. Bull. 67* [2] (1988).

[17] P. McIntire, *Nondestructive Testing Handbook*, Am. Soc. for Nondestructive Testing Publishers, Columbus, OH, pp. 227–232 (1985).

[18] C.E. Hoge and S. Thomas, "Some Considerations of the Gold-Silicon Die Bond Based on Surface Chemical Analysis," 18th Annual Proceedings of Reliability Physics, IEEE (1980).

Rapid Non-Destructive Testing of Ceramic Multilayer Capacitors by a Resonance Method

O. Boser, P. Kellawon and R. Geyer
Philips Laboratories
North American Philips Corporation
Briarcliff Manor, N.Y. 10510

ABSTRACT

A rapid non-destructive test method for ceramic multilayer capacitors made from piezoelectric materials such as barium titanate or lead containing materials is described and evaluated. The test method is based on the internal excitation of standing acoustic waves in the capacitors. The standing waves are severely dampened by defects such as delaminations and pores. An undampened resonance is a good indication of a defect free ceramic multilayer capacitor. This finding was used in a non-destructive test set-up to evaluate about 1,000 capacitors. The test set-up has the potential to test over 100,000 capacitors an hour. Through metallographic (cross section) examination a sorting accuracy of 2% false accepts and 8% false rejects was determined for the high speed test set-up.

INTRODUCTION

Ceramic multilayer capacitors are composite structures made from alternating layers of metal (electrodes) and ceramic (dielectric material). As can be imagined, joining such different materials into a monolithic structure presents problems especially since the sintering of ceramic requires a $1,200^{\circ}C$ to $1,500^{\circ}C$ heat treatment. The electrodes are generally much thinner than the dielectric and the former can act as crack initiation sites if the ceramic multilayer capacitor has not been heat treated carefully. Therefore it is possible to have delaminations between electrode and ceramic that can not be detected by capacitance measurements. Currently delaminations are detected in a destructive fashion i.e. the production is sampled periodically and the ceramic multilayer capacitors are sectioned and inspected. If the samples tested do not show delaminations the whole batch is declared good.

Non-destructive testing of ceramic capacitors is difficult, because the testing method has to be rapid, efficient and low cost. Different test methods have been proposed: Neutron irradiation [1], scanning acoustic microscopy [2] and acoustic emission measured as a function of applied mechanical stress [3] or electrical stress [4]. All test methods have shown some success but none has proved economically and technically efficient enough for full scale application. In the following, a new non-destructive test method [5] for capacitors made from piezoelectric materials will be discussed that is based on an acoustic resonance method and might fulfill the above criteria.

ELECTROMECHANICAL RESONANCES IN CERAMIC CAPACITORS

It is well known that $BaTiO_3$ shows piezoelectric effects and therefore can be used to fabricate ultrasonic transducers. On the other hand $BaTiO_3$ is also used as dielectric material in ceramic capacitors. Although designed for a different purpose, ceramic capacitors should show electromechanical resonances similar to the ones observed in piezoelectric transducers. It seems that observations of electromechanical resonances in ceramic capacitors have only been reported recently [6].

Unlike transducers, ceramic capacitors do not show a remnant polarization after application of a bias field. However, during the application of a bias voltage a polarization is built up. Therefore electromechanical resonances due to piezoelectric effects should appear and be observed with a d.c. bias applied.

To determine the resonance behavior of the capacitors the impedance of the ceramic multilayer capacitors was measured with a Hewlett-Packard low frequency impedance analyzer that has a frequency range from 5 Hz to 13 MHz. The impedance

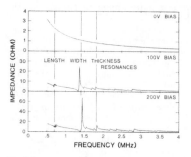 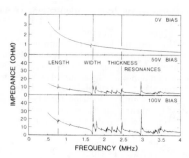

Fig.(1) Impedance vs frequency for (a) X7R-1206 capacitor, 1.25 *mm* thick, (100 *nF*) with 0, 100 and 200 *V* bias and (b) Z5U-1206 capacitor, 1.0 *mm* thick, (100 *nF*) with 0, 50 and 100 *V* bias.

analyzer was controlled by an HP computer that can sweep over the entire range of resonances and print out the results.

The impedance analyzer has a built-in dc bias of up to 35 *V*. For measurements with a larger bias a sample holder was constructed according to the HP manual [7]. This sample holder allowed the use of bias voltages of up to 200 *V*. The set-up was calibrated in the shorted and open condition at the upper frequency of the swept frequency range.

For the study of the mechanical resonances in ceramic capacitors the impedance for one size capacitor made from 2 different materials was measured from 0.5 *MHz* to 4 *MHz* with and without a dc bias. The capacitor size is 1206 (.12 x .06 x .05 *inch* or 3.2 x 1.6 x 1.25 *mm*). The materials were X7R [8] that has a small temperature coefficient and Z5U [8] that has a much larger temperature coefficient. As can be seen from Fig.(1) the resonance effects increase with increasing bias voltage. The resonance effects are particularly pronounced for the Z5U capacitor, Fig.(1b). Except for the length, width and thickness resonance it was not possible to associate any of the additional resonances with a dimension in the capacitor.

The results, Fig.(1), can be analyzed under the assumption that the observed fluctuations in impedance are due to electromechanical resonances excited in ceramic capacitors through piezoelectric effects. It is difficult to identify the exact mode of the resonances. However, it seems reasonable to associate the resonance at the lowest frequency with the longest dimension in the capacitor, usually the length dimension. Assuming a compressive standing wave the wavelength is equal to twice the length dimension. A simple equation relates the wavelength, the frequency f and the sound velocity v:

$$v = 2lf \qquad (1)$$

From this equation the sound velocity can be determined as about 5,000 *m/sec* which is in good agreement with the established value for the sound velocity in $BaTiO_3$, [9]. The most pronounced resonance can be associated with the width dimension. It shows the largest swing in impedance going from the resonance to the anti-resonance. The thickness resonance is difficult to determine, Fig.(1). It seems to be severely dampened through the metal ceramic interface.

The varying dampening for the different resonances is due to the construction of the capacitor. The thickness resonance is dampened by the electrode dielectric material layers. The length resonance is dampened by the terminations at either end of the capacitor. Only the width resonance is relatively undampened which leads to the large swing in impedance between resonance and anti-resonance and is a sign for a high quality factor Q [10].

The additional resonances observed, can not be associated with any dimension in the ceramic capacitor. This is not surprising since the resonance modes in rectangular parallelepipeds are numerous and very difficult to trace mathematically, [11]. In spite of these difficulties it is possible to recognize a certain pattern between the width and the thickness resonance and a pattern above the thickness resonance, see Fig.(1). The pattern seems to be independent of the material and independent of the size of the capacitor.

Since no permanent polarization can be built up the polarization is always parallel to the bias field and the signal field. This arrangement can produce longitudinal- or expander waves only, [10]. In addition, it can be seen that the length and width resonance have the same character, that is the wave propagates perpendicular to the bias- and signal field. The symmetry is different for the thickness resonance where the wave propagation is parallel to the bias- and signal field. This should result in slightly different sound velocities for the length/width resonance on one hand and the thickness resonance on the other. However, within the error limits of the measurements no difference in sound velocity can be established between the thickness mode and the length/width mode.

NON-DESTRUCTIVE TEST METHOD AND TEST SAMPLES

It is well known [12], [13], [14] that mechanical resonances are affected by flaws and density variations in the structures under test. Yet the dampening of resonances is not often used in non-destructive testing probably because the dimensions of the objects under test are rather large (> 10 cm). Hence the resonance frequencies are rather low and measuring the dampening of the resonance is not very discriminatory. In this situation pulse echo methods are much more suitable. However, in the case of multilayer capacitors the pulse echo method is difficult to apply because of the interfaces formed by the electrodes and the small dimensions of the capacitors (< 10 mm).

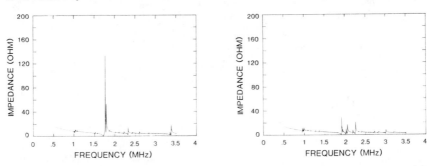

Fig.(2) Impedance vs frequency for Z5U-1005 capacitor (20 nF) with 50 V bias (a) Delamination free capacitor (b) Capacitor with delamination.

Fig.(3) Micrographs of capacitors tested electrically in Fig.(2). (a) Capacitor free of delaminations. (b) Capacitor with delamination.

Due to the fortunate circumstance of internal excitation of resonances in these small pieces, the dampening of the resonances by defects seems to be the almost ideal test method. Delaminations and pores in ceramic multilayer capacitors represent such defects, that can dampen or even inhibit the appearance of the acoustic resonances depending on the size of the delamination.

To test this hypothesis Z5U capacitors with nominal dimensions of .10 x .05 x .05 *inch* (2.5 x 1.25 x 1.25 *mm*) i. e. size 1005, were used with a capacitance value of 20 *nF*. The test batch of ceramic capacitors showed an exceptionally large incidence of delaminations. The impedance of the Z5U capacitors of size 1005 was measured with the above described test set-up in the frequency range between 0.5 *MHz* and 4 *MHz* with a 50 *V* bias. The impedance versus frequency curve of a good capacitor is shown in Fig.(2a).

It is possible to estimate the expected length and width resonance with the help of the sound velocity derived earlier. One obtains for the length and width resonance 1 *MHz* and 2 *MHz*, respectively. However, since the cross section of the capacitor is square other modes different from the ones discussed above can appear. Different modes propagate at different sound velocities. In the capacitors tested here the width and the thickness resonances occur at the same frequency (nominally 2 *MHz*) due to the square cross section. The coincidence of the two standing waves perpendicular to each other, leads to a new vibrational mode that has a resonance frequency of 1.8 *MHz*, slightly lower than the one predicted for single standing waves.

In Fig.(2b) the impedance versus frequency curve is shown for a delaminated capacitor. The difference in the impedance at anti-resonance between the good and the delaminated capacitor is quite pronounced. This difference in impedance seems to be a good measure to determine the quality of a ceramic capacitor. The impedance of 140 Ω measured in Fig.(2a) is a representative value for good capacitors with fluctuations of about 10 %. The "bad" capacitors had a consistently lower impedance not higher than 50 Ω.

To confirm the suggestion that the change in impedance at resonance is caused by delaminations, the capacitors were cut and polished. Fig.(3) shows the micrographs of the capacitors and it confirms that the capacitor tested in Fig.(2a) is good and the one tested in Fig.(2b) is delaminated. The example shown in Fig.(2b) is rather obvious. No detailed study of the detection limit was made. However, the detection limit derived from the metallographic inspection falls somewhere between 0.25 *mm* to 0.5 *mm* crack length. From experiments on other capacitors it became obvious that the impedance at anti-resonance is very sensitive to porosity.

OSCILLATOR TEST SET-UP

To accelerate the testing of ceramic multilayer capacitors an oscillator test circuit was proposed that can be combined with the breakdown and poling test set-up. The test circuit was designed as a marginally stable oscillator circuit, where the anti-resonance frequency of the capacitor determines the frequency of oscillation and the magnitude of the anti-resonance impedance determines whether oscillations occur: If the impedance is high the capacitor is good and oscillations occur, however, if delaminations are present the impedance is low and there will be no oscillations. The circuit is shown in Fig.(4). The op-amp, and R_3 and R_4 form a non-inverting amplifier with a gain given in Eq.(2).

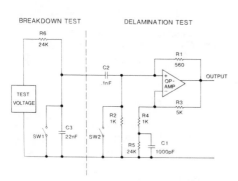

Fig.(4) Circuit diagram of automated test oscillator.

$$Gain = (R_3 + R_4) / R_4 \qquad (2)$$

The resistor R_5 lowers the gain of the amplifier at low frequencies to prevent spurious oscillations, and C_1 bypasses R_5 at the anti-resonance frequency of the capacitor under test (C_3). The capacitor C_2 is a blocking capacitor and prevents the poling voltage from reaching the op-amp. At the anti-resonance frequency the resistor R_1 and the impedance Z_{ar} divide the output signal of the amplifier. If Z_{ar} is the impedance of the capacitor under test at the anti-resonance frequency one obtains the following ratio.

$$Z_{ar}/(Z_{ar} + R_1) \qquad (3)$$

This divided signal is applied to the input of the amplifier. The condition for oscillation is that the gain of the amplifier and the voltage divider is greater than unity.

HIGH SPEED NON-DESTRUCTIVE TESTING AND RESULTS

In a non-destructive test the capacitor to be tested (C_3) is inserted in the test set-up with both switches closed. The test starts by opening switch #1 to apply the poling voltage to the capacitor. A period of 5 *msec* was found to be enough to let all transients die out in Z5U capacitors. Then switch #2 was opened to start the oscillations. It was left open for 12 *msec*. Then switch #2 and switch #1 are closed again. It shows that a non-destructive test can be performed in less that 20 *msec* not counting insertion time. If large batches are inserted at a time and the testing is controlled electronically the insertion time can be kept at a minimum not adding more than about 5 *msec* per capacitor.

For the actual high speed test the critical impedance Z_{ar} to start oscillations in good capacitors was set at 127 Ω (effective value 112 Ω). This value has to be compared with the impedance measured at anti-resonance for good capacitors of better than 120 Ω. The relatively low value of 112 Ω was chosen to bias the test in favor of false rejects and against false accepts.

A batch of 954 capacitors of size 1005 was tested. Of these, 244 tested free of delaminations and 710 had delaminations. From each of these two groups 50 capacitors were chosen at random for destructive testing. Metallographic preparation techniques at different heights in the capacitor gave the results summarized in Table (I).

TABLE (I)
Summary of results from metallographic testing of 100 capacitors.

Correct accepts	49	Correct rejects	46
False accepts	1	False rejects	4

As can be seen the biased high speed non-destructive test has decreased the number of false accepts at the expense of the number of false rejects.

CONCLUSION

The test results show that a ceramic multilayer capacitor can be non-destructively tested in about 25 *msec*. This translates into over 100,000 capacitors an hour. In addition, the non-destructive test set-up can be integrated with the standard voltage breakdown test reducing capital investment to a minimum. Under the above described circumstances it was possible to classify 95 % of the capacitors correctly. The test was biased to favor false rejects over false accepts. This bias translates into 98 % of correct accepts.

ACKNOWLEDGMENT

We want to thank Dr. S. Long, Mepco/Centralab, Los Angeles, CA for the samples and the support of this investigation.

REFERENCES

(1) G. F. Kiernan, "Comparison of Screening Techniques for Ceramic Capacitors," Proc. of Symposium held at Marshall Space Flight Center, NASA Conference Publication # 2186, Marshall Space Flight Center, Alabama, 1981

(2) C. L. Vorres, D. E. Yuhas and L. W. Kessler "Non-Destructive Evaluation of Ceramic Chip Capacitors by Means of the Scanning Laser Acoustic Microscope," 2nd Capacitor and Resistor Technology Symposium, pp. F3-1 - 6, Components Technology Inst., Inc, Suite 1122-K, 303 Williams Ave., Huntsville, Alabama, 35801, 1982

(3) S. R. Kahn and R. W. Checkaneck, "Acoustic Emission Testing of Multilayer Ceramic Capacitors," IEEE Transactions on Components, Hybrids and Manufacturing Technology, **6** , p. 517 - 526, 1983

(4) N. H. Chan and B. S. Rawal, "An Electrically Excited Acoustic Emission Test Technique for Screening Multilayer Ceramic Capacitors" Proceedings of the Electronic Components Conf., Los Angeles, CA, 1988, vol. 38, pp. 502 - 506, publ. by IEEE and EIA, New York

(5) O. Boser, US Patent #4,644,259, February 1987

(6) O. Boser, "Electromagnetic Resonances in Ceramic Capacitors and Evaluation of the Piezoelectric Materials' Properties," Advanced Ceramic Materials, **2** , pp. 167 - 172, 1987

(7) "Operations and Service Manual for Model #4192A Impedance Analyzer," Yokogawa - Hewlett - Packard Ltd, Tokyo, Japan, March 1983

(8) "EIA Standards Manual for Ceramic Dielectric Capacitors, Classes I, II, III and IV" RS-198-C Revision of RS-198-B, November 1983, Electrical Industries Association, Engineering Dept. 2001 Eye St. Washington, DC 20006.

(9) Landoldt-Bornstein, Vol. 3, Editors: K.-H. Hellwege and A. M. Hellwege, Springer Verlag, Heidelberg, 1969

(10) Belincourt, D. A., D. R. Curran and H. Jaffe, "Piezoelectric and Piezomagnetic Materials and Their Function in Transducers" Physical Acoustics, Vol. I, Part A, Editor: W. A. Masson Academic Press, New York, 1964, p. 169 - 270

(11) R. Holland and E. P. EerNisse, "Design of Resonant Piezoelectric Devices" Research Monograph #56, The MIT Press, Cambridge, Mass, 1969.

(12) V. M. Baranov and E. M. Kudryavtsev, "Application of the Ultrasonic Resonance Method in the Inspection of Small Products," Soviet J. Non-Destructive Testing, **15** , pp. 750 - 755, 1980

(13) L. R. Testardi, S. J. Norton and T. Hsieh, "Determination of Inhomogeneities of Elastic Modulus and Density for One-Dimensional Structures Using Acoustic Dimensional Resonances," J. Applied Physics, **56** , pp. 2681 -2685, 1984

(14) L. R. Testardi, S. J. Norton and T. Hsieh, "Acoustic Dimensional Resonance Tomography: Some Examples in One-Dimensional Systems," J. Applied Physics, **59** , pp. 55 - 58, 1985

DETECTING AREAS OF HIGH CHLORIDE CONCENTRATION IN BRIDGE DECKS

K. LIM AND D.L. GRESS
Department of Civil Engineering, University of New Hampshire
Durham, New Hampshire 03824
K. MASER
Department of Civil Engineering, Massachusetts Institute of Technology
Cambridge, Massachusetts 02138.

ABSTRACT

The presence of chloride ions in New England bridge decks has caused severe damage and has contributed greatly in reducing service life. The source of the chloride ions is mainly from deicing salts applied for snow and ice removal to enable safer driving during the winter. The chloride ion acts as an accelerator in the corrosion of reinforcing bars causing expansion and therefore deterioration of the bridge deck concrete.

At present the method of determining the percentage of chloride present in bridge decks is a very time consuming, expensive and destructive process. The need for a fast nondestructive procedure to determine the amount of chloride in a concrete bridge deck is apparent. This paper introduces a means of utilizing ground penetrating radar as a viable nondestructive means of detecting areas of high chloride concentration in concrete bridge decks. The technique is based on the attenuation of a radar signal emitted from a radar unit in the presence of chloride.

INTRODUCTION

PROBLEM DEFINITION

Since the mid 1950's, the use of deicing salts on the roads in New England during the winter accelerated, due to increased traffic as well as concern for maintaining safe roads free of dangerous ice and snow. Cracks present in concrete bridge decks provide easy access for moisture and deicing salt penetration. Cracking is the result of many variables including plastic shrinkage, drying shrinkage, thermal stress and loading stress at nearly ages and during the bridge deck's service life. Penetration of moisture and chloride ions leads to the corrosion of the reinforcing bars which causes internal expansion of the bars [1]. The production of corrosion products on the reinforcing bars causes horizontal cracks in the concrete which in turn causes spalling and or delaminations. These failures lead to substantial loss in strength of the bridge deck and can eventually lead to total destruction if left unchecked.

The corrosion process starts when chloride ions, moisture and oxygen are present at the surface of a reinforcing bar. The presence of these three elements provide the ideal corrosion environment. Although corrosion can and does occur without the presence of chloride, it becomes accelerated in its presence. Knowing the level of chloride present is of great importance in predicting if corrosion is active. Measurement of the amount of chloride present in the bridge therefore yields valuable information on whether corrosion is occurring and also provides an indication of whether total bridge replacement is necessary [2].

The detection of areas of high concentration of chloride in a concrete deck is a slow and tedious process involving drilling or coring into the bridge deck in order to obtain samples for testing.

TESTING PROCEDURE

A GSSI (Geophysical Survey System Incorporate) radar unit was provided to the University of New Hampshire by the Public Service of New Hampshire. Modifications were made to the GSSI so as to be able to obtain more information from the radar signal produced. The GSSI unit was equipped with two output ports, one to emit the waveform and the other to emit the start of a scan pulse. The modifications allowed the central processing unit of the radar system to be connected to a microcomputer by way of an oscilloscope which allowed for a more detailed analysis of each individual waveform. A Gulf Applied System radar unit, from Federal Highway Administration, was also used in this research.

The waveform is refracted and reflected when it travels through changing media. It is the reflected waveform that is detected by the receiver in the antenna and analyzed by the processor. The strength of the reflected waveform is measured in volts. Some of the strength of the waveform is lost when the waveform travels through different media. It is the loss in strength of the waveform that was used to correlate with zones of high chloride content. The voltage output was recorded as a function of time, in nanoseconds.

Actual bridge slabs were brought to the University of New Hampshire laboratory from three different bridges, Route 1 at Portsmouth, New Hampshire, Interstate 93 over Route 128 and Interstate 93 in New Hampton, New Hampshire. The top surface of each slab was marked with a one foot grid to allow for easy identification and location notation.

The radar antenna was placed on the specified test location of the bridge slab and the resulting radar waveform was then transferred to the computer by way of the oscilloscope and stored for later analysis. The antenna was then repositioned for the next observation. The antenna was positioned parallel to the direction of the bridge traffic thus polarizing the longitudinal reinforcing bars. The program adopted for evaluating the radar waveform was Waveform Basic. It provides the user an easy means of determining the values of the different peaks required in the analysis.

Once all the waveforms were collected, the concrete sampling process to determine the chloride and moisture content of the dust samples was done in accordance to FHWA-KS-RD 75-2.

The dust samples were tested for moisture content and chloride content in accordance to ASTM 566-78 and AASHTO T260 respectively. The chloride test deviated from the standard procedure only in that the sample was placed in a centrifuge at 2000 g's at room temperature instead of being filtered as called for by the AASHTO standards. Centrifuging was found to be faster and also provided a cleaner sample which produced better results.

A total of four slabs were tested. Two slabs from Route 1 and two from Interstate 93 in New Hampshire. Each slab had a different range in moisture and chloride content. The original concrete mixtures in all four slabs were similar in design and made with gravel aggregate.

Once all the data had been collected, the data along with its location were placed into a contour program called Plotcall manufactured by Golden Graphics Systems, Golden, Colorado. The program develops contour lines from the input data. Data points were not collected at the edge of the slab due to the effect of the slabs edge on the return signal of the radar waveform. Areas of high chloride concentration were easily identified by the peaks and valleys of the topographical maps.

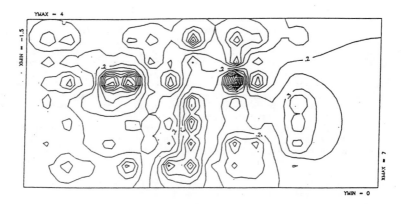

FIGURE 1A: TOPOGRAPHICAL PLOT OF D/C VALUES.

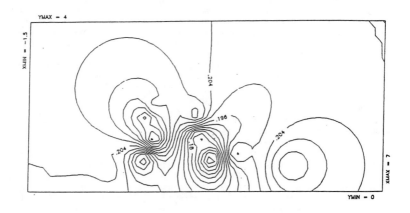

FIGURE 1B: TOPOGRAPHICAL PLOT OF CHLORIDE CONTENTS.

TOPOGRAPHICAL PLOTS PRODUCED WITH THE
GULF APPLIED RADAR SYSTEM

EXPERIMENTAL

DATA AND DISCUSSION

The object of this project was to be able to determine areas of high chloride content in concrete bridge decks using an existing radar system.

It has been shown that chloride is capable of affecting the attenuation of the radar signal. This attenuation occurs in areas of high concentration of chloride due to a greater loss of radar energy, caused by an increase in the material's conductivity [3]. The presence of chloride in the zone of question results in a weaker reflected waveform. As the amount of chloride increases in the bridge deck, the relative permittivity of the bridge deck decreases [3]. The return waveform from the surface of the reinforcing bar (D peak) was divided by the return waveform from the asphalt concrete interface (C peak) to provide a D/C ratio. This normalized variable should in concept be relatable to the chloride content present. Topographical plots of the chloride content were developed along with values of the D/C ratio in an effort to get a visual understanding of the data. It was hoped that areas containing high concentrations of chloride would result in relatively low values for the absolute value of the D/C ratios and vice versa.

It was found in general that areas of high concentration of chloride occur in areas where the D/C ratio is low. The D/C ratios obtained are relative for each individual slab. Figure 1 was developed using a Gulf Applied System radar obtained from the Federal Highway while Figure 2 was developed by using the GSSI radar system.

In the plot obtained by using the Gulf Applied System radar, the areas of high chloride concentration were more accurately defined by the D/C ratio plot. The plot obtained from the GSSI radar system did not identify the high chloride concentrated areas as accurately as the Gulf Applied System radar unit. The "ringing" application on the return pulse from the GSSI radar unit was suggested as the cause for the distortion of the signal [3]. It is suggested that identifying areas of high chloride concentration be limited to general areas instead of a point source. In reality, this is typical of a chloride distribution as it is very variable to the x, y and z directions in a given bridge deck. This "ringing" is not apparent in the Gulf Applied System radar.

The average moisture content of the slabs tested using the Gulf Applied System radar was found to be 3.54 percent, while the moisture content for the slabs tested using the GSSI radar unit was 3.89 percent. Although the difference appears to be trivial, it does in fact represent almost a 10 percent increase in moisture which could probably be masking the influence of the chloride attenuation. The dielectric constant of water is 81 and its presence affects the attenuation of the radar signal drastically.

It should also be noted that the antenna of the radar unit emits the waveform at an angle and not perpendicular to the slab, thereby covering a larger area instead of a specific point. It is assumed however that the sample utilized in determining the chloride content of the location be a representative sample of the test area in question.

It was found that the antenna must be placed at least 18 inches from the edge of the slab to reliably determine areas of high chloride concentration. This assures the elimination of edge effects which result form the waveform leaving the edge of the bridge deck. When the waveform leaves through the edge of the bridge deck it essentially passes through two different medias thereby causing a refracted and reflected wave to occur. If the antenna is placed close enough to the edge of the bridge deck, the reflected waveform can be mistakenly used in the analysis.

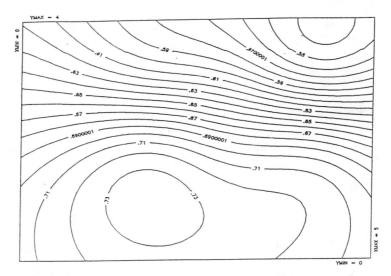

FIGURE 2A: TOPOGRAPHICAL PLOT OF D/C VALUES.

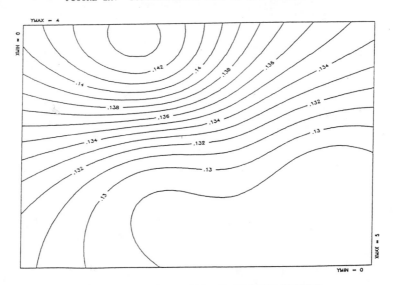

FIGURE 2B: TOPOGRAPHICAL PLOT OF CHLORIDE CONTENTS.

TOPOGRAPHICAL PLOTS PRODUCED WITH THE
GSSI RADAR SYSTEM

CONCLUSION

Based on the research performed on the concrete slabs, the following conclusions were reached:
1) The method is a viable nondestructive testing means of locating areas of high chloride concentration.
2) Both the Gulf Applied System and the GSSI radar system can be used however the Gulf Applied System appears to be more sensitive at this time.

ACKNOWLEDGMENTS

This research was sponsored in part by the New England Consortium. The support of the New England Consortium and the loaning of equipment from the Public Service of New Hampshire is very much appreciated.

REFERENCES

1. Ayyub, B.M. and White, R. "Detection of Delamination and Cavities in Concrete", SHARR, FHWA/MD-87/05, (1987) p. 1.

2. NCHRP Report NR57CL "Durability of Concrete Bridge Decks", (1979) p. 11.

3. Maser, K.R. "Detection of Progressive Deterioration in Bridge Decks Using Ground Penetrating Radar", Proceedings of an ASCE/EM Division Specialty Conference in Boston, MA. Oct. (1986).

PART VI

Metals and Metallic
Bonded Structures

NONDESTRUCTIVE DEPTH PROFILING OF A DIFFUSE INTERFACE

ANTHONY N. SINCLAIR, PHINEAS DICKSTEIN, AND MICHAEL A. GRAF
University of Toronto, Dept. of Mechanical Engineering, 5 King's College Road, Toronto, Canada M5S-1A4

ABSTRACT

A numerical solution of the one-dimensional wave equation is used to find the characteristics of wave propagation in a non-homogeneous medium. The solution is used to determine the magnitude and phase of the reflection coefficient at a diffuse interface. The result is found to be strongly dependent on sonic frequency. Comparison is made between theoretical calculations and measurements of the reflection coefficient at a copper-to-nickel diffusion bond.

INTRODUCTION

Ultrasonics is well-suited to the detection of a sharp boundary between two dissimilar materials. In this capacity, there are established procedures for finding delaminations in a bonded structure or slag inside a weld. Under ideal conditions, the magnitude and phase of the ultrasonic echo returned by these foreign bodies can be predicted by wave theory: the wave equation for a homogeneous material is solved for each of the material constituents, and then appropriate boundary conditions are applied at the sharp interfaces linking the various materials. For the case of a planar wave incident at 90° on a flat surface, the reflection coefficient R is easily shown to be:

$$R = \frac{Z_2 - Z_1}{Z_2 + Z_1} \tag{1}$$

where Z_i is the mechanical impedance of material i. Similarly, the reflection coefficient can also be calculated for a multi-layered medium by a linear superposition of the waves reflected by the multiple layers.

Complications arise, however, if the interface between two media is diffuse, i.e., if there is a *gradual* transition from one material to another over a finite region of thickness L. On a macroscopic scale, such interfaces are found in diffusion welds and in materials whose surfaces have been subjected to radiation or ion bombardment. On an atomic scale, virtually all interfaces are diffuse, although this factor is usually ignored if L is much less than the wavelength λ.

In this work, both theoretical and experimental studies were conducted to find the reflection coefficient of ultrasound at a diffusion weld. A range of sound frequencies was selected for the tests, such that the dependence of R on λ would be readily apparent.

THEORY

For the case of a plane compression wave, the classical wave equation can be reduced to a one-dimensional formulation of Newton's Law of motion combined with Hooke's Law:

$$\rho(x)\frac{\partial^2 u(x,t)}{\partial t^2} = \frac{\partial}{\partial x}\left[\tilde{E}(x)\frac{\partial u(x,t)}{\partial x}\right] \tag{2}$$

where $\tilde{E}(x)$ is the effective stiffness of the material at location x under conditions of uniaxial strain. The spatial and temporal dependence of the displacement u can be separated to give $u(x,t) = X(x)\tau(t)$. A second order ordinary differential equation can then be written for $X(x)$:

$$X_{xx} + \frac{\omega^2 \rho}{\tilde{E}}X + \frac{\tilde{E}_x}{\tilde{E}}X_x = 0 \tag{3}$$

To consider the reflection coefficient at a diffuse layer linking two dissimilar materials, the geometry of Fig. 1 is adopted. An incident plane wave traveling in the positive x direction in medium 1 is incident at $90\,°$ on the interface. Two parameters are required to specify the characteristics of sound propagation in the isotropic, non-attenuating media 1 and 2. By specifying the density ρ_i and stiffness $\tilde{E}_i$ for material $i = 1,2$, the speed of sound for compression waves c_i and mechanical impedance Z_i are uniquely determined for these media. In the interfacial region of Fig. 1, all acoustic parameters vary in a continuous manner between their specified values in the two bulk materials.

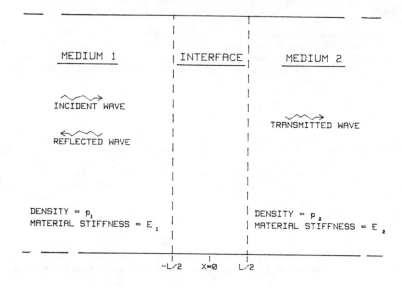

Figure 1: Geometry of Specimen with Diffuse Interface

Irrespective of the precise functional form of $\rho(x)$ or $\tilde{E}(x)$ within the interfacial zone, the general solution of Eq. (3) can be expressed as:

$$X(x) = \begin{cases} A_1 e^{ik_1x} + A_2 e^{-ik_1x} & , \qquad x \leq -L/2 \\ B_1\Psi_1(x) + B_2\Psi_2(x) & , \qquad -L/2 \leq x \leq L/2 \\ C_1 e^{ik_2x} & , \qquad x \geq L/2 \end{cases} \tag{4}$$

where wavenumber k_i of medium i is defined in terms of frequency f by $k_i = 2\pi f/c_i$. The reflection coefficient is given by $R = A_2/A_1$. In general, the parameters A_1 and A_2 may be complex to incorporate a phase factor.

Brekhovskikh [1] and Levine [2] both attempted to find analytical solutions to Eq. (3) in order to determine values for A_1 and A_2. Although solutions were found for a limited class of interfacial profiles, they did not in general constitute a good representation of actual material profiles found in engineering components. A numerical solution to Eq. (3) is therefore needed to accommodate an arbitrary makeup of the interfacial zone.

To determine $X(x)$ for a specified wave frequency, the parameter C_1 of Eq. (4) is arbitrarily set to 1. This immediately gives the solution $X(x)$ for $x \geq L/2$, i.e., the transmitted wave. A numerical scheme is now adopted to determine the values of A_1 and A_2 that correspond to this particular choice of C_1; the reflection coefficient can then be determined [3] from Eq. (4).

A fourth order Runge-Kutta method [4] is used to solve for $X(x)$ across the interfacial region and into material 1 where $x \leq L/2$. The solution is based on Eq. (3), using the known solution of $X(x)$ for $x \geq L/2$ to obtain the initial conditions $X(x_o)$ and $X'(x_o)$ at the starting point x_o of the numerical solution.

It is noteworthy that a solution is required for both the real and imaginary parts of $X(x)$. This is because there are two unknowns, A_1 and A_2 for which values are sought. These two unknowns can then be determined from the first line of Eq. (4). In general, the solution will be a function of wave frequency ω. Results are most conveniently shown for specific bulk materials and interfacial profiles; the wide spectrum of possible material profiles makes it difficult to present results in non-dimensionalized form.

Figure 2 shows the magnitude of the reflection coefficient $|R|$ as a function of frequency for a plane wave in nickel-200 incident on a a diffuse interface with copper. For this case, it is assumed that the density and stiffness of the material vary in a continuous linear manner across the interfacial zone. Results are shown for L ranging from 0.1 to 1 mm. Bulk material properties are shown in Table I.

material	$\tilde{E}$ (Pa)	ρ (kg/m^3)	Z $(kg/m^2 \cdot s)$	c_L (m/s)
Nickel-200	2.82×10^{11}	8881	5.0×10^7	5630
Copper	1.97×10^{11}	8936	4.2×10^7	4700

Table I: Mechanical Properties of Test Materials

The following trends are noted in Figure 2. For $\lambda \gg L$, the sound behaves as if the interface were infinitely thin and the reflection coefficient approaches the value given by Eq. (1), equal to -0.087 for a wave traveling from nickel into copper. If the wavelength is of the same order as L, then $|R|$ is seen to go into an oscillating pattern, with the amplitude of oscillation decreasing with increasing frequency. For $\lambda \ll L$, the interface becomes totally transparent with transmission coefficient $T = 1$, and $R = 0$.

Brekhovskikh [1] also observed the oscillating pattern in $|R|$ for intermediate frequencies and concluded it was due to some sort of "resonance" effect. Physically, the wave is not reflected at any specific point within the interface but is rather being continuously reflected as the wave crosses from one bulk medium into the other. At certain frequencies, these reflected components add to zero due to their different phases, resulting in the "nodes" in the reflection coefficient seen in Figure 2.

Figure 3 shows the phase of the reflection coefficient calculated under the same conditions as Figure 2. A direct correlation between the two Figures is clearly visible. For very large wavelengths, the phase is equal to $-180°$, as predicted by Eq. (1) for the case where $Z_2 < Z_1$. For higher frequencies, an oscillating pattern is seen with the phase going to zero whenever $|R|$ goes to zero.

EXPERIMENT

Samples of half-inch (1.22 cm) diameter nickel-200 and copper rod were cut into 4 cm lengths. Nickel-copper pairs were then friction-welded together. The ends of each specimen were then trimmed to yield samples approximately 2 cm in length, with the friction weld located close to the midpoint. Interfacial echo signals were then obtained using the experimental set-up of Figure 4 and pulsed excitation. Probes ranged from 2.25 to 15 MHz central frequency with bandwidth at -6 dB between 40% and 70%. These signals were needed for calibration purposes; the reflection coefficient at the weld interface for these freshly welded samples is given by Eq. (1).

The samples were then heat-treated for 48 hours at 1273 K in order to generate a diffusion zone on the order of 100 μm across. The profile was numerically determined by a solution of the diffusion equation, and later verified by destructive analysis. Unlike the profile used to generate Figures 2 and 3, this interface had no sharply defined boundary, but instead the mechanical parameters on both sides of the interface approached those of the bulk media in an asymptotic manner.

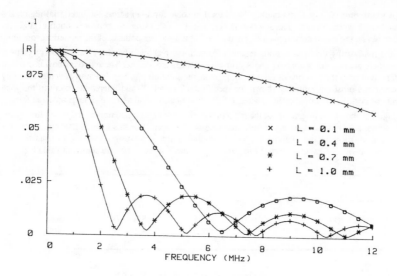

Figure 2: Magnitude of reflection coefficient at a linear nickel-copper interface of thickness *L*.

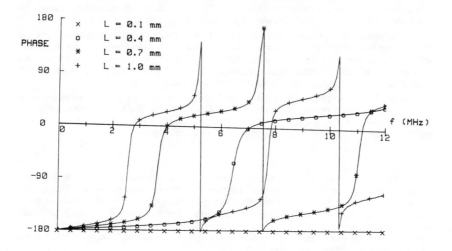

Figure 3: Phase of reflection coefficient at a linear nickel-copper interface of thickness *L*.

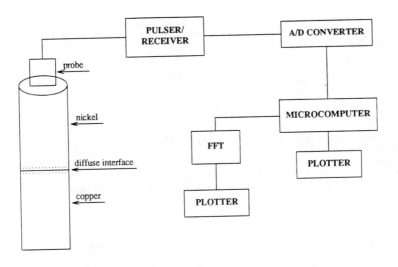

Figure 4: Experimental Configuration.

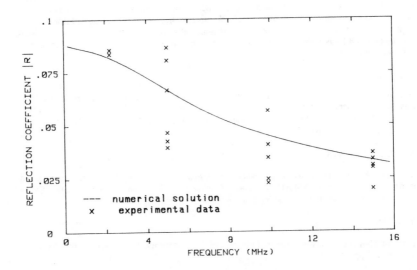

Figure 5: Magnitude of reflection coefficient at nickel-copper diffusion bond. Heat treatment was for 48 hours at 1273 K.

A second set of interfacial echo signals was then obtained using the identical experimental set-up as before. The magnitude of the reflection coefficient at any frequency f_o could then be determined from the Fast Fourier Transform $y(f)$ of the echo signals:

$$|R(f_o)|_{heat\ treated} = \frac{Z_2 - Z_1}{Z_2 + Z_1} \times \left[\frac{\int\limits_{f_o-\Delta f}^{f_o+\Delta f} y^2(f)\, df\ |_{heat\ treated}}{\int\limits_{f_o-\Delta f}^{f_o+\Delta f} y^2(f)\, df\ |_{before\ heat\ treatment}} \right]^{\frac{1}{2}} \tag{5}$$

Results of these measurements are shown in Figure 5 along with the theoretical curve calculated by numerical solution of the wave equation. Data that showed a high sensitivity to the exact location of the probe on the specimen were not included in Figure 5, as such data indicated a lack of uniformity in the friction weld. Note that Eq. (5) will yield good results only if applied over a narrow frequency interval $2\Delta f$ at a central frequency f_o close to the central frequency of the probe, where the signal energy is strong.

ANALYSIS

Considerable scatter is seen in the experimental results of Figure 5. Destructive analysis of the specimens later showed that the heat treatment cycle had led to a congregation of oxides at the nickel-copper surface. This is believed to be the major reason for the observed variation in experimental results, and would also be a contributing factor to the deviation between experiment and theory. Despite this factor, a rough correlation between experimental and theoretical results is observed.

The dependence of $|R|$ on frequency for the copper nickel samples is seen to be markedly different than that shown in Figure 2 for the linear interfacial profile. This is an indication that an inversion scheme might be developed whereby both a characteristic width L and functional profile of the interfacial zone might be derived from the frequency dependence of the reflection coefficient. In most industrial cases, it is the parameter L that is of key interest; its value may be quickly estimated from the frequency at which $|R|$ shows a sharp dip from the asymptotic value given by Eq. 1. A more precise estimate can be obtained by comparing the experimental results with theoretical curves generated using a range of values of L.

CONCLUSION

The reflection coefficient R has been numerically determined for a plane wave incident on a diffuse boundary layer. A marked dependence of R on frequency and layer thickness was observed. Comparison of theoretical results with those of experiments was complicated by the difficulty in producing experimental specimens that were free from defects.

ACKNOWLEDGEMENTS

This study has been supported by equipment and operating grants from the Natural Sciences and Engineering Research Council (NSERC) of Canada. Gratitude is expressed to Mr. Robert Batt of Stelco Research for the friction welding of experimental specimens.

REFERENCES

1. L.M. Brekhovskikh, Waves in Layered Media, 2nd ed. (Academic Press, New York, 1980), pp. 161-180 and pp. 188-192.

2. H. Levine, Unidirectional Wave Motions, (North-Holland, Amsterdam, 1978), pp. 323-335.

3. A.N. Sinclair and M. Graf, submitted to Research in Nondestructive Evaluation (1988).

4. T.E. Shoup, Applied Numerical Methods for the Microcomputer, (Prentice Hall, Englewood Cliffs, New Jersey, 1984), pp. 116-119.

ULTRASONIC CHARACTERIZATION OF THIN ADHESIVE BONDS

P. DICKSTEIN [*] , A.N. SINCLAIR[*] , E. SEGAL[**] , Y. SEGAL[**]

[*]   Department of Mechanical Engineering, 5 King's College Road, Toronto, Canada M5S-1A4

[**] Department of Nuclear Engineering, Technion-Israel Institute of Technology, Haifa, Israel-32000

ABSTRACT

A series of nondestructive and destructive tests were performed on samples of aluminum-to-aluminum bonded plates. The specimens featured a range nominal thickness of the adhesive layer from 0.0 up to 0.5 mm. It was found that a commercial bondtester was inappropriate for assessing the thickness of the adhesive in this range. However, the frequency peak of an ultrasonic echo signal from the adhesive layer was a reliable indicator of bond thickness. It is proposed that nondestructive assessments of the bond thickness could serve as an indicator of the longterm resistance of the adhesive to water ingress and bond degradation.

INTRODUCTION

The thickness of the adhesive layer in a structural adhesive joint is an important parameter in determining both the initial strength of the joint, as well as its long term durability characteristics. Bickerman [1] reviews several works dealing with the strength of adhesive joints of different configurations and the thickness B of the adhesive layer. For B ranging from 0 to 0.5 mm, no consistent pattern could be found linking the destructive shear strength of the bond to its precise thickness. Above 0.5 mm, however, a monotonic decrease in strength with increasing B has been observed by a majority of researchers, e.g., MacIver and Thompson [2] who examined bond thicknesses ranging from 0.25 to 2.5 mm.

Bickerman argues that when the thickness of the adhesive is increased, the probability of an internal imperfection or weak point also increases. Anderson et al [3], however, suggest several other factors which may contribute to the correlation between bond strength and B. In a tension-loaded butt joint, for example, the adherends will tend to impose a condition of triaxial constraint on the adhesive, and thereby reduce the shear stress. The extent of triaxial constraint is dependent on the bond thickness, and therefore a relationship is established between B and the bond strength. Another consideration is that the residual stresses in the adhesive due to the cure and thermal shrinkage are a function of adhesive thickness. Bulk material properties may also be linked to B; Knollman [4] observed a pronounced variation of the shear-modulus as a function of the adhesive thickness.

The strength of an adhesive bond immediately after manufacture is not the sole parameter of interest. There also exists the question of the in-service durability of a joint. Lloyd and Wadhwani [5] immersed bonded specimens with B equal to 0.02 and 0.2 mm in water at $60^{o}C$ and monitored periodically the degradation of the bond due to water ingress. Although the moisture take-up was not within the adhesive itself but instead occurred at the adhesive-adherend interface, the reduction in the bond's quality was observed first with the specimens with the thicker adhesive layer. This indicates that the durability of adhesive joints is partially dependent on the thickness of the bond line.

The inspection of adhesive joints with B in the interval 0.0 to 0.5 *mm* is the subject addressed in this work. This range of adhesive thicknesses is characteristics of joints in the aviation industry. This study employs a destructive shear test, plus two nondestructive techniques: an ultrasonic pulse-echo test, and an assessment of the resonant frequency characteristics of the joints using a commercial ultrasonic bondtester.

EXPERIMENT

Six aluminum-to-aluminum bonded specimens were prepared, following the ASTM D1002-72 standard for the strength properties of adhesives in shear by tension loading (metal to metal). The two aluminum alloy adherends were Al-2043 T3, each of 1.62 *mm* thickness.

Each bonded specimen had a different bond thickness B, controlled by foils placed between the two adherends during manufacture. Nominal thicknesses of the bonds were 0.0, 0.05, 0.1, 0.2, 0.3, and 0.5 *mm*. Prior to bonding, the adherends underwent a chromic-acid anodization according to the MIL-A-8625D standard. The anodization treatment causes an oxidation layer of 15000 Å thick to be formed consisting of highly uniform tightly-packed columns, each about 400 Å in diameter [6]. Three types of tests were then conducted on the specimens.

Ultrasonic Pulse-Echo Test

The specimens were immersed in an ultrasonic immersion tank with a Trienco model 705 X-Y scanning bridge and a C-scan recorder. A transducer of nominal central frequency 2.25 *MHz* was driven by a Panametrics 5052 PRX100 pulser receiver. Gated ultrasound echo signals from the adhesive layer were digitized by a Tektronix 7912 A/D programmable digitizer, with a sampling rate of 1 *GHz*; ten such echoes were recorded for each of the six specimens. The digitized signals were transferred to a Digital PDP 11/34 mini computer for processing and storage.

Bondtester Inspection

Each of the six specimens was inspected by the Fokker[a] model 70L ultrasonic bondtester, according to the manufacturer's directions. Use of the bondtester is based on the model of a test specimen as two adherends joined by a spring and damper in a parallel configuration. By sweeping over a wide frequency range, the bondtester finds the mechanical resonant frequency of the specimen, which is then compared to that of a reference sample; a marked variation in resonant frequency is normally an indication of a defective bond. For the case of the six specimens under consideration here, there should be a direct correlation between the bond thickness and the effective stiffness of the model spring joining the two adherends.

Destructive Test

Following the nondestructive tests, each of the six bonded plate specimens was sliced to give seven test strips of width 1 inch (2.54 *cm*). The two ends of each strip were then clamped

---

[a] Fokker B.V., Amsterdam, The Netherlands

in vice grips of a loadframe, and the load was increased until failure of the adhesive. From the specimen geometry, the failure mode was predominantly shear.

RESULTS AND DATA ANALYSIS

Ultrasonic Test

The large amount of data acquired with the pulse-echo testing necessitated a method for compressing the information into a form amenable for analysis and storage. For each of the digitized echo signals, a feature vector f was defined:

$$f = \{f_1, f_2, \cdots f_n\} \tag{1}$$

Each element of the feature vector represents a parameter derived from the ultrasonic signal from either its time or frequency domain representation. The elements of f were selected based on empirical evidence from other researchers as to the signal features that correlate well with bond quality. Eight features were calculated for each signal, including moments of the signal representation in both the time and frequency domains, parameters obtained from Higher Order Crossing (HOC) analysis [7], and the magnitude of the highest peak in the power spectrum.

The majority of the eight features showed no significant correlation with bond thickness, though, previous work has indicated the sensitivity of these features to key parameters of an adhesive bond [8].

The feature f_1, representing the magnitude of the highest peak of the spectrum, was found to be the only one of the eight selected features that was sensitive to the thickness of the adhesive layer. Complementary results have been obtained by Biggiero et al [9,10] who examined specimens of uniform adhesive thickness, but a range of adhesives constituents.

Figure 1 shows a bar chart of the ten f_1 values corresponding to each bonded plate specimen. The ten measurements, from which the values of the ten f_1 values were derived, were taken at different successive locations. The chart clearly shows the dependence of f_1 on the adhesive thickness. In general, the value of f_1 shows little variation within each specimen. This indicates a relatively uniform bond thickness.

A two-dimensional view of the dependence of f_1 on the thickness of the adhesive layer is given in Figure 2. A monotonic decrease in f_1 is obtained when the thickness of the adhesive layer is increased. The fitting curve for the values of f_1 is of the form:

$$f_1(B) = f_1(o) \, e^{-\alpha B} \tag{2}$$

For the case of figure 2, $\alpha \approx 2$ and $f_1(o) \approx 2.5$. The parameter α can be considered as a pseudo-attenuation coefficient of f_1; it can be expressed in Nepers or decibels per unit length.

Before adopting the parameter f_1 as an indicator of adhesive thickness, it is necessary to investigate the sensitivity of f_1 to other manufacturing parameters. One such parameter that was investigated is the pretreatment procedure. A second series of bonded plates was prepared, but this time the surface pretreatment of the adherends consisted only of acetone cleaning, rather than the chromic anodization sequence used before. (It has been established that the acetone cleaning treatment by itself does not generate the thick uniform layer of oxide needed

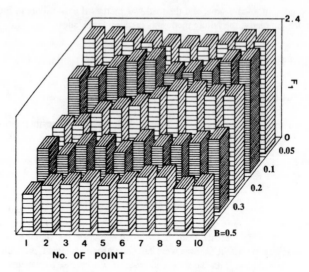

Figure 1: f_1 Values from Bonds with Different Thickness of The Adhesive Layer B .

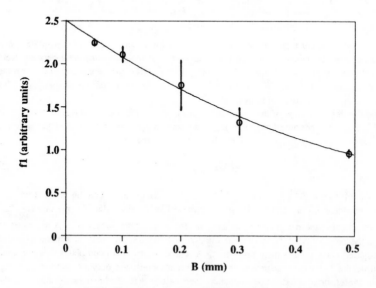

Figure 2: The Feature f_1 Vs. The Thickness of The Adhesive Layer B

on the adherends for long-term durability of the bond [11-13]). A comparison of the frequency spectra of signals obtained from plates corresponding to the two different surface pre-treatments was carried out. As expected, signals in this frequency range have little sensitivity to the surface pre-treatment, and therefore the dependence of f_1 on adhesive thickness is essentially unchanged from before.

Bondtester Results

The results of the Fokker bondtester assessment of the plates are shown in Figure 3. The bondtester could detect no statistically significant difference among the plates for which $B < 0.5$ mm. The reason can be seen from the dynamic model of the test specimen. For bond thicknesses less than 0.5 mm, the stiffness of the adhesive (which is proportional to $1/B$) is extremely large; therefore the overall stiffness of the specimen is dominated by the adherends. The resonant frequency under these circumstances is insensitive to bond thickness; the thickness therefore cannot be assessed by the bondtester.

The bondtester did indicate a significant deviation in bond quality in the plate for which $B = 0.5$ mm. The adhesive in this sample formed a bubble-like structure rather than a homogeneous layer [14]. This porosity led to a "weak link" between the two adherends which significantly reduced the overall stiffness of the sample, and its resonant frequency.

Destructive Shear Test Results

Results of the destructive tests on the test strips cut from the bonded plates are shown in Figure 4. As observed by other researchers and indicated by the bondtester, there was no statistically significant difference in strength among the strips for which $B < 0.5$ mm. For these specimens, the failure was of a cohesive nature. The specimens with $B = 0.5$ mm were notably weaker than the rest, and the breaking mode was a mixture of cohesive and adhesive failure [14].

These results indicate that the bondtester satisfactorily assessed the strength of these samples. Its failure to note any variation in the thickness of the bond line when $B < 0.5$ mm has negative implications for a durability assessment of such joints, as the rate of degradation of the bond due to water ingress is dependent on this parameter [5].

SUMMARY AND CONCLUSIONS

A variety of features were calculated from ultrasonic echo signals reflected by the adhesive layer in aluminum-to-aluminum bonded plate specimens. The sensitivity of the magnitude of the peak maximum of the spectrum to the thickness of thin adhesive layers in the range of 0.0 to 0.5 mm has been demonstrated. This range of adhesive thicknesses is widely applied in the aeronautical industry. It has been shown experimentally that the adhesive's thickness has implications on the long-term durability of the joint, and should therefore be monitored during the manufacturing process. An ultrasonic commercial bondtester is not suited to this task, as the resonant frequency of the specimens is not sensitive to the thickness of the adhesive in this range.

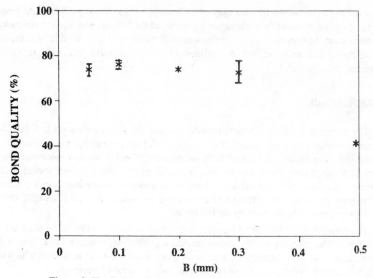

Figure 3: The Fokker Relative Quality of The Bonds Vs. B.

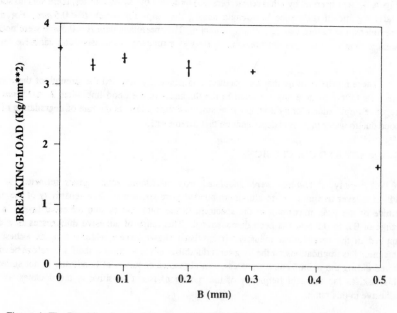

Figure 4: The Breaking-Load of The Bonds Vs. The Thickness of The Adhesive Layer B

REFERENCES

1. J.J. Bickerman, The Science of Adhesive Joints , 2nd ed. (Academic Press, New York and London, 1968).

2. G.M. MacIver and D.P. Thompson II, in: Adhesives in Manufacturing , edited by L. Schneberger, (Marcel Dekker Inc., New York and Basel, 1983).

3. G.P. Anderson, S.J. Bennett and K.L. DeVries, Analysis and Testing of Adhesive Bonds , (Academic Press, 1977).

4. G.C. Knollman, Int. J. Adhes. $\underline{5}$, (3), 137-141, (1985).

5. E.A. Lloyd and D.S. Wadhwani, in: Adhesion 4 , edited by K.W. Allen, (Applied Science Publishers Ltd., London, 1980).

6. J.D. Venables, D.K. McNamara, J.M. Chen, T.S. Sun and R.L. Hopping in: 10th Nat. SAMPE Tech. Conf., 362, 1978.

7. B. Kedem, Proc. IEEE, $\underline{74}$, 11, (1986).

8. P. Dickstein, D.Sc Thesis, Technion - Israel Institute of Technology, 1988.

9. G. Biggiero, G. Canella and A. Moschini, NDT International, $\underline{1983}$, 4.

10. G. Biggiero and G. Canella, Int. J. Adhes., $\underline{5}$, 3 (1985).

11. W.J. Russell and C.A.L. Westerdahl, in: Corrosion Control by Coatings , edited by H. Leidheiser Jr., (Science Press, Princeton, 1979).

12. W.G. Brockmann, in: Structural Adhesives in Engineering , (Mechanical Engineering Publications Ltd., London 1988).

13. S.H. Hartshorn, in: Structural Adhesives, Chemistry and Technology , edited by S.H. Hartshorn , (Plenum Press, 1986).

14. P. Dickstein, E. Segal and Y. Segal, in: Nondestructive Testing , edited by J.M. Fraely and R.V. Nichols, (Plenum Press, 1987).

CORRELATING MICROSTRUCTURE
with
BACKSCATTERED ULTRASONIC ENERGY

John Mittleman and David W. Mohr

Naval Coastal Systems Center
Panama City, FL 32407

ABSTRACT
High frequency ultrasonic energy interacts with metallurgical microstructure to produce scattered energy fields that are useful in characterizing the material. By using a focused transducer whose focal spot size and ultrasonic wavelength are both comparable to the scale of grain structure in high purity copper specimens, the authors have shown a strong correlation between microstructure and the pattern of backscattered ultrasonic signals. A systematic series of experiments logged the ultrasonic returns from cold-rolled copper with the direction of the ultrasonic beam both parallel to and transverse to the rolling direction. Fourteen samples were annealed to various degrees, capturing several stages of recovery and recrystallization.

Use of backscattered ultrasonic energy rather than specularly reflected energy allows isolation of grain boundary scattering in an otherwise quiet acoustic environment. Clear differences between returns of backscattered shear waves from parallel and transverse beam orientations relative to the rolling direction gradually disappear as the annealing process re-establishes a field of equiaxial crystals. Backscattered surface waves also show variations that appear to correspond with the relaxation of residual stresses before the onset of recrystallization, and preferred orientation after recrystallization.

INTRODUCTION
The relationship between microstructure and ultrasonic signals has been studied by a great number of researchers. Hecht [1] observed the frequency spectra of normal incidence returns as well as signals backscattered from waves incident at the Rayleigh angle for austenitic steel sheets. The excitation of backscattered Rayleigh leaky waves is described by Adler and de Billy [2,3]. Klinman related grain size and yield strength to ultrasonic attenuation and its frequency dependence [4,5]. Fracture toughness was related to ultrasonic parameters at higher frequencies by Vary [6-8], and the relationship between grain size and attenuation was also reported by Serabian [9]. Numerous other researchers have also observed and reported correlations between microstructure and ultrasonic waveform features, primarily attenuation; the theoretical basis for these correlations is presented by Green [10]. Scattering of surface waves was known to the Krautkramers [11] who cite correlations between grain size and acoustic velocity noted as early as 1960.

In this study the authors have observed high frequency (50 MHz), focused ultrasonic energy backscattered from polished copper specimens which had been cold-rolled and annealed to produce a range of grain sizes and structures. Backscattered signals, which exist in an extremely quiet acoustic environment,

Mat. Res. Soc. Symp. Proc. Vol. 142. ⊂1989 Materials Research Society

compete only with random electrical noise and with stationary transducer noise, but not with echoes produced at the specimen's surfaces. Relatively high signal-to-noise ratios are achieved in this manner.

PREPARATION OF SAMPLES

Copper specimens used in this study were all prepared from a single 0.75 inch diameter bar of oxygen-free high conductivity (OHFC) copper. A short section of the bar was cold-rolled to a thickness of 0.10 inches (86.7% reduction), using multiple passes, each of 0.005" reduction. Deformation was by elongation; increase in width of the bar was only to 0.825".

Coupons approximately one inch long were taken from the rolled bar and subjected to a variety of heat treatments:

```
No anneal
5 min. @ 100 C      1 hr. @ 100 C      10 min. @ 650 C
5 min. @ 200 C      1 hr. @ 200 C      10 min. @ 800 C
5 min. @ 300 C      1 hr. @ 300 C      10 min. @ >1000 C
5 min. @ 400 C      1 hr. @ 400 C      (sufficient to
5 min. @ 500 C      1 hr. @ 500 C       melt the specimen)
```

The rolling plane of each specimen was polished and examined by ultrasound, optical microscopy, penetration hardness testing, and X-ray diffraction methods.

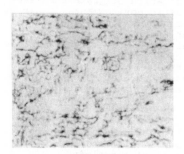

1a. Unannealed

1b. Annealed at 400 C

1c. Annealed at 650 C

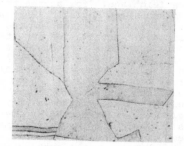

1d. Annealed at 800 C

Figure 1. Copper Specimens Etched
           with Marble's Reagent

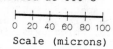

0 20 40 60 80 100
Scale (microns)

METALLOGRAPHIC EXAMINATION OF THE SPECIMENS

Polished copper samples were etched with Marble's reagent and examined in an optical microscope with reflected light (Metals Handbook [12]). Grain sizes were measured on photomicrographs of known magnification. Selected photomicrographs are shown in Figure 1.

A transverse cut through the starting copper stock revealed an annealed microstructure whose average grain size is 200 microns. Examination of the cold-rolled specimens on their rolling planes revealed typically deformed microstructure with grains elongated parallel to the rolling direction. Maximum size in the rolling plane is 75 x 1000 microns. Grain boundaries are irregular; slip planes are abundant. Examination of specimens cut perpendicular to the rolling plane showed a maximum grain thickness of 20 microns. Optical examination of samples annealed to temperatures of 100 and 200 degrees Centigrade revealed no discernable differences from the unannealed material. Samples annealed at 300 C and higher showed complete recrystallization, with the formation of equiaxed, twinned grains having sharp, planar boundaries. For a given temperature of anneal, the differing anneal times yielded no optically discernable changes in microstructure. Penetration hardness, measured on the Rockwell "F" scale, showed a sharp decrease between 200 C and 300 C, evidencing the effects of recrystallization (Figure 2).

Grain size increased monotonically with temperature of anneal, as shown in Table I. The sample which was melted in the presence of air showed a columnar microstructure of the Cu-CuO eutectic with individual columns of copper up to 200 microns across. CuO particles reached a maximum size of 10 microns along growth-front boundaries.

TABLE I: ANNEALING TEMPERATURE versus GRAIN SIZE

Annealing Temperature	Approximate Grain Size
300 C	15 microns
400 C	25 microns
500 C	40 microns
650 C	75 microns
800 C	200 microns

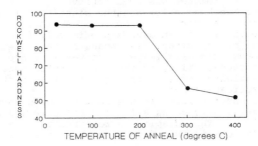

HARDNESS vs. TEMPERATURE of ANNEAL
Rockwell Hardness, "F" Scale

Figure 2. Penetration Hardness Measurements

X-Ray Diffraction Results
Ratio of Intensities (D111/D200)

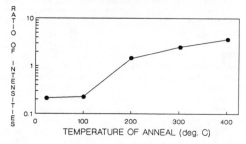

Figure 3. X-Ray Diffraction Results

X-RAY DIFFRACTION ANALYSIS

Samples annealed for five minutes at 100, 200, 300, and 400 degrees Centigrade, as well as a coupon of the unannealed, cold-rolled copper, were investigated by routine XRD analysis [13]. Freshly etched samples were mounted so that X-rays impinged on the rolling plane.

Increases in the temperature of anneal above 100 C correlate to a strong increase in the intensity of the (111) reflection and a moderate diminution in the intensity of the (200) and (311) reflections (Figure 3). No discernable difference in relative intensities of X-ray peaks was found between the unannealed sample and that annealed at 100 C. Apparently, relaxation of residual stresses and incipient recrystallization occurred between 100 C and 200 C, but a temperature between 200 C and 300 C was required to allow optical resolution of the neoblasts.

ULTRASONIC EXAMINATION OF THE SPECIMENS

Each of the copper specimens was examined ultrasonically, using a 50 MHz (nominal) focused immersion transducer with a 0.5" (nominal) focal length, and the general experimental configuration shown in Figure 4. The point of focus of the beam was positioned on the sample's surface for all runs. The angle of the ultrasonic beam axis relative to the specimen's surface was chosen to be above the first critical angle (18.4 deg.) or second critical angle (41.0 deg.) to excite shear or surface waves, respectively. Scans were actually run at 29 degrees and 42 degrees.

A motorized micrometer was used to move the transducer perpendicular to the beam axis, and the signal was recorded approximately every 40 microns of travel. A single scan comprises 256 waveforms taken over approximately 10 millimeters of travel across the specimen. Two scans of each specimen were made with the beam axis parallel or transverse to the rolling direction. The repeatability of the experimental runs was confirmed by examining the stationarity of variance statistics derived from replicate scans.

At each transducer location the backscattered waveform represents signal amplitude as a function of time (relative to the initial pulse). This display is referred to as an "A-scan". Systematic variations in this signal occur as the transducer is

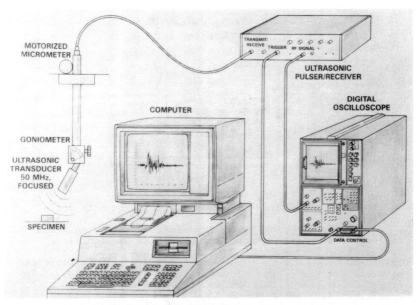

Figure 4. Experimental Set-up

scanned across the surface of each specimen. The spatial
coherence of each ensemble of signals is revealed by use of a
"stacked A-scan" display in which consecutive A-scans are
plotted as parallel lines, with amplitude coded on a gray scale.
The result is much like a top view of the sequence of A-scans,
with axes of ultrasonic travel time versus position across the
specimen (Figure 5). High amplitude returns from local areas of
the specimen show up in the stacked A-scans as dark patches, and
areas that are acoustically quiet show up as light areas. In
this nonparametric presentation of the data one can observe
systematic variations in the pattern of the images as the
temperature of anneal is varied.

CORRELATIONS BETWEEN ULTRASONIC SIGNALS AND MICROSTRUCTURE
    Shear wave backscattering images shown in Figure 6 reveal
that the cold-rolled structure responds very differently to
shear waves introduced transverse to and parallel to the rolling
direction. This anisotropic response is visible in samples
annealed at 100 C and 200 C, but is not evident for samples
annealed at and above 300 C (where recrystallization was first
observable). Between 500 C and 650 C the approximate size of
grains becomes comparable to the ultrasonic wavelength, and at
this point the strength of ultrasonic returns in the later
portions of the waveforms decreases sharply. It is hypothesized
that the more efficient scattering from larger grain boundaries
leaves less energy to propagate deeper into the material.
Finally, the very large grains found in the melted material
provide few scatterers. Since these are seen at large distances
from the focal spot as well as close to it, it appears that the
internal structure of the material is responsible for the
observed signals.

STACKED A-SCANS
Axes and Approximate Scale

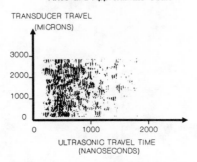

Figure 5. Stacked A-Scan Axes and Approximate Scale

Surface wave backscattering images demonstrated a somewhat different behavior. Anisotropic images (with significant beam orientation dependences) were produced in samples annealed both below and above 300 C. This behavior is thought to correlate with the preferred crystal orientations visible in the microphotographs. Additionally, the signals generated with the beam axis parallel to the rolling direction showed sensitivity to microstructural changes that occur before recrystallization; this may correlate with XRD results that show significant recovery between 100 C and 200 C. Significant differences between shear and surface wave image patterns may result from differences in the particle trajectories characteristic of each mode of propagation.

SUMMARY AND CONCLUSIONS

Backscattered energy produced by exciting polished copper specimens with a focused, 50 MHz ultrasonic beam was observed. The specimens were characterized by optical microscopy, X-ray diffraction and penetration hardness testing. Both shear waves and surface wave ultrasonic images corresponding to deformed microstructures are highly anisotropic. Shear wave images of recrystallized structures appear to be independent of the ultrasonic beam's orientation relative to the rolling direction. There is some evidence that the surface waves reveal the relaxation of residual stresses introduced by cold-working, and respond to crystallographic orientations more strongly than do shear waves.

This investigation has been largely phenomenological; a satisfactory theoretical scattering model has not yet been identified. However, it is clear that there is a strong dependence of backscattered ultrasonic signals on local microstructural details such as grain size and grain boundary stiffness. Signal processing techniques capable of extracting this dependence are under development.

This method, requiring access to only a small surface area, may eventually develop into an assessment technique for in-service materials. It is hoped that other researchers will replicate these experiments, improving on the experimental and analytical techniques employed.

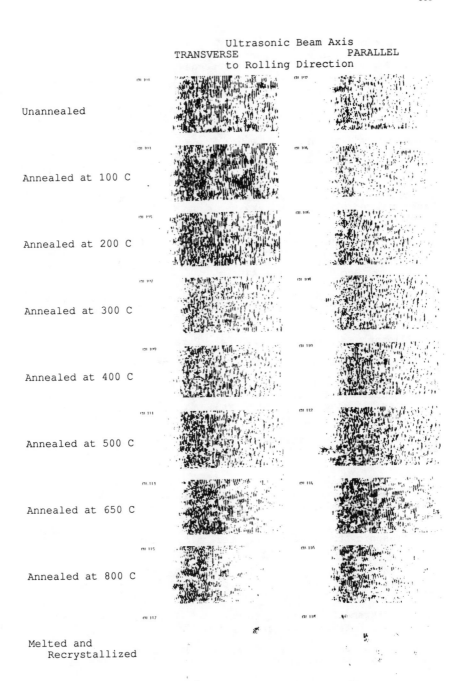

Figure 6. Shear Wave Backscattering Images

364

ACKNOWLEDGEMENTS
    This research was supported by the Naval Coastal Systems
Center's Independent Research/Independent Exploratory
Development program. The authors would like to extend special
thanks to Dr. L. Flax and Dr. M. McNeil for their support and
direction.

REFERENCES

[1] Hecht, A., R. Thiel, E. Neumann and E. Mundry,
"Nondestructive Determination of Grain Size in Austenitic Sheet
by Ultrasonic Backscattering", Mat. Eval., Vol. 39, Sep 81, pp
934-938

[2] de Billy, Michel, and Laszlo Adler, "Parameters Affecting
Backscattered Ultrasonic Leaky-Rayleigh Waves from Liquid-Solid
Interfaces", J. Ac.Soc. Am. Vol. 72, No. 3, Sep 82, pp 1018-1020

[3] Adler, Laszlo, Michel de Billy, and Gerard J. Quentin,
"Excitation of Ultrasonic Rayleigh Leaky Waves at Liquid-Solid
Interface for General Angle of Incidence", J. Appl. Phys., Vol.
53, No. 12, Dec 82, pp 8756-8758

[4] Klinman, R. and E.T. Stephenson, "Ultrasonic Prediction of
Grain Size and Mechanical Properties in Plain Carbon Steel",
Mat. Eval., Vol. 39, Nov 81, pp 1116-1120

[5] Klinman, R., G.R. Webster,, F.J. Marsh, and E.T. Stephenson,
"Ultrasonic Prediction of Grain Size, Strength, and Toughness in
Plain Carbon Steel", Mat. Eval., Vol. 38, Oct 80, pp 26-32

[6] Vary, A. and D.R. Hull, "Interrelation of Material
Microstructure, Ultrasonic Factors, and Fracture Toughness of a
Two-Phase Titanium Alloy", Mat. Eval., Vol. 41, March 83, pp
309-314

[7] Vary, A., "Correlations among Ultrasonic Propagation Factors
and Fracture Toughness Properties of Metallic Materials",
Mat. Eval., Vol. 36, Jun 78, pp 55-64

[8] Vary, A. "Concepts for Interrelating Ultrasonic Attenuation,
Microstructure, and Fracture Toughness in Polycrystalline
Solids", Mat. Eval., Vol. 46, Apr 88, pp 642-649

[9] Serabian, S. and R.S. Williams, "Experimental Determination
of Ultrasonic Attenuation Characteristics Using the Roney
Generalized Theory", Mat. Eval., Vol. 36, Jul 78, pp 55-62

[10] Green, Robert E., Jr. "Ultrasonic Investigation of
Mechanical Properties", Treatise on Materials Science and
Technology (Volume 3), Academic Press, New York, 1973

[11] Krautkramer, J. and H. Krautkramer, "Ultrasonic Testing of
Materials", Springer-Verlag, NY, 1969

[12] American Society for Metals, "Metals Handbook:
Metallography and Microstructures (Volume 9)", ASM, Metals Park,
OH, 1985

[13] Barrett, C.S., "Structure of Metals: Crystallographic
Methods, Principles, and Data", McGraw Hill, NY, 1952

NONDESTRUCTIVE CHARACTERIZATION OF BERYLLIUM EFFECTS ON Ni-Cr DENTAL ALLOY ELASTIC PROPERTIES AND MICROSTRUCTURE: ULTRASONICS, X-RAY DIFFRACTOMETRY, SCANNING ELECTRON MICROSCOPY AND WAVELENGTH DISPERSIVE SPECTROMETRY

SURENDRA SINGH, J. LAWRENCE KATZ[*] and B.S. ROSENBLATT[*]

NDE and Sensor Technology, Timken Research, Canton, OH, USA.
*Department of Biomedical Engineering, RPI, Troy, NY, USA.

ABSTRACT

Knowledge of structure-properties relationship is a key factor in the development and improvement of new and existing metal alloys through manipulation in their chemical-compositions. In this study, the elastic properties and microstructure of cast Ni-Cr-Be and Ni-Cr dental alloys were studied. The elastic properties, i.e., Young's, shear and bulk moduli and Poisson's ratios, were determined using measurements on the ultrasonic velocities and densities. Both the shear and the longitudinal (dilatational) velocities were measured using an ultrasonic pulse-through-transmission method; density was measured using a buoyant force method. In microstructure, crystallinity, porosity, particle-size and quantitative elemental compositions were studied using x-ray diffractometry (XRD), scanning electron microscopy (SEM) and wavelength dispersive spectrometry (WDS) respectively. These results show that: (1) the addition of Be increased significantly the alloy's elastic moduli and Poisson's ratio; and (2) the presence of Be in Ni-Cr alloy also significantly modified its microstructure by producing a second binary phase, Ni-Be, in eutectic areas.

INTRODUCTION

Attempts have been made to improve the properties of base alloys by modifying their structure either through changes in chemical-compositions or in processing procedures [1-4]. For example, chromium is added to the base metal Ni to provide corrosion resistance [4]. Similarly, beryllium is added to Ni-Cr alloys to reduce the solidification temperature, to increase castability and to improve the bond strength between the alloy substrate and the ceramic veneer [1-3]. The disadvantage of beryllium is its well-known toxic effects, in particular by inhalation of fumes and dust [5]. In addition, the presence of Be in Ni-Cr has been reported to produce a second binary phase, Ni-Be, in eutectic areas. This particular phase has been reported to be more prone to corrosion in bio-environment [5,6]. Although the beryllium effect on microstructure provides some guidelines for interpretation of results, it fails to provide a scientific basis for physical-properties prediction. Thus, there is need to study the properties and microstructure of these alloys. In the present study, the elastic properties of alloys Ni-Cr-Be and Ni-Cr have been determined using ultrasonic wave-propagation methods. Microstructure has also been studied by examining crystallinity, particle-size and porosity and chemical compositions using XRD, SEM and WDS respectively. The objectives of this work were: (1) to determine the elastic properties and microstructure; (2) to examine whether or not the structural changes are reflected in the elastic properties; and (3) to use these results for the optimization of properties by incorporating certain modifications in processing of the materials understudy.

MATERIALS AND METHODS

Specimen Preparation

The non-precious dental alloys, along with their chemical compositions (approximate weight %), are listed in Table 1.

TABLE 1

Alloys	Ni	Cr	Mo	Nb	Al	Si	Be	Ti
Ni-Cr-Be	78	12.5	4.9	--	2.3		1.9	0.4
Ni-Cr	77	11.5	3.5	3	2	3	--	0.5

These alloys were induction heated and were cast using silica crucibles. The cooling time from 1315 degrees C to room temperature was almost one hour. No carbon was added during casting. Eight specimens, four from each alloy, were used in this study. Samples were prepared in cylindrical shape with diameter = 9.90 mm and thickness = 12.80, adopting a standard polishing procedure.

Ultrasonic Elastic Properties Measurements

Young's, shear and bulk moduli and Poisson's ratios were obtained for each alloy by measuring longitudinal (V_ℓ, in m/s) and shear (V_t, in m/s) velocities using ultrasonic wave propagation techniques and densities (ρ, in kg/m³) using a buoyant force method.

The formulae used in these calculations were:

Shear modulus, $G = \rho V_t^2$ (1)

Bulk modulus, $K = \rho (V_\ell^2 - \frac{4}{3} V_t^2)$ (2)

Once G and K were determined, Y and υ could be calculated from the standard equations for isotropic materials:

Young's modulus, $Y = \dfrac{9\,KG}{(3K + G)}$ (3)

Poisson's ratio, $\upsilon = \dfrac{(3K - 2G)}{2(3K + G)}$ (4)

Both the longitudinal and the shear velocities were determined using a pulse-through transmission technique (Figure 1) and by computing the ratio of the specimen thickness to the time-delay as reported earlier by the author (7). Transmitting transducer was excited using a low pulse-width ($t_w = 0.5$ μS) with pulse-repetition rate 10 kHz and with termination of 50 ohms at a resonant frequency of 1 MHz. The flight-time was measured within an error of +/- 5 nS. This measurement was based on averaging hundred waveforms and digitizing each waveform using 1024 points per waveform.

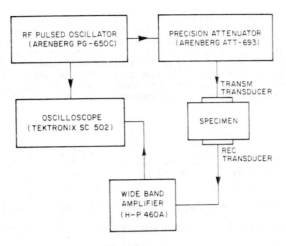

Figure 1

Fifteen independent observations were made on each of four specimens of each type of alloy. The density was measured using buoyant force method (7)

MICROSTRUCTURE

X-Ray Diffractometry (Crystallinity)

X-ray diffractometry was used on specimens obtained in fine powdered form with particle sizes of approximately 45 micrometer (325 sieves). This range was selected to avoid orientation effects in the specimen and subsequent error in reading diffraction angle due to broadening of peaks (8). Diffraction patterns were obtained using an x-ray diffractometer (ADP 3520, PW 1710, Phillips) in an angle range from 10^{0} to 95^{0}.

Microscopy (Porosity and Particle-size)

Specimen microstructure was examined by obtaining SEM micrographs of polished and etched specimen surfaces using a scanning electron microscope (JEOL, JSM- 840 II). These SEM micrographs were obtained at magnifications of x 5,000 and were taken at 20 kV accelerating voltage, 10^{-10} ampere probe-current and working distance in the range from 6 to 48 mm and a tilt angle of 10-20 (9).

Wavelength Dispersive Spectrometry (WDS)

Wavelength dispersive spectrometry was performed using a JEOL 733 Superprobe at a probe-current 22.14 nA and voltage 15 kV.

RESULTS AND DISCUSSIONS

Elastic Properties

The ultrasonic elastic properties, e.g., Young's, shear and bulk moduli and Poisson's ratios at 1 MHz, for these materials, are given in Table 2. These results show an average value for the sixty observations made on the four specimens from each group of alloys.

The Young's moduli of the alloys were in good agreement with the measured values in the literature (10, 11). From Table 2, it is clear that the elastic properties for the Ni-Cr-Be alloy were significantly higher that those of the Ni-Cr alloy (p < 0.001). The significant increase in the elastic properties of Ni-Cr-Be may be attributed to the presence of Be (1.9 weight %). It is clear that ultrasonic velocity in beryllium (longer wavelength material) is almost two times higher than in Ni-based alloys and this is the reason that the elastic properties are increased in Ni-based Ni-Cr-Be alloys (12).

Table 2

Materials	Shear Modulus G (GPa)	Bulk Modulus K (GPa)	Young's Modulus Y (GPa)	Poisson's Ratio
Ni-Cr-Be	79.52 (6.16)	191.0 (16.5)	209 (14)	0.32 (0.02)
Ni-Cr	75.89 (5.43)	150.9 (9.88)	195 (11)	0.28 (0.02)
Ni-Cr-Be vs. Ni-Cr	p < 0.001	p < 0.001	p < 0.01	p < 0.001

MICROSTRUCTURE

X-Ray Diffractometry

Diffractograms in Figure 2 show that these materials were crystalline. The Ni-Cr alloy had a structure consisting of a single binary f.c.c. phase. However, the addition of Be to Ni-Cr alloys produced a second binary phase, Ni-Be, in addition to the f.c.c. Ni-Cr phase in Ni-Cr-Be alloy. This Ni-Be phase had the CsCl (B_2) structure which is simple cubic and not b.c.c. as in a recent study (5,6) which mistakenly called the CsCl arrangement b.c.c. rather than simple cubic (13). The lattice parameters for these phases in each of the two alloys were calculated from the x-ray patterns:

(1) Ni-Cr, f.c.c., a = 0.3557 nm
(2) Ni-Cr, f.c.c., a = 0.3562 nm and
Ni-Be, (CsCl (B_2)), simple cubic a = 0.2620 nm

Thus it is clear from the x-ray analysis study that the presence of Be produced an additional binary phase, simple cubic Ni-Be, in the Ni-Cr-Be alloy.

Scanning Electron Microscopy

SEM micrographs at x 5,000 for polished and etched specimens are shown in Figures 3a,b. From Figure 3b, it is clear that in the Ni-Cr-Be alloy the Ni-Be phase is found in the eutectic areas. In view of the recent reports by Hero et al. (5,6) that the Ni-Be phase is more prone to corrosion in a bio-environment, this finding of the presence of the Ni-Be phase is important when considering certain clinical applications in which corrosion may be a possibility.